THIRD EDITION

Modern Control Systems

THIRD EDITION

Modern Control Systems

RICHARD C. DORF

UNIVERSITY OF CALIFORNIA, DAVIS

ADDISON-WESLEY PUBLISHING COMPANY

READING, MASSACHUSETTS

MENLO PARK, CALIFORNIA

LONDON

AMSTERDAM

DON MILLS, ONTARIO

SYDNEY

This book is in the
Addison-Wesley Series in Electrical Engineering

Library of Congress Cataloging in Publication Data

Dorf, Richard C
    Modern control systems.

    Includes bibliographical references and index.
    1. Control systems.   I.   Title.
TJ216.D67   1980   629.8′3   79-16320

Reprinted with corrections, April 1983

ISBN 0-201-01258-8
GHIJKLMN-DO-89876543

Man cannot inherit the past; he has to recreate it.

To Christine and Renée
as they seek to create

Preface

Man cannot inherit the past; he has to recreate it.* The most important and productive approach to learning is for the reader to rediscover and recreate anew the answers and methods of the past. Thus, the ideal is to present the student with a series of problems and questions and point to some of the answers that have been obtained over the past decades. The traditional method of confronting the student not with the problem but with the finished solution means depriving him or her of all excitement, to shut off the creative impulse, to reduce the adventure of human-kind to a dusty heap of theorems. The issue, then, is to present some of the unanswered and important problems which we continue to confront. For it may be asserted that what we have truly learned and understood we discovered ourselves.

The purpose of this book is to present the structure of feedback control theory and to provide a sequence of exciting discoveries as one proceeds through the text and problems. In that this book is able to assist the student in discovering feedback control system theory and practice, it will have succeeded.

The book is organized around the concepts of control system theory as they have been developed in the frequency- and time-domain. A real attempt has been made to make the selection of topics, as well as the systems discussed in the examples and problems, modern in the best sense. Therefore, one will find a discussion of sensitivity, performance indices, state variables, and computer control systems, to name a few. However, a valiant attempt has been made to retain the classical topics of control theory which have proven to be so very useful in practice.

The text is written in an integrated form so that one should proceed from the first to the last chapter. However, it is not necessary to include all the sections of a given chapter in any given course, and there appears to be quite a large number of combinations of sequences of the sections for study. The book is designed for an introductory undergraduate course in control systems for engineering students. There is very little demarcation between electrical, mechanical, chemical, and industrial engineering in control system practice; therefore, this text is written with-

---

* A. Koestler, *The Act of Creation,* Hutchinson, London, 1964; p. 266.

out any conscious bias toward one discipline. Thus, it is hoped that this book will equally be useful for all engineering disciplines and, perhaps, assist in illustrating the unity of control engineering. The problems and examples are chosen from all fields, and the examples of the sociological, biological, ecological, and economic control systems are intended to provide the reader with an awareness of the general applicability of control theory to many facets of life.

The book is primarily concerned with linear, constant parameter control systems. This is a deliberate limitation, since the author believes that for an introduction to control systems, it is wisest to initially consider linear systems. Nevertheless, several nonlinear systems are introduced and discussed where it is appropriate.

Chapter 1 provides an introduction and basic history of control theory. Chapter 2 is concerned with developing mathematical models of these systems. With the models available, we are able to describe the characteristics of feedback control systems in Chapter 3 and illustrate why feedback is introduced in a control system. In Chapter 4 we examine the performance of control systems, and in Chapter 5 we investigate the stability of feedback systems. Chapter 6 is concerned with the s-plane representation of the characteristic equation of a system and the root locus. Chapters 7 and 8 treat the frequency response of a system and the investigation of stability using the Nyquist criterion. Chapter 9 develops the time-domain concepts in terms of the state variables of a system. Chapter 10 describes and develops several approaches to designing and compensating a control system. Finally, Chapter 11 discusses digital computer control systems.

This book is suitable for an introductory course in control systems. The text, in its first and second editions, has been used for a senior level course for engineering students at over one hundred colleges and universities. Also, it has been used for a course for engineering graduate students with no previous background in control system theory.

The text presumes a reasonable familiarity with the Laplace transformation and transfer functions as developed in a first course in linear system analysis or network analysis. These concepts are discussed in Chapter 2 and are used to develop mathematical models for control system components. Answers to selected problems are provided at the end of the book.

This material has been developed with the assistance of many individuals to whom I wish to express my sincere appreciation. Among those to whom I owe a particular debt of gratitude are Professors L. Gould, G. Thaler, and S. Weissenberger, and Mr. S. Mori. I wish to acknowledge the unflagging assistance of my secretaries, Mrs. M. McKenna, Mrs. M. Mahaffey, Mrs. B. Moore, and Mrs. P. Needle. Finally, I can only partially acknowledge the encouragement and patience of my wife, Joy, who helped to make this book possible.

Davis, California   R.C.D.
January 1980

Contents

9 Time-Domain Analysis of Control Systems

10 The Design and Compensation of Feedback Control Systems

11 Digital Control Systems

1 / Introduction to Control Systems

1.1 INTRODUCTION

Engineering is concerned with understanding and controlling the materials and forces of nature for the benefit of mankind. Control system engineers are concerned with understanding and controlling segments of their environment, often called *systems,* in order to provide useful economic products for society. The twin goals of understanding and control are complementary since, in order to control more effectively, the systems under control must be understood and modeled. Furthermore, control engineering often must consider the control of poorly understood systems such as chemical process systems. The present challenge to control engineers is the modeling and control of modern, complex, interrelated systems such as traffic-control systems, chemical processes, and economic regulation systems. However, simultaneously, the fortunate engineer has the opportunity to control many very useful and interesting industrial automation systems. Perhaps the most characteristic quality of control engineering is the opportunity to control machines, and industrial and economic processes for the benefit of society.

Control engineering is based on the foundations of feedback theory and linear system analysis, and integrates the concepts of network theory and communication theory. Therefore, control engineering is not limited to any engineering discipline but is equally applicable for aeronautical, chemical, mechanical, environmental, civil, and electrical engineering. For example, quite often a control system includes electrical, mechanical, and chemical components. Furthermore, as the understanding of the dynamics of business, social, and political systems increases, the ability to control these systems will increase also.

A *control system* is an interconnection of components forming a system configuration which will provide a desired system response. The basis for analysis of a system is the foundation provided by linear system theory which assumes a cause-

1

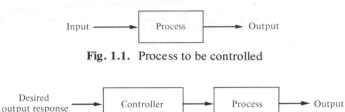

Fig. 1.1. Process to be controlled

Fig. 1.2. Open-loop control system.

effect relationship for the components of a system. Therefore, a component or process to be controlled can be represented by a block as shown in Fig. 1.1. The input-output relation represents the cause and effect relationship of the process, which in turn represents a processing of the input signal to provide an output signal variable, often with a power amplification. An *open-loop* control system utilizes a controller or control actuator in order to obtain the desired response as shown in Fig. 1.2.

In contrast to an open-loop control system, a closed-loop control system utilizes an additional measure of the actual output in order to compare the actual output with the desired output response. A simple *closed-loop feedback control system* is shown in Fig. 1.3. A standard definition of a feedback control system is as follows: A feedback control system is a control system which tends to maintain a prescribed relationship of one system variable to another by comparing functions of these variables and using the difference as a means of control.

A feedback control system often uses a function of a prescribed relationship between the output and reference input to control the process. Often the difference between the output of the process under control and the reference input is amplified and used to control the process so that the difference is continually reduced. The feedback concept has been the foundation for control system analysis and design.

' Due to the increasing complexity of the system under control and the interest in achieving optimum performance, the importance of control system engineering has grown in this decade. Furthermore, as the systems become more complex, the interrelationship of many controlled variables must be considered in the control scheme. A block diagram depicting a *multivariable control system* is shown in Fig. 1.4. A humorous example of a closed-loop feedback system is shown in Fig. 1.5.

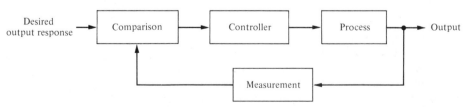

Fig. 1.3. Closed-loop feedback control system.

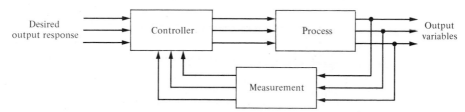

Fig. 1.4. Multivariable control system.

1.2 HISTORY OF AUTOMATIC CONTROL

The use of feedback in order to control a system has had a fascinating history. The first applications of feedback control rest in the development of float regulator mechanisms in Greece in the period 300 to 1 B.C. [1, 2]. The waterclock of Ktesibios used a float regulator (refer to Problem 1.11). An oil lamp devised by Philon in approximately 250 B.C. used a float regulator in an oil lamp for maintaining a constant level of fuel oil. Heron of Alexandria, who lived in the first century A.D., published a book entitled *Pneumatica* which outlined several forms of water level mechanisms using float regulators [1].

The first feedback system to be invented in modern Europe was the temperature regulator of Cornelis Drebbel (1572–1633) of Holland [1]. Dennis Papin [1647–1712] invented the first pressure regulator for steam boilers in 1681. Papin's pressure regulator was a form of safety regulator similar to a pressure-cooker valve.

The first automatic feedback controller used in an industrial process is generally agreed to be James Watt's flyball governor developed in 1769 for controlling the speed of a steam engine [1, 2]. The all-mechanical device, as shown in Fig. 1.6, measured the speed of the output shaft and utilized the movement of the flyball with

Fig. 1.5. Rube Goldberg's elaborate creations were almost all closed-loop feedback systems. Goldberg called this simply, "Be Your Own Dentist." (© Rube Goldberg, permission granted by King Features Syndicate, Inc., 1979.)

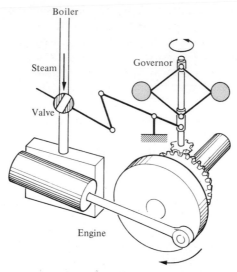

Fig. 1.6. Watt flyball governor.

speed to control the valve and therefore the amount of steam entering the engine. As the speed increases, the ball weights rise and move away from the shaft axis thus closing the valve. The flyweights require power from the engine in order to turn and therefore make the speed measurement less accurate.

The first historical feedback system claimed by the Soviet Union is the water level float regulator said to have been invented by I. Polzunov in 1765 [4]. The level regulator system is shown in Fig. 1.7. The float detects the water level and controls the valve which covers the water inlet in the boiler.

The period preceding 1868 was characterized by the development of automatic control systems by intuitive invention. Efforts to increase the accuracy of the con-

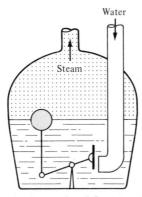

Fig. 1.7. Water-level float regulator.

trol system led to slower attenuation of the transient oscillations and even to unstable systems. It then became imperative to develop a theory of automatic control. J. C. Maxwell formulated a mathematical theory related to control theory using a differential equation model of a governor [5]. Maxwell's study was concerned with the effect various system parameters had on the system performance. During the same period, I. A. Vyshnegradskii formulated a mathematical theory of regulators [6].

Prior to World War II, control theory and practice developed in the United States of America and Western Europe in a different manner than in the U.S.S.R. and Eastern Europe. One main impetus for the use of feedback in the U.S.A. was the development of the telephone system and electronic feedback amplifiers by Bode, Nyquist, and Black at the Bell Telephone Laboratories [7, 8, 9, 10, 12]. The frequency domain was used primarily to describe the operation of the feedback amplifiers in terms of bandwidth and other frequency variables. In contrast, the eminent mathematicians and applied mechanicians in Russia inspired and dominated the field of control theory. Therefore, the Russian theory tended to utilize a time-domain formulation using differential equations.

A large impetus to the theory and practice of automatic control occurred during World War II when it became necessary to design and construct automatic airplane pilots, gun-positioning systems, radar antenna control systems, and other military systems based on the feedback control approach. The complexity and expected performance of these military systems necessitated an extension of the available control techniques and fostered interest in control systems and the development of new insights and methods. Prior to 1940, for most cases, the design of control systems was an art involving a trial and error approach. During the decade of the 1940's, mathematical and analytical methods increased in number and utility, and control engineering became an engineering discipline in its own right [10, 11, 12].

Frequency-domain techniques continued to dominate the field of control following World War II with the increased use of the Laplace transform and the complex frequency plane. During the 1950s, the emphasis in control engineering theory was on the development and use of the s-plane methods and, particularly, the root locus approach. Furthermore, during the 1950s, the utilization of both analog and digital computers for control components became possible. These new controlling elements possessed an ability to calculate rapidly and accurately which was formerly not available to the control engineer. There are now over thirty thousand digital process control computers installed in the United States [13, 14]. These computers are employed especially for process control systems in which many variables are measured and controlled simultaneously by the computer.

With the advent of Sputnik and the space age, another new impetus was imparted to control engineering. It became necessary to design complex, highly accurate control systems for missiles and space probes. Furthermore, the necessity to minimize the weight of satellites and to control them very accurately has spawned the important field of optimal control. Due to these requirements, the time-domain methods due to Liapunov, Minorsky, and others have met with great interest in the last decade [15]. Furthermore, new theories of optimal control have been developed by L. S. Pontryagin in Russia and R. Bellman in the U.S.A. It now appears that control engineering must consider both the time-domain and the frequency-domain approaches simultaneously in the analysis and design of control systems.

1.3 CONTROL ENGINEERING PRACTICE

Control engineering is concerned with the analysis and design of goal-oriented systems. Therefore, the mechanization of goal-oriented policies has grown into a hierarchy of goal-oriented control systems. Modern control theory is concerned with systems with the self-organizing, adaptive, learning, and optimum qualities. This interest has aroused even greater excitement among control engineers.

The control of an industrial process (manufacturing, production, etc.) by automatic rather than human means is often called *automation*. Automation is prevalent in the chemical, electric power, paper, automobile, and steel industries, among others. The concept of automation is central to our industrial society. Automatic machines are used to increase the production of a plant per worker in order to offset rising wages and inflationary costs. Thus, industries are concerned with the productivity per worker of their plant. *Productivity* is defined as the ratio of physical output to physical input. In this case we are referring to labor productivity, which is real output per hour of work. In a study conducted by the U.S. Commerce Department it was determined that labor productivity grew at an average annual rate of 2.8% from 1948 to 1978 [20]. In order to continue these productivity gains, expenditures for factory automation in the United States are expected to double from 1.5 billion dollars in 1978 to 3.0 billion dollars in 1984 [13, 14]. World-wide, expenditures for process control and manufacturing plant control are expected to grow from 4.4 billion dollars in 1976 to 9.5 billion dollars in 1986 [22]. The U.S. manufacturers currently supply approximately one-half of worldwide control equipment.

There are about 120,000 control engineers in the United States and over 100,000 control engineers in the Soviet Union. In the United States alone, the control industry does a business of over eight billion dollars per year! The theory, practice, and application of automatic control is a large, exciting, and extremely useful engineering discipline. One can readily understand the motivation for a study of modern control systems.

1.4 EXAMPLES OF MODERN CONTROL SYSTEMS

Feedback control is a fundamental fact of modern industry and society. Driving an automobile is a pleasant task when the auto responds rapidly to the driver's com-

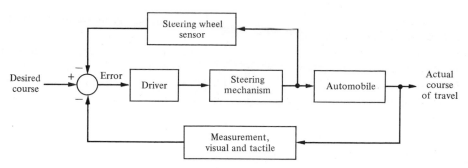

Fig. 1.8. Automobile steering control system.

mands. Many cars have power steering and brakes which utilize hydraulic amplifiers for amplification of the force to the brakes or the steering wheel. A simple
block diagram of an automobile steering control system is shown in Fig. 1.8. The
desired course is compared with a measurement of the actual course in order to
generate a measure of the error. This measurement is obtained by visual and tactile
(body movement) feedback. There is an additional feedback from the feel of the
steering wheel by the hand (sensor). This feedback system is a familiar version of
the steering control system in an ocean liner or the flight controls in a large airplane.
All these systems operate in a closed-loop sequence as shown in Fig. 1.9. The actual
and desired outputs are compared and a measure of the difference is used to drive
the power amplifier. The power amplifier causes the actuator to modulate the process in order to reduce the error. The sequence is such that if the ship is heading
incorrectly to the right, the rudder is actuated in order to direct the ship to the left.
The system shown in Fig. 1.9 is a *negative feedback* control system, since the
output is subtracted from the input and the difference is used as the input signal to
the power amplifier.

A basic manually controlled closed-loop system for regulating the level of fluid
in a tank is shown in Fig. 1.10. The input is a reference level of fluid that the operator is instructed to maintain. (This reference is memorized by the operator.) The
power amplifier is the operator and the sensor is visual. The operator compares the
actual level with the desired level and opens or closes the valve (actuator) to maintain the desired level.

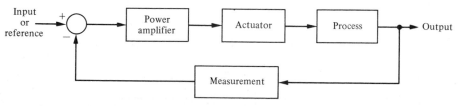

Fig. 1.9. Basic closed-loop control system.

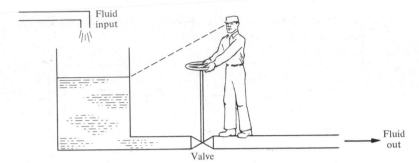

Fig. 1.10. A manual control system for regulating the level of fluid in a tank by adjusting the output valve. The operator views the level of fluid through a port in the side of the tank.

Other very familiar control systems have the same basic elements as the system shown in Fig. 1.9. A refrigerator has a temperature setting or desired temperature, a thermostat to measure the actual temperature and the error, and a compressor motor for power amplification. Other examples in the home are the oven, furnace, and water heater. In industry, there are speed controls, process temperature and pressure controls, position, thickness, composition, and quality controls among many others [26, 27, 28].

In order to provide a mass transportation system for modern urban areas, a large, complex, high-speed system is necessary. Several automatic train systems are being designed for trains to run at 2-minute intervals at high speeds. Automatic control is necessary in order to maintain a constant flow of trains and for comfortable deceleration and braking conditions at stations. The block diagram of a train control system is shown in Fig. 1.11(a) [11]. A measurement of the distance from the station and the speed of the train is used to determine the error signal and therefore the braking signal. The central control room of the 75-mile Bay Area Rapid Transit system is shown in Fig. 1.11(b). BART uses a computer control system for control of the trains.

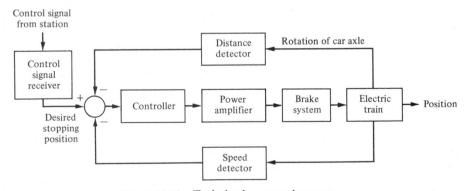

Fig. 1.11(a). Train-brake-control system.

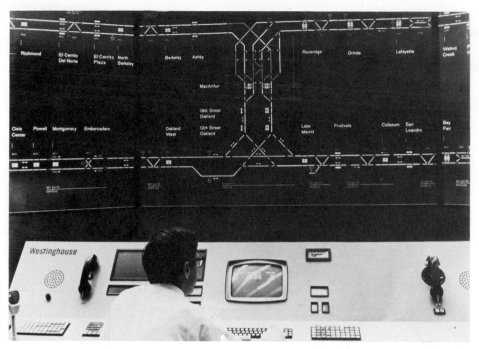

Fig. 1.11(b). The central control room of the Bay Area Rapid Transit system. (Photo courtesy of Westinghouse Corp.)

There has been considerable discussion recently concerning the gap between practice and theory in control engineering. However, it would be natural that theory precedes the applications in many fields of control engineering. Nonetheless, it is interesting to note that in the electric power industry, the largest industry in the United States, the gap is relatively insignificant [29, 30, 31]. The electric power industry is primarily interested in energy conversion, control, and distribution. It is critical that computer control be increasingly applied to the power industry in order to improve the efficiency of use of energy resources. Also, the control of power plants for minimum waste emission has become increasingly important [24]. The modern large capacity plants which exceed several hundreds of megawatts require automatic control systems which account for the interrelationship of the process variables and the optimum power production. It is common to have as many as ninety or more manipulated variables under coordinated control. A simplified model showing several of the important control variables of a large boiler-generator system is shown in Fig. 1.12. This is an example of the importance of measuring many variables, such as pressure and oxygen, in order to provide information to the computer for control calculations. It is estimated that more than thirty thousand computer control systems have been installed in the United States [13, 14]. The diagram

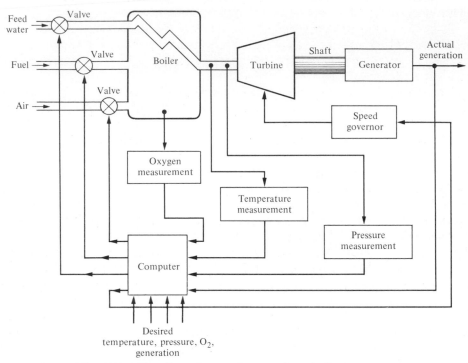

Fig. 1.12. Coordinated control system for a boiler-generator.

of a computer control system is shown in Fig. 1.13. The electric power industry has utilized the modern aspects of control engineering for significant and interesting applications. It appears that in the process industry, the factor that maintains the applications gap is the lack of instrumentation to measure all the important process variables, including the quality and composition of the product. As these instruments become available, the applications of modern control theory to industrial systems should increase measurably.

Another very important industry, the metallurgical industry, has had considerable success in automatically controlling its processes. In fact, in many cases, the control applications are beyond the theory [28]. For example, a hot strip steel mill

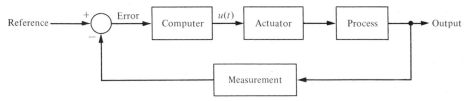

Fig. 1.13. A computer control system.

which involves a 100-million-dollar investment is controlled for temperature, strip width, thickness, and quality.

Rapidly rising energy costs coupled with threats of energy curtailment are resulting in new efforts for efficient automatic energy management. Computer controls are used to control energy use in industry and stabilize and connect loads evenly to gain fuel economy [32, 33].

There has been considerable interest recently in applying the feedback control concepts to automatic warehousing and inventory control [23, 28]. Furthermore, automatic control of agricultural systems (farms) is meeting increased interest. Automatically controlled silos and tractors have been developed and tested. Automatic control of wind turbine generators, solar heating and cooling, and automobile engine performance are important modern examples [34, 35].

Also, there have been many applications of control system theory to biomedical experimentation, diagnosis, prosthetics, and biological control systems [36, 37]. The control systems under consideration range from the cellular level to the central nervous system, and include temperature regulation, neurological, respiratory, and cardiovascular control. Most physiological control systems are closed-loop systems. However, we find not one controller but rather control loop within control loop forming a hierarchy of systems. The modeling of the structure of biological processes confronts the analyst with a high-order model and a complex structure. Prosthetic devices that aid the 46 million handicapped individuals in the U.S. are designed to provide automatically controlled aids to the disabled [38, 39, 40, 41]. An artificial hand that uses force feedback signals and is controlled by the amputee's bioelectric control signals, which are called electromyographic signals, is shown in Fig. 1.14.

Finally, it has become of interest and value to attempt to model the feedback processes prevalent in the social, economic, and political spheres. This approach is undeveloped at present but appears to have a reasonable future. Society, of course, is comprised of many feedback systems and regulatory bodies such as the Interstate Commerce Commission and the Federal Reserve Board which are controllers exerting the necessary forces on society in order to maintain a desired output [42]. A simple lumped model of the national income feedback control system is shown in Fig. 1.15. This type of model helps the analyst to understand the effects of government control—granted its existence—and the dynamic effects of government spending. Of course, many other loops not shown also exist, since, theoretically, government spending cannot exceed the tax collected without a deficit, which is itself a control loop containing the Internal Revenue Service and the Congress. Of course, in a communist country the loop due to consumers is deemphasized and the government control is emphasized. In that case, the measurement block must be accurate and must respond rapidly; both are very difficult characteristics to realize from a bureaucratic system. This type of political or social feedback model, while usually nonrigorous, does impart information and understanding.

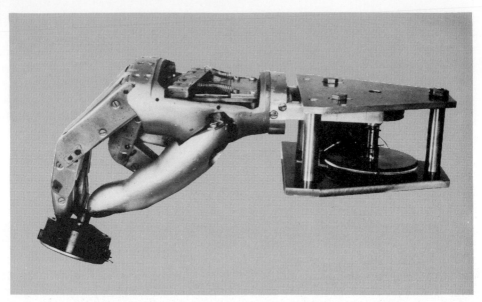

Fig. 1.14. An artificial hand developed by Professor Tomovic is operated by electro-myographic signals from the amputee's stump. An amputee can grip a glass without breaking it because the hand has force feedback. (Photo courtesy of Professor R. Tomovic, University of Belgrade, Yugoslavia.)

Feedback control systems are used extensively in industrial applications. An industrial manipulator controlled by a human operator is shown in Fig. 1.16. There are thousands of industrial robots currently in use. Manipulators can pick up objects weighing hundreds of pounds and position them with an accuracy of one-tenth of

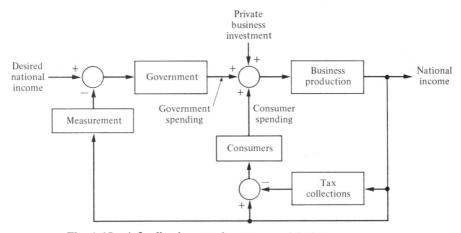

Fig. 1.15. A feedback control system model of the economy.

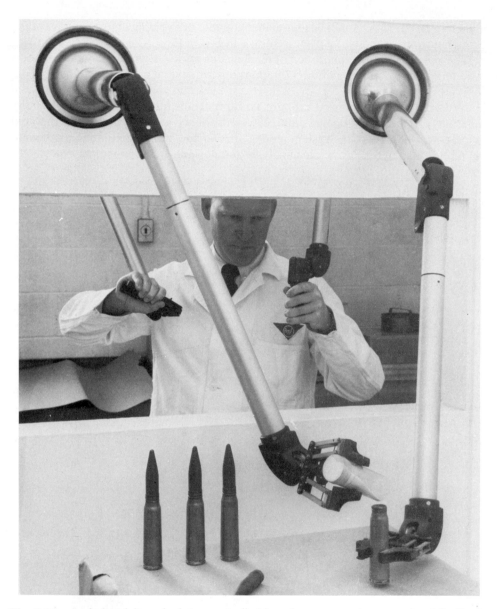

Fig. 1.16. An industrial manipulator controlled by a human operator uses visual feedback. This is the Mini-Manip, manufactured by AMF Co. (Photo courtesy of American Machine and Foundry Co.)

an inch or better [44]. A mobile automaton capable of avoiding objects and traveling through a room or industrial plant is shown in Fig. 1.17.

The potential future application of feedback control systems and models appears to be unlimited. Estimates of the U.S. markets for several control systems applications are given in Table 1.1 [21, 22]. It appears that the theory and practice of modern control systems have a bright and important future, and certainly justify the study of modern automatic control system theory and application. In the next chapter, we shall study the system models further to obtain quantitative mathematical models useful for engineering analysis and design.

Fig. 1.17. The automaton vehicle used by SRI in the application of artificial intelligence principles to the development of integrated robot systems. The vehicle is propelled by electric motors and carries a television camera and optical range finder in the movable "head." The vehicle responds to commands from a computer. The sensors are the bump detector, the TV camera, and the range finder. [43] (Courtesy of Stanford Research Institute.)

Table 1.1

Application	(Millions of dollars)*			
	1972	1973	1976	1980
Motor controls (speed, position)	90.3	100.5	112	150
Numerical controls	43.4	47.3	76	100
Thickness controls (steel, paper)	45.4	57.8	99	180
Process controls (oil, chemical)	318.5	357.2	449	700
Pollution monitoring and control	14.0	17.0	26	75
Nuclear reactor control	9.3	11.1	19	25

*U.S. Market estimates for several control system applications. The examples given in parentheses are not all-inclusive of the applications.

PROBLEMS

1.1. Draw a schematic block diagram of a home heating system. Identify the function of each element of the thermostatically controlled heating system.

1.2. Control systems have, in the past, used a human operator as part of a closed-loop control system. Draw the block diagram of the valve control system shown in Fig. P1.2.

1.3. In a chemical process control system, it is valuable to control the chemical composition of the product. In order to control the composition, a measurement of the composition may be obtained by using an infrared stream analyzer as shown in Fig. P1.3. The valve on the additive stream may be controlled. Complete the control feedback loop and draw a block diagram describing the operation of the control loop.

1.4. The accurate control of a nuclear reactor is important for power system generators. Assuming the number of neutrons present is proportional to the power level, an ionization chamber is used to measure the power level. The current, i, is proportional to the power level. The position of the graphite control rods moderates the power level. Complete the control system of the nuclear reactor shown in Fig. P1.4 and draw the block diagram describing the operation of the control loop.

1.5. A light-seeking control system is shown in Fig. P1.5. The output shaft, driven by the

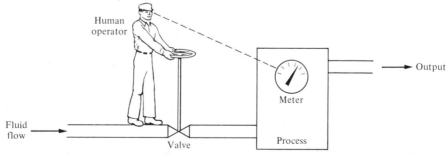

Figure P1.2

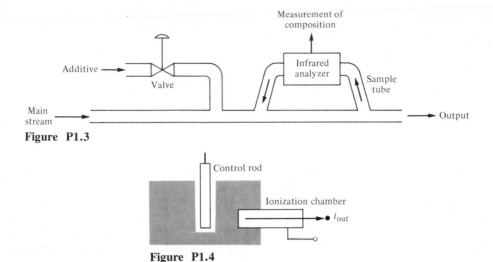

Figure P1.3

Figure P1.4

motor through a worm reduction gear, has a bracket attached on which are mounted two photocells. Complete the closed-loop system in order that the system follows the light source.

1.6. Feedback systems are not always negative feedback systems in nature. Economic inflation, which is evidenced by continually rising prices, is a positive feedback system. A positive feedback control system, as shown in Fig. P1.6, *adds* the feedback signal to the input signal and the resulting signal is used as the input to the process. A simple model of the price-wage inflationary spiral is shown in Fig. P1.6. Add additional feedback loops, such as legislative control or control of the tax rate, in order to stabilize the system. It is assumed that an increase in workers' salaries, after some time delay, results in an increase in prices. Under what conditions could prices be stabilized by falsifying or delaying the availability of cost of living data? How did the President's wage and price economic guideline program of 1979–1980 affect the feedback system?

1.7. The story is told about the Sergeant who stopped at the jewelry store every morning at 9 o'clock and compared and reset his watch with the chronometer in the window. Finally, one day the Sergeant went into the store and complimented the owner on the accuracy of the chronometer.

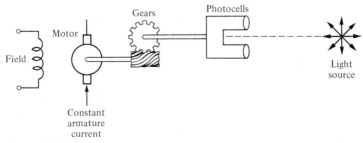

Figure P1.5

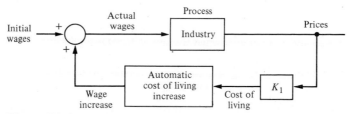

Figure P1.6

"Is it set according to time signals from Arlington?" asked the Sergeant.

"No," said the owner, "I set it by the 5 o'clock cannon fired from the fort. Tell me, Sergeant, why do you stop every day and check your watch?"

The Sergeant replied, "I'm the gunner at the fort!"

Is the feedback prevalent in this case positive or negative? The jeweler's chronometer loses one minute each 24-hour period and the Sergeant's watch loses one minute during each 8 hours. What is the net time error of the cannon at the fort after 15 days?

1.8. The student-teacher learning process is inherently a feedback process intended to reduce the system error to a minimum. The desired output is the knowledge being studied and the student may be considered the process. With the aid of Fig. 1.3, construct a feedback model of the learning process and identify each block of the system.

1.9. Models of physiological control systems are valuable aids to the medical profession. A model of the heart rate control system is shown in Fig. P1.9 [45]. This model includes the processing of the nerve signals by the brain. The heart rate control system is, in fact, a multivariable system and the variables x, y, w, v, z, and u are vector variables. In other words, the variable x represents many heart variables x_1, x_2, ..., x_n. Examine the model of the heart rate control system and add or delete the blocks, if necessary. Determine a control system model of one of the following physiological control systems:

1. respiratory control system,
2. adrenalin control system,
3. human arm control system,
4. eye control system,
5. pancreas and the blood-sugar level control system,
6. circulatory system.

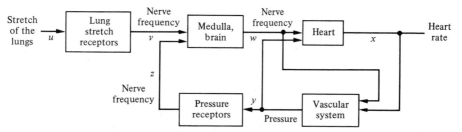

Figure P1.9

1.10. The role of air traffic control systems is increasing as airplane traffic increases at busy airports. A tragic example of a traffic control mishap was the collision of the Pacific Southwest Airways 727 and a privately owned Cessna at San Diego airport in October 1978 [47]. Engineers are developing flight control systems, air traffic control systems and collision avoidance systems [48]. Investigate these and other systems designed to improve air traffic safety; select one and draw a simple block diagram of its operation.

1.11. Automatic control of water level using a float level was used in the Middle East for a water clock [1, 11]. The water clock, shown in Fig. P1.11, was used from sometime before Christ until the 17th century. Discuss the operation of the water clock and establish how the float provides a feedback control which maintains the accuracy of the clock.

1.12. An automatic turning gear for windmills was invented by Meikle about 1750 [1, 11]. The fantail gear shown in Fig. P1.12 automatically turns the windmill into the wind. The fantail windmill at right angle to the mainsail was used to turn the turret. The gear ratio was of the order of 3000 to 1. Discuss the operation of the windmill and establish the feedback operation that maintains the main sails into the wind.

1.13. The Environmental Protection Agency is developing pollution emission standards to be applied to power plants across the nation. With established standards of waste discharges to the air and water, a power plant will have to be monitored by environmental sensors. Assume the plant has air and water quality measurement devices. Devise a power plant control system which will monitor and maintain operation of the plant within the quality standards. You may assume that the plant uses a fossil-fuel to generate electricity [24].

1.14. Adam Smith (1723–1790) discussed the issue of free competition between the participants of an economy in his book *Wealth of Nations*. It may be said that Smith employed social feedback mechanisms to explain his theories [46]. Smith suggests that (1) the available workers as a whole compare the various possible employments and enter that one which

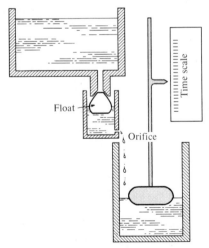

Fig. P1.11. (From Newton, Gould, and Kaiser, *Analytical Design of Linear Feedback Controls*. Wiley, New York, 1957, with permission.)

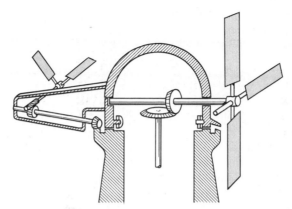

Fig. P1.12. (From Newton, Gould, and Kaiser, *Analytical Design of Linear Feedback Controls*. Wiley, New York, 1957, with permission.)

offers the greatest rewards; and (2) in any employment the rewards diminish as the number of competing workers rises. Let r = total of rewards averaged over all trades; c = total of rewards in a particular trade; q = influx of workers into the specific trade. Draw a feedback loop to represent this system.

1.15. The law of supply and demand is a basic law in a free economy. This basic law may be represented as a feedback system with the output of the system as the actual market or selling price of a specific item. The law states that the market demand for the item decreases as its price increases. Also the law states that a stable market price is achieved if and only if the supply is equal to the demand. Draw a block diagram which includes the following four blocks: supplier, demander, pricer, market. The desired command input is a market price change equal to zero ($r = 0$).

REFERENCES

1. O. Mayr, *The Origins of Feedback Control,* M.I.T. Press, Cambridge, Mass., 1970.

2. O. Mayr, "The Origins of Feedback Control," *Scientific American,* **223,** 4, October 1970; pp. 110–118.

3. O. Mayr, *Feedback Mechanisms in the Historical Collections of the National Museum of History and Technology,* Smithsonian Institution Press, Washington, D.C., 1971.

4. E. P. Popov, *The Dynamics of Automatic Control Systems,* Gostekhizdat, Moscow, 1956; Addison-Wesley, Reading, Mass., 1962.

5. J. C. Maxwell, "On Governors," *Proc. of the Royal Society of London,* 16, 1868; in *Selected Papers on Mathematical Trends in Control Theory*. Dover, New York, 1964; pp. 270–283.

6. I. A. Vyshnegradskii, "On Controllers of Direct Action," *Izv. SPB Tekhnolog. Inst.,* 1877.

7. H. W. Bode, "Feedback—The History of an Idea," in *Selected Papers on Mathematical Trends in Control Theory*. Dover, New York, 1964; pp. 106–123.

8. H. S. Black, "Inventing the Negative Feedback Amplifier," *IEEE Spectrum,* December 1977; pp. 55–60.

9. J. E. Brittain, *Turning Points in American Electrical History,* IEEE Press, New York, 1977, Sect. II-E.

10. G. J. Thaler, *Automatic Control: Classical Linear Theory,* Dowden, Hutchinson, and Ross, Inc., Stroudsburg, Pa., 1974.

11. G. Newton, L. Gould, and J. Kaiser, *Analytical Design of Linear Feedback Controls,* Wiley, New York, 1957.

12. M. D. Fagen, *A History of Engineering and Science on the Bell System,* Bell Telephone Laboratories, 1978; Ch. 3.

13. L. B. Evans, "Impact of the Electronics Revolution on Industrial Process Control," *Science,* March 18, 1977; pp. 1146–1151.

14. R. Sugarman, "Electrotechnology to the Rescue," *IEEE Spectrum,* October 1978; pp. 53–73.

15. A. P. Sage and C. C. White, *Optimum Systems Control,* 2nd ed., Prentice-Hall, Englewood Cliffs, N.J., 1977.

16. R. H. Maskrey and W. J. Thayer, "A Brief History of Electrohydraulic Servomechanisms," *ASME J. of Dynamic Systems, Measurement and Control,* June 1978; pp. 110–116.

17. T. E. Fortmann and K. L. Hitz, *An Introduction to Linear Control Systems,* Marcel Dekker, Inc., New York, 1977.

18. C. N. Day, "Modern Control Applications to Manual Control—Historical Perspective and Future Direction," *Proceedings of IEEE Conference on Decision and Control,* December 1977; pp. 226–229.

19. C. R. Walker, *Toward the Automotive Factory,* Greenwood Press, Westport, Conn., 1977.

20. R. Allan, "Electronics to Boost Productivity," *IEEE Spectrum,* January 1978; pp. 45–48.

21. "13% Growth to Push Work Equipment Sales Past $136 Billion," *Electronics,* January 4, 1979; pp. 105–113.

22. "World Process Control Market to be $9.5 Billion by 1986," *Control Engineering,* August 1978; p. 9.

23. P. E. Caines, "Society, Research Policy and Systems Science," *Proceedings of the IEEE Conference on Decision and Control,* December 1977; pp. 455–458.

24. G. J. Vachtsevanos, "Simulation Studies in the Energy-Environment Interface," *Proceedings of the IEEE Conference on Decision and Control,* December 1977; pp. 1004–1008.

25. N. Rosenberg, *Technology and American Economic Growth,* Sharpe, Inc., White Plains, N.Y., 1977.

26. E. J. Kompass, "Control in the Paper Industry up to the Minute," *Control Engineering,* December 1978; pp. 44–46.

27. D. Tesar, "Mission-Oriented Research for Light Machinery," *Science,* September 1978; pp. 880–886.

28. C. F. Carter, Jr., "An Overview of Manufacturing Problems and Control Solutions," *Proceedings of the 1976 JACC American Society of Mechanical Engineers,* New York, 1976; pp. 12–16.

29. F. C. Schweppe, "Power Systems 2000," *IEEE Spectrum,* July 1978; pp. 42–53.

30. F. D. Boardman, "Control and Operation of the UK Electricity Supply System," *Proceedings of the IEEE,* January 1978; pp. 61–65.

31. B. Frogner and H. S. Rao, "Control of Nuclear Power Plants," *IEEE Transactions on Automatic Control,* June 1978; pp. 405–415.

32. A. Kaya, "Industrial Energy Control," *IEEE Spectrum,* July 1978; pp. 48–53.

33. C. H. Cho and J. Lipe, "Managing Energy With Computers," *Instruments and Control Systems,* April 1978; pp. 35–38.

34. L. B. Anderson and H. E. Rauch, "Application of Optimization Techniques to Solar Heating and Cooling," *J. of Energy,* February 1977; pp. 18–24.

35. T. O. Jones, "Some Recent and Future Automotive Electronic Developments," *Science,* March 18, 1977; pp. 1156–1160.

36. E. Satinoff, "Neural Organization and Evolution of Thermal Regulation in Mammals," *Science,* July 1978; pp. 16–22.

37. L. R. Young, "Man's Internal Navigation System," *Technology Review,* May 1978; pp. 41–45.

38. M. Pines, "Modern Bioengineers Reinvent Human Anatomy With Spare Parts," *Smithsonian,* November 1978; pp. 50–57.

39. S. Solomon, "Spare Parts for Humans," *Forbes,* May 29, 1978; pp. 52–54.

40. W. J. Spencer, "For Diabetics: An Electronic Pancreas," *IEEE Spectrum,* June 1978; pp. 38–42.

41. M. D. Zimmerman, "Rehabilitation Engineering," *Machine Design,* November 9, 1978; pp. 24–29.

42. J. D. Pitchford and S. J. Turnovsky, *Application of Control Theory to Economic Analysis,* Elsevier, New York, 1977.

43. R. C. Dorf, *Introduction to Computers and Computer Science,* Boyd and Fraser Publishing Co., San Francisco, 2nd ed., 1977; Chaps. 13, 14.

44. R. Malone, *The Robot Book,* Harcourt, Brace, Jovanovich, New York, 1978.

45. R. C. Dorf and J. Unmack, "A Time-Domain Model of the Heart Rate Control System," *Proceedings of the San Diego Symposium for Biomedical Engineering,* 1965; pp. 43–47.

46. O. Mayr, "Adam Smith and the Concept of the Feedback System," *Technology and Culture,* **12,** 1, January 1971; pp. 1–22.

47. K. Labich, "Collision Course," *Newsweek,* October 9, 1978; pp. 48–53.

48. P. R. Kurzhals, "New Directions in Civil Avionics," *Astronautics and Aeronautics,* March 1978; pp. 38–40.

2 / Mathematical Models of Systems

2.1 INTRODUCTION

In order to understand and control complex systems, one must obtain quantitative *mathematical models* of these systems. Therefore it is necessary to analyze the relationships between the system variables and to obtain a mathematical model. Since the systems under consideration are dynamic in nature, the descriptive equations are usually *differential equations*. Furthermore, if these equations can be *linearized*, then the *Laplace transform* can be utilized in order to simplify the method of solution. In practice, the complexity of systems and the ignorance of all the relevant factors necessitate the introduction of *assumptions* concerning the system operation. Therefore we shall often find it useful to consider the physical system, delineate some necessary assumptions, and linearize the system. Then, by using the physical laws describing the linear equivalent system, we can obtain a set of linear differential equations. Finally, utilizing mathematical tools, such as the Laplace transform, we obtain a solution describing the operation of the system. In summary, the approach to dynamic system problems can be listed as follows:

1. define the system and its components;
2. formulate the mathematical model and list the necessary assumptions;
3. write the differential equations describing the model;
4. solve the equations for the desired output variables;
5. examine the solutions and the assumptions; and then
6. reanalyze or design.

2.2 DIFFERENTIAL EQUATIONS OF PHYSICAL SYSTEMS

The differential equations describing the dynamic performance of a physical system are obtained by utilizing the physical laws of the process [1, 39, 40]. This approach

Table 2.1 Summary of Through- and Across-Variables for Physical Systems .

System	Variable Through Element	Integrated Through Variable	Variable Across Element	Integrated Across Variable
Electrical	Current, i	Charge, q	Voltage difference, v_{21}	Flux linkage, λ_{21}
Mechanical translational	Force, F	Translational momentum, P	Velocity difference, v_{21}	Displacement difference, y_{21}
Mechanical rotational	Torque, T	Angular momentum, h	Angular velocity difference, ω_{21}	Angular displacement difference, θ_{21}
Fluid	Fluid volumetric rate of flow, Q	Volume, V	Pressure difference, P_{21}	Pressure momentum, γ_{21}
Thermal	Heat flow rate, q	Heat energy, H	Temperature difference, $\mathfrak{I}_{21}$	

applies equally well to mechanical [2], electrical [3], fluid, and thermo-dynamic systems [4]. A summary of the variables of dynamic systems is given in Table 2.1 [5]. We prefer to use the International System of units (SI) in contrast to the British system of units. The International System of units is given in Table 2.2. The conversion of other systems of units to SI units is facilitated by Table 2.3. A summary of the describing equations for lumped, linear, dynamic elements is given in Table 2.4 [5]. The equations in Table 2.4 are idealized descriptions and only approximate the actual conditions (for example, when a linear, lumped approximation is used for a distributed element).

Table 2.2 The International System of Units (SI)

	Unit	Symbol
Basic Units		
Length	meter	m
Mass	kilogram	kg
Time	second	s
Temperature	kelvin	K
Electric current	ampere	A
Derived Units		
Velocity	meters per second	m/s
Area	square meter	m^2
Force	newton	$N = kgm/s^2$
Torque	kilogram-meter	kgm
Pressure	pascal	Pa
Energy	joule	$J = Nm$
Power	watt	$W = J/s$

Table 2.3 Conversion Factors for Converting to SI Units

From	Multiply by	To obtain
Length		
inches	25.4	millimeters
feet	30.48	centimeters
Speed		
miles per hour	0.4470	meters per second
Mass		
pounds	0.4536	kilograms
Force		
pounds-force	4.448	newtons
Torque		
foot-pounds	0.1383	kilogram-meters
Power		
horsepower	746	watts
Energy		
British thermal unit	1055	joules
kilowatt-hour	3.6×10^6	joules

Nomenclature

Through-variable: F = force, T = torque, i = current, Q = fluid volumetric flow rate, q = heat flow rate.

Across-variable: v = translational velocity, ω = angular velocity, v = voltage, P = pressure, $\mathfrak{I}$ = temperature.

Inductive storage: L = inductance, $1/k$ = reciprocal translational or rotational stiffness, I = fluid inertance.

Capacitive storage: C = capacitance, M = mass, J = moment of inertia, C_f = fluid capacitance, C_t = thermal capacitance.

Energy dissipators: R = resistance, f = viscous friction, R_f = fluid resistance, R_t = thermal resistance.

The symbol $v(t)$ is used for both voltage in electrical circuits and velocity in translational mechanical systems, and is distinguished within the context of each differential equation. For mechanical systems, one utilizes Newton's laws, and for electrical systems, Kirchhoff's voltage laws. For example, the simple spring-mass-damper mechanical system shown in Fig. 2.1 is described by Newton's second law of motion. Therefore, we obtain

$$M \frac{d^2y(t)}{dt^2} + f \frac{dy(t)}{dt} + Ky(t) = r(t), \tag{2.1}$$

where K is the spring constant of the ideal spring and f is the friction constant. Equation 2.1 is a linear constant coefficient differential equation of second order.

Table 2.4 Summary of Describing Differential Equations for Ideal Elements

Type of element	Physical element	Describing equation	Energy E or power $\mathcal{P}$	Symbol
Inductive storage	Electrical inductance	$v_{21} = L\dfrac{di}{dt}$	$E = \dfrac{1}{2}Li^{.2}$	
	Translational spring	$v_{21} = \dfrac{1}{K}\dfrac{dF}{dt}$	$E = \dfrac{1}{2}\dfrac{F^2}{K}$	
	Rotational spring	$\omega_{21} = \dfrac{1}{K}\dfrac{dT}{dt}$	$E = \dfrac{1}{2}\dfrac{T^2}{K}$	
	Fluid inertia	$P_{21} = I\dfrac{dQ}{dt}$	$E = \dfrac{1}{2}IQ^2$	
Capacitive storage	Electrical capacitance	$i = C\dfrac{dv_{21}}{dt}$	$E = \dfrac{1}{2}Cv_{21}^2$	
	Translational mass	$F = M\dfrac{dv_2}{dt}$	$E = \dfrac{1}{2}Mv_2^2$	
	Rotational mass	$T = J\dfrac{d\omega_2}{dt}$	$E = \dfrac{1}{2}J\omega_2^2$	
	Fluid capacitance	$Q = C_f\dfrac{dP_{21}}{dt}$	$E = \dfrac{1}{2}C_fP_{21}^2$	
	Thermal capacitance	$q = C_t\dfrac{d\mathfrak{I}_2}{dt}$	$E = C_t\mathfrak{I}_2$	
Energy dissipators	Electrical resistance	$i = \dfrac{1}{R}v_{21}$	$\mathcal{P} = \dfrac{1}{R}v_{21}^2$	
	Translational damper	$F = fv_{21}$	$\mathcal{P} = fv_{21}^2$	
	Rotational damper	$T = f\omega_{21}$	$\mathcal{P} = f\omega_{21}^2$	
	Fluid resistance	$Q = \dfrac{1}{R_f}P_{21}$	$\mathcal{P} = \dfrac{1}{R_f}P_{21}^2$	
	Thermal resistance	$q = \dfrac{1}{R_t}\mathfrak{I}_{21}$	$\mathcal{P} = \dfrac{1}{R_t}\mathfrak{I}_{21}$	

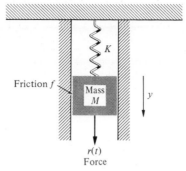

Fig. 2.1. Spring-mass-damper system.

Alternatively, one may describe the electrical *RLC* circuit of Fig. 2.2 by utilizing Kirchhoff's current law. Then we obtain the following integrodifferential equation:

$$\frac{v(t)}{R} + C\frac{dv(t)}{dt} + \frac{1}{L}\int_0^t v(t)\,dt = r(t). \tag{2.2}$$

The solution of the differential equation describing the process may be obtained by classical methods such as the use of integrating factors and the method of undetermined coefficients [1]. For example, when the mass is displaced initially a distance $y(t) = y(0)$ and released, the dynamic response of an *underdamped* system is represented by an equation of the form

$$y(t) = K_1 e^{-\alpha_1 t}\sin(\beta_1 t + \theta_1). \tag{2.3}$$

A similar solution is obtained for the voltage of the *RLC* circuit when the circuit is subjected to a constant current $r(t) = I$. Then the voltage is

$$v(t) = K_2 e^{-\alpha_2 t}\cos(\beta_2 t + \theta_2). \tag{2.4}$$

A voltage curve typical of an underdamped *RLC* circuit is shown in Fig. 2.3.

In order to further reveal the close similarity between the differential equations for the mechanical and electrical systems, we shall rewrite Eq. (2.1) in terms of velocity,

$$v(t) = \frac{dy(t)}{dt}$$

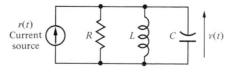

Fig. 2.2. *RLC* circuit.

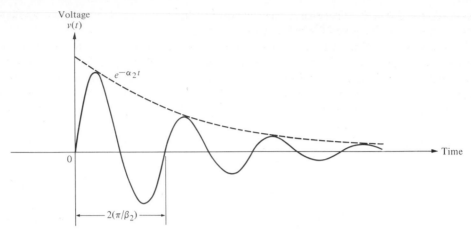

Fig. 2.3. Typical voltage curve for underdamped RLC circuit.

Then we have

$$M \frac{dv(t)}{dt} + fv(t) + K \int_0^t v(t) \, dt = r(t). \tag{2.5}$$

One immediately notes the equivalence of Eqs. (2.5) and (2.2) where velocity $v(t)$ and voltage $v(t)$ are equivalent variables, usually called *analogous* variables, and the systems are analogous systems. Therefore the solution for velocity is similar to Eq. (2.4) and the curve for an underdamped system is shown in Fig. 2.3. The concept of analogous systems is a very useful and powerful technique for system modeling. The voltage-velocity analogy, often called the force-current analogy, is a natural analogy since it relates the analogous through- and across-variables of the electrical and mechanical systems. However, another analogy which relates the velocity and current variables is often used, and is called the force-voltage analogy.

Analogous systems with similar solutions exist for electrical, mechanical, thermal, and fluid systems. The existence of analogous systems and solutions provides the analyst with the ability to extend the solution of one system to all analogous systems with the same describing differential equations. Therefore, what one learns about the analysis and design of electrical systems is immediately extended to an understanding of fluid, thermal, and mechanical systems.

2.3 LINEAR APPROXIMATIONS OF PHYSICAL SYSTEMS

A great majority of physical systems are linear within some range of the variables. However, all systems ultimately become nonlinear as the variables are increased without limit. For example, the spring-mass-damper system of Fig. 2.1 is linear and described by Eq. (2.1) so long as the mass is subjected to small deflections $y(t)$. However, if $y(t)$ were continually increased, eventually the spring would be over-extended and break. Therefore, the question of linearity and the range of applicability must be considered for each system.

A system is defined as linear in terms of the system excitation and response. In the case of the electrical network, the excitation is the input current $r(t)$ and the response is the voltage $v(t)$. In general, a *necessary condition* for a linear system can be determined in terms of an excitation $x(t)$ and a response $y(t)$. When the system at rest is subjected to an excitation $x_1(t)$, it provides a response $y_1(t)$. Furthermore, when the system is subjected to an excitation $x_2(t)$, it provides a corresponding response $y_2(t)$. For a linear system, it is *necessary* that the excitation $x_1(t)$ + $x_2(t)$ result in a response $y_1(t)$ + $y_2(t)$. This is usually called the *principle of superposition*.

Furthermore, it is necessary that the magnitude scale factor is preserved in a linear system. Again, consider a system with an input x which results in an output y. Then it is necessary that the response of a linear system to a constant multiple β of an input x is equal to the response to the input multiplied by the same constant so that the output is equal to βy. This is called the property of *homogenity*. A system is linear if and only if the properties of superposition and homogenity are satisfied.

A system characterized by the relation $y = x^2$ is not linear since the superposition property is not satisfied. A system which is represented by the relation $y = mx$ + b is not linear, since it does not satisfy the homogenity property. However, this device may be considered linear about an operating point x_0, y_0 for small changes Δx and Δy. When $x = x_0 + \Delta x$ and $y = y_0 + \Delta y$, we have

$$y = mx + b$$

or

$$y_0 + \Delta y = mx_0 + m\,\Delta x + b$$

and therefore $\Delta y = m\,\Delta x$, which satisfies the necessary conditions.

The linearity of many mechanical and electrical elements can be assumed over a reasonably large range of the variables [7]. This is not usually the case for thermal and fluid elements, which are more frequently nonlinear in character. Fortunately, however, one can often linearize nonlinear elements assuming small-signal conditions. This is the normal approach used to obtain a linear equivalent circuit for electronic circuits and transistors. Consider a general element with an excitation (through-) variable $x(t)$ and a response (across-) variable $y(t)$. Several examples of dynamic system variables are given in Table 2.1. The relationship of the two variables is written as

$$y(t) = g(x(t)), \tag{2.6}$$

where $g(x(t))$ indicates $y(t)$ is a function of $x(t)$. The relationship might be shown graphically, as in Fig. 2.4. The normal operating point is designated by x_0. Since the curve (function) is continuous over the range of interest, a *Taylor series* expansion

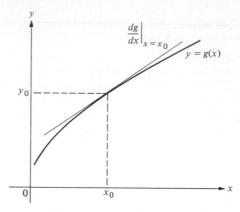

Fig. 2.4. A graphical representation of a nonlinear element.

about the operating point may be utilized. Then we have

$$y = g(x) = g(x_0) + \frac{dg}{dx}\Big|_{x=x_0} \frac{(x - x_0)}{1!} + \frac{d^2g}{dx^2}\Big|_{x=x_0} \frac{(x - x_0)^2}{2!} + \cdots. \quad (2.7)$$

The slope at the operating point,

$$\frac{dg}{dx}\Big|_{x=x_0},$$

is a good approximation to the curve over a small range of $(x - x_0)$, the deviation from the operating point. Then, as a reasonable approximation, Eq. (2.7) becomes

$$y = g(x_0) + \frac{dg}{dx}\Big|_{x=x_0} (x - x_0) = y_0 + m(x - x_0), \quad (2.8)$$

where m is the slope at the operating point. Finally, Eq. (2.8) can be rewritten as the linear equation

$$(y - y_0) = m(x - x_0)$$

or

$$\Delta y = m \, \Delta x. \quad (2.9)$$

This linear approximation is as accurate as the assumption of small signals is applicable to the specific problem.

If the dependent variable y depends upon several excitation variables, x_1, x_2, ..., x_n, then the functional relationship is written as

$$y = g(x_1, x_2, \ldots, x_n). \quad (2.10)$$

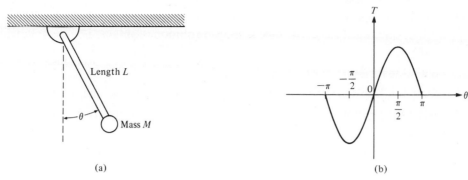

Fig. 2.5. Pendulum oscillator.

The Taylor series expansion about the operating point x_{1_0}, x_{2_0}, . . . , x_{n_0} is useful for a linear approximation to the nonlinear function. When the higher-order terms are neglected, the linear approximation is written as

$$y = g(x_{1_0}, x_{2_0}, \ldots, x_{n_0}) + \left.\frac{\partial g}{\partial x_1}\right|_{x=x_0} (x_1 - x_{1_0}) + \left.\frac{\partial g}{\partial x_2}\right|_{x=x_0} (x_2 - x_{2_0})$$

$$+ \cdots + \left.\frac{\partial g}{\partial x_n}\right|_{x=x_0} (x_n - x_{n_0}). \quad (2.11)$$

where x_0 is the operating point. An example will clearly illustrate the utility of this method.

Example 2.1.   Consider the pendulum oscillator shown in Fig. 2.5(a). The torque on the mass is
$$T = MgL \sin \theta, \quad (2.12)$$

where g is the gravity constant. The equilibrium condition for the mass is $\theta_0 = 0°$. The nonlinear relation between T and θ is shown graphically in Fig. 2.5(b). The first derivative evaluated at equilibrium provides the linear approximation which is

$$T = MgL \left.\frac{\partial \sin \theta}{\partial \theta}\right|_{\theta=\theta_0} (\theta - \theta_0)$$
$$= MgL(\cos 0°)(\theta - 0°)$$
$$= MgL\theta. \quad (2.13)$$

This approximation is reasonably accurate for $-\dfrac{\pi}{4} \leq \theta \leq \dfrac{\pi}{4}$.

2.4 THE LAPLACE TRANSFORM

The ability to obtain linear approximations of physical systems allows the analyst to consider the use of the *Laplace transformation*. The Laplace transform method substitutes the relatively easily solved algebraic equations for the more difficult differential equations [1, 3]. The time response solution is obtained by the following operations:

1. obtain the differential equations;
2. obtain the Laplace transformation of the differential equations;
3. solve the resulting algebraic transform of the variable of interest.

The Laplace transform exists for linear differential equations for which the transformation integral converges. Therefore, in order that $f(t)$ be transformable, it is sufficient that

$$\int_0^\infty |f(t)| e^{-\sigma_1 t} \, dt < \infty$$

for some real, positive σ_1 [1]. If the magnitude of $f(t)$ is $|f(t)| < Me^{\alpha t}$ for all positive t, the integral will converge for $\sigma_1 > \alpha$. The region of convergence is there-fore given by $\infty > \sigma_1 > \alpha$, and σ_1 is known as the abscissa of absolute convergence. Signals that are physically possible always have a Laplace transform. The Laplace transformation for a function of time, $f(t)$, is

$$F(s) = \int_0^\infty f(t)e^{-st} \, dt = \mathcal{L}\{f(t)\}. \tag{2.14}$$

The *inverse Laplace transform* is written as

$$f(t) = \frac{1}{2\pi j} \int_{\sigma-j\infty}^{\sigma+j\infty} F(s)e^{+st} \, ds. \tag{2.15}$$

The transformation integrals have been used to derive tables of Laplace transforms which are ordinarily used for the great majority of problems. A table of Laplace transforms is provided in Appendix A. Some important Laplace transform pairs are given in Table 2.5.

Alternatively, the Laplace variable s can be considered to be the differential operator so that

$$s \equiv \frac{d}{dt}. \tag{2.16}$$

Then we also have the integral operator

$$\frac{1}{s} \equiv \int_{0+}^t dt. \tag{2.17}$$

The inverse Laplace transformation is usually obtained by using the Heaviside partial fraction expansion. This approach is particularly useful for systems analysis and design, since the effect of each characteristic root or eigenvalue may be clearly observed.

Table 2.5 Important Laplace Transform Pairs

$f(t)$	$F(s)$
Step function, $u(t)$	$\dfrac{1}{s}$
e^{-at}	$\dfrac{1}{s + a}$
$\sin \omega t$	$\dfrac{\omega}{s^2 + \omega^2}$
$\cos \omega t$	$\dfrac{s}{s^2 + \omega^2}$
$e^{-at}f(t)$	$F(s + a)$
t^n	$\dfrac{n!}{s^{n+1}}$
$f^{(k)}(t) = \dfrac{d^k f(t)}{dt^k}$	$s^k F(s) - s^{k-1}f(0^+) - s^{k-2}f'(0^+)$ $- \cdots - f^{(k-1)}(0^+)$
$\displaystyle\int_{-\infty}^{t} f(t)\,dt$	$\dfrac{F(s)}{s} + \dfrac{\int_{-\infty}^{0} f\,dt}{s}$
Impulse function, $\delta(t)$	1

In order to illustrate the usefulness of the Laplace transformation and the steps involved in the system analysis, reconsider the spring-mass-damper system described by Eq. (2.1), which is

$$M \frac{d^2 y}{dt^2} + f \frac{dy}{dt} + Ky = r(t). \tag{2.18}$$

We wish to obtain the response, y, as a function of time. The Laplace transform of Eq. (2.18) is

$$M \left(s^2 Y(s) - sy(0^+) - \frac{dy(0^+)}{dt} \right) + f(sY(s) - y(0^+)) + KY(s) = R(s). \tag{2.19}$$

When

$$r(t) = 0, \qquad y(0^+) = y_0, \qquad \left.\frac{dy}{dt}\right|_{t=0+} = 0,$$

we have

$$Ms^2 Y(s) - Msy_0 + fsY(s) - fy_0 + KY(s) = 0. \tag{2.20}$$

Solving for $Y(s)$, we obtain

$$Y(s) = \frac{(Ms + f)y_0}{Ms^2 + fs + K} = \frac{p(s)}{q(s)}. \tag{2.21}$$

The denominator polynomial $q(s)$, when set equal to zero, is called the *characteristic equation,* since the roots of this equation determine the character of the time response. The roots of this characteristic equation are also called the *poles* or *singularities* of the system. The roots of the numerator polynomial $p(s)$ are called the *zeros* of the system; for example, $s = -f/M$. Poles and zeros are critical frequencies. At the poles the function $Y(s)$ becomes infinite; while at the zeros, the function becomes zero. The complex frequency s-*plane* plot of the poles and zeros graphically portrays the character of the natural transient response of the system.

For a specific case, consider the system when $K/M = 2$ and $f/M = 3$. Then Eq. (2.21) becomes

$$Y(s) = \frac{(s + 3)y_0}{(s + 1)(s + 2)}.$$ (2.22)

The poles and zeros of $Y(s)$ are shown on the s-plane in Fig. 2.6.

Expanding Eq. (2.22) in a partial fraction expansion, we obtain

$$Y(s) = \frac{k_1}{s + 1} + \frac{k_2}{s + 2},$$ (2.23)

where k_1 and k_2 are the coefficients of the expansion. The coefficients, k_i, are called *residues* and are evaluated by multiplying through by the denominator factor of Eq. (2.22) corresponding to k_i and setting s equal to the root. Evaluating k_1 when $y_0 = 1$, we have

$$k_1 = \frac{(s - s_1)p(s)}{q(s)}\Bigg|_{s=s_1}$$
$$= \frac{(s + 1)(s + 3)}{(s + 1)(s + 2)}\Bigg|_{s_1=-1} = 2$$ (2.24)

and $k_2 = -1$. Alternatively, the residues of $Y(s)$ at the respective poles may be evaluated graphically on the s-plane plot, since Eq. (2.24) may be written as

$$k_1 = \frac{s + 3}{s + 2}\Bigg|_{s=s_1=-1}$$
$$= \frac{s_1 + 3}{s_1 + 2}\Bigg|_{s_1=-1} = 2.$$ (2.25)

The graphical representation of Eq. (2.25) is shown in Fig. 2.7. The graphical method of evaluating the residues is particularly valuable when the order of the characteristic equation is high and several poles are complex conjugate pairs.

The inverse Laplace transform of Eq. (2.22) is then

$$y(t) = \mathcal{L}^{-1}\left\{\frac{2}{s + 1}\right\} + \mathcal{L}^{-1}\left\{\frac{-1}{s + 2}\right\}.$$ (2.26)

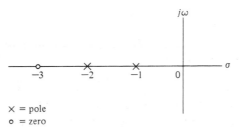

X = pole
o = zero

Fig. 2.6. An s-plane pole and zero plot.

Using Table 2.5, we find that

$$y(t) = 2e^{-t} - 1e^{-2t}. \tag{2.27}$$

Finally, it is usually desired to determine the *steady-state* or *final value* of the response of $y(t)$. For example, the final or steady-state rest position of the spring-mass-damper system should be calculated. The final value can be determined from the relation

$$\lim_{t \to \infty} y(t) = \lim_{s \to 0} s Y(s), \tag{2.28}$$

where a simple pole of $Y(s)$ at the origin is permitted, but poles on the imaginary axis and in the right half-plane and higher-order poles at the origin are excluded. Therefore, for the specific case of the spring, mass, and damper, we find that

$$\lim_{t \to \infty} y(t) = \lim_{s \to 0} s Y(s) = 0. \tag{2.29}$$

Hence the final position for the mass is the normal equilibrium position $y = 0$.

In order to clearly illustrate the salient points of the Laplace transform method, let us reconsider the mass-spring-damper system for the underdamped case. The equation for $Y(s)$ may be rewritten as

$$\begin{aligned}
Y(s) &= \frac{(s + f/M)(y_0)}{(s^2 + (f/M)s + K/M)} \\
&= \frac{(s + 2\zeta\omega_n)(y_0)}{s^2 + 2\zeta\omega_n s + \omega_n^2},
\end{aligned} \tag{2.30}$$

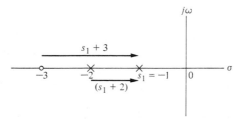

Fig. 2.7. Graphical evaluation of the residues.

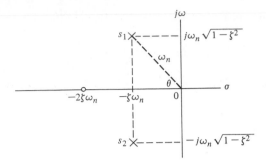

Fig. 2.8. An s-plane plot of the poles and zeros of $Y(s)$.

where ζ is the dimensionless *damping ratio* and ω_n is the *natural frequency* of the system. The roots of the characteristic equation are

$$s_1, s_2 = -\zeta\omega_n \pm \omega_n\sqrt{\zeta^2 - 1}, \tag{2.31}$$

where, in this case, $\omega_n = \sqrt{K/M}$ and $\zeta = f/(2\sqrt{KM})$. When $\zeta > 1$, the roots are real; and when $\zeta < 1$, the roots are complex and conjugates. When $\zeta = 1$, the roots are repeated and real and the condition is called *critical damping*.

When $\zeta < 1$, the response is underdamped and

$$s_{1,2} = -\zeta\omega_n \pm j\omega_n\sqrt{1 - \zeta^2}. \tag{2.32}$$

The s-plane plot of the poles and zeros of $Y(s)$ is shown in Fig. 2.8, where $\theta = \cos^{-1}\zeta$. As ζ varies with ω_n constant, the complex conjugate roots follow a circular locus as shown in Fig. 2.9. The transient response is increasingly oscillatory as the roots approach the imaginary axis when ζ approaches zero.

The inverse Laplace transform can be evaluated using the graphical residue

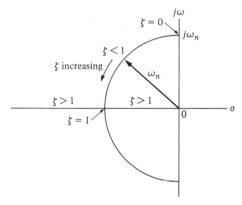

Fig. 2.9. The locus of roots as ζ varies with ω_n constant.

evaluation. The partial fraction expansion of Eq. (2.30) is

$$Y(s) = \frac{k_1}{(s - s_1)} + \frac{k_2}{(s - s_2)}. \tag{2.33}$$

Since s_2 is the complex conjugate of s_1, the residue k_2 is the complex conjugate of k_1 so that we obtain

$$Y(s) = \frac{k_1}{(s - s_1)} + \frac{k_1^*}{(s - s_1^*)},$$

where the asterisk indicates the conjugate relation. The residue k_1 is evaluated from Fig. 2.10 as

$$k_1 = \frac{(y_0)(s_1 + 2\zeta\omega_n)}{(s_1 - s_1^*)}$$
$$= \frac{(y_0)M_1 e^{j\theta}}{M_2 e^{j\pi/2}}, \tag{2.34}$$

where M_1 is the magnitude of $(s_1 + 2\zeta\omega_n)$ and M_2 is the magnitude of $(s_1 - s_1^*)$. In this case, we obtain

$$k_1 = \frac{(y_0)(\omega_n e^{j\theta})}{(2\omega_n\sqrt{1 - \zeta^2}\ e^{j\pi/2})}$$
$$= \frac{(y_0)}{2\sqrt{1 - \zeta^2}}\ e^{j(\theta - \pi/2)}, \tag{2.35}$$

where $\theta = \cos^{-1} \zeta$. Therefore,

$$k_2 = \frac{(y_0)}{2\sqrt{1 - \zeta^2}}\ e^{j(\pi/2 - \theta)} \tag{2.36}$$

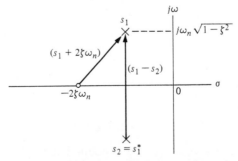

Fig. 2.10. Evaluation of the residue k_1.

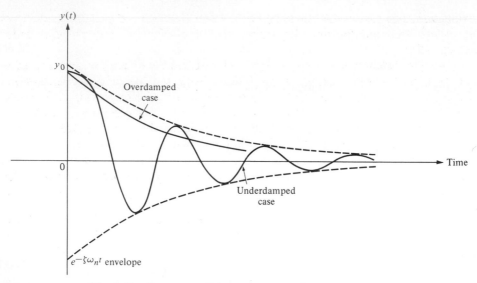

Fig. 2.11. Response of the spring-mass-damper system.

Finally, we find that

$$y(t) = k_1 e^{s_1 t} + k_2 e^{s_2 t}$$

$$= \frac{y_0}{2\sqrt{1 - \zeta^2}} (e^{j(\theta - \pi/2)} e^{-\zeta \omega_n t} e^{j\omega_n \sqrt{1-\zeta^2}t} + e^{j(\pi/2 - \theta)} e^{-\zeta \omega_n t} e^{-j\omega_n \sqrt{1-\zeta^2}t})$$

$$= \frac{y_0}{\sqrt{1 - \zeta^2}} e^{-\zeta \omega_n t} \sin (\omega_n \sqrt{1 - \zeta^2}\, t + \theta). \tag{2.37}$$

The transient response of the overdamped ($\zeta > 1$) and underdamped ($\zeta < 1$) cases are shown in Fig. 2.11.

The direct and clear relationship between the s-plane location of the poles and the form of the transient response is readily interpreted from the s-plane pole-zero plots. Furthermore, the magnitude of the response of each root, represented by the residue, is clearly visualized by examining the graphical residues on the s-plane. The Laplace transformation and the s-plane approach is a very useful technique for system analysis and design where emphasis is placed on the transient and steady-state performance. In fact, since the study of control systems is concerned primarily with the transient and steady-state performance of dynamic systems, we have real cause to appreciate the value of the Laplace transform techniques.

2.5 THE TRANSFER FUNCTION OF LINEAR SYSTEMS

The *transfer function* of a linear system is defined as the ratio of the Laplace transform of the output variable to the Laplace transform of the input variable, with all initial conditions assumed to be zero. The transfer function of a system (or element) represents the relationship describing the dynamics of the system under consideration.

A transfer function may only be defined for a linear, stationary (constant parameter) system. A nonstationary system, often called a time-varying system, has one or more time-varying parameters, and the Laplace transformation may not be utilized. Furthermore, a transfer function is an input-output description of the behavior of a system. Thus, the transfer function description does not include any information concerning the internal structure of the system and its behavior.

The transfer function of the spring-mass-damper system is obtained from the original describing equation, Eq. (2.19), rewritten with zero initial conditions as follows:

$$Ms^2 Y(s) + fs Y(s) + KY(s) = R(s). \tag{2.38}$$

Then the transfer function is

$$\frac{\text{Output}}{\text{Input}} = G(s) = \frac{Y(s)}{R(s)} = \frac{1}{Ms^2 + fs + K}. \tag{2.39}$$

The transfer function of the RC network shown in Fig. 2.12 is obtained by writing the Kirchhoff voltage equation, which yields

$$V_1(s) = \left(R + \frac{1}{Cs} \right) I(s). \tag{2.40}$$

The output voltage is

$$V_2(s) = I(s) \left(\frac{1}{Cs} \right). \tag{2.41}$$

Therefore, solving Eq. (2.40) for $I(s)$ and substituting in Eq. (2.41), we have

$$V_2(s) = \frac{(1/Cs)V_1(s)}{R + 1/Cs}. \tag{2.42}$$

Then the transfer function is obtained as the ratio $V_2(s)/V_1(s)$, which is

$$G(s) = \frac{V_2(s)}{V_1(s)} = \frac{1}{RCs + 1}$$

$$= \frac{1}{\tau s + 1}$$

$$= \frac{(1/\tau)}{s + 1/\tau}, \tag{2.43}$$

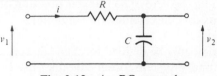

Fig. 2.12. An RC network.

where $\tau = RC$, the *time constant* of the network. Equation (2.43) could be imme-diately obtained if one observes that the circuit is a voltage divider, where

$$\frac{V_2(s)}{V_1(s)} = \frac{Z_2(s)}{Z_1(s) + Z_2(s)} \tag{2.44}$$

and $Z_1(s) = R$, $Z_2 = 1/Cs$.

A multiloop electrical circuit or an analogous multiple mass mechanical system results in a set of simultaneous equations in the Laplace variable. It is usually more convenient to solve the simultaneous equations by using matrices and determinants [1, 3, 16]. An introduction to matrices and determinants is provided in Appendix C [31].

Example 2.2.   Consider the mechanical system shown in Fig. 2.13(a) and its elec-trical circuit analog shown in Fig. 2.13(b). The electrical circuit analog is a force-current analog as outlined in Table 2.1. The velocities, $v_1(t)$ and $v_2(t)$, of the mechanical system are directly analogous to the node voltage $v_1(t)$ and $v_2(t)$ of the electrical circuit. The simultaneous equations, assuming the initial conditions are zero, are

$$M_1 s V_1(s) + (f_1 + f_2) V_1(s) - f_1 V_2(s) = R(s), \tag{2.45}$$

$$M_2 s V_2(s) + f_1(V_2(s) - V_1(s)) + K \frac{V_2(s)}{s} = 0. \tag{2.46}$$

Rearranging Eqs. (2.45) and (2.46) we obtain

$$(M_1 s + (f_1 + f_2)) V_1(s) + (-f_1) V_2(s) = R(s), \tag{2.47}$$

$$(-f_1) V_1(s) + \left(M_2 s + f_1 + \frac{K}{s} \right) V_2(s) = 0, \tag{2.48}$$

or, in matrix form, we have

$$\begin{bmatrix} (M_1 s + f_1 + f_2) & (-f_1) \\ (-f_1) & \left(M_2 s + f_1 + \dfrac{K}{s} \right) \end{bmatrix} \begin{bmatrix} V_1(s) \\ V_2(s) \end{bmatrix} = \begin{bmatrix} R(s) \\ 0 \end{bmatrix}. \tag{2.49}$$

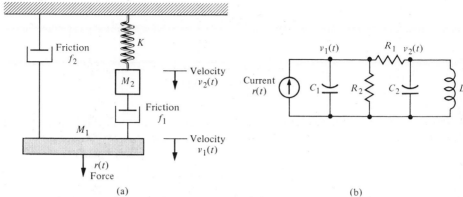

(a)   (b)

Fig. 2.13. (a) Two-mass mechanical system. (b) Two-node electric circuit analog $C_1 = M_1$, $C_2 = M_2$, $L = 1/K$, $R_1 = 1 f_1$, $R_2 = 1/f_2$.

Assuming the velocity of M_1 is the output variable, we solve for $V_1(s)$ by matrix inversion or Cramer's rule to obtain [1, 3]

$$V_1(s) = \frac{(M_2 s + f_1 + (K/s))R(s)}{(M_1 s + f_1 + f_2)(M_2 s + f_1 + (K/s)) - f_1^2}. \tag{2.50}$$

Then the transfer function of the mechanical (or electrical) system is

$$G(s) = \frac{V_1(s)}{R(s)} = \frac{(M_2 s + f_1 + (K/s))}{(M_1 s + f_1 + f_2)(M_2 s + f_1 + (K/s)) - f_1^2}$$

$$= \frac{(M_2 s^2 + f_1 s + K)}{(M_1 s + f_1 + f_2)(M_2 s^2 + f_1 s + K) - f_1^2 s}. \tag{2.51}$$

If the transfer function in terms of the position $x_1(t)$ is desired, then we have

$$\frac{X_1(s)}{R(s)} = \frac{V_1(s)}{sR(s)} = \frac{G(s)}{s}. \tag{2.52}$$

As an example, let us obtain the transfer function of the very important electrical control component, the dc-motor [7].

Example 2.3. The dc-motor is a power actuator device which delivers energy to a load as shown in Fig. 2.14.

The transfer function of the dc-motor will be developed for a linear approximation to an actual motor, and second-order effects, such as hysteresis and the voltage drop across the brushes, will be neglected. The input voltage may be applied to the field or armature terminals. The air-gap flux of the motor is proportional to the field current, provided the field is unsaturated, so that

$$\phi = K_f i_f. \tag{2.53}$$

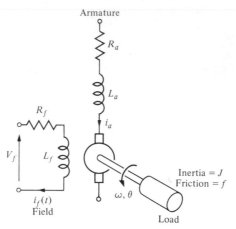

Fig. 2.14. A dc-motor.

The torque developed by the motor is assumed to be related linearly to ϕ and the armature current as follows:

$$T_m = K_1\phi i_a(t) = K_1 K_f i_f(t) i_a(t). \tag{2.54}$$

It is clear from Eq. (2.54) that in order to have a linear element one current must be maintained constant while the other current becomes the input current. First, we shall consider the field current controlled motor which provides a substantial power amplification. Then we have in Laplace transform notation

$$T_m(s) = (K_1 K_f I_a)I_f(s) = K_m I_f(s), \tag{2.55}$$

where $i_a = I_a$ is a constant armature current and K_m is defined as the motor constant. The field current is related to the field voltage as

$$V_f(s) = (R_f + L_f s)I_f(s). \tag{2.56}$$

The motor torque $T_m(s)$ is equal to the torque delivered to the load. This relation may be expressed as

$$T_m(s) = T_L(s) + T_d(s), \tag{2.57}$$

where $T_L(s)$ is the load torque and $T_d(s)$ is the disturbance torque which is often negligible. However, the disturbance torque often must be considered in systems subjected to external forces such as antenna wind-gust forces. The load torque for rotating inertia as shown in Fig. 2.14 is written as

$$T_L(s) = Js^2\theta(s) + fs\theta(s). \tag{2.58}$$

Rearranging Eqs. (2.55), (2.56), and (2.57), we have

$$T_L(s) = T_m(s) - T_d(s), \tag{2.59}$$

$$T_m(s) = K_m I_f(s), \tag{2.60}$$

$$I_f(s) = \frac{V_f(s)}{R_f + L_f s}. \tag{2.61}$$

Therefore, the transfer function of the motor-load combination is

$$\frac{\theta(s)}{V_f(s)} = \frac{K_m}{s(Js + f)(L_f s + R_f)}$$

$$= \frac{K_m/JL_f}{s(s + f/J)(s + R_f/L_f)}. \tag{2.62}$$

The block diagram model of the field controlled dc-motor is shown in Fig. 2.15. Alternatively, the transfer function may be written in terms of the time constants of the motor as

$$\frac{\theta(s)}{V_f(s)} = G(s) = \frac{K_m/fR_f}{s(\tau_f s + 1)(\tau_L s + 1)}, \tag{2.63}$$

where $\tau_f = L_f/R_f$ and $\tau_L = J/f$. Typically, one finds that $\tau_L > \tau_f$ and often the field time constant may be neglected.

The *armature controlled dc-motor* utilizes a constant field current, and therefore the motor torque is

$$T_m(s) = (K_1 K_f I_f) I_a(s) = K_m I_a(s). \tag{2.64}$$

The armature current is related to the input voltage applied to the armature as

$$V_a(s) = (R_a + L_a s) I_a(s) + V_b(s), \tag{2.65}$$

where $V_b(s)$ is the back electromotive-force voltage proportional to the motor speed. Therefore we have

$$V_b(s) = K_b \omega(s), \tag{2.66}$$

and the armature current is

$$I_a(s) = \frac{V_a(s) - K_b \omega(s)}{(R_a + L_a s)}. \tag{2.67}$$

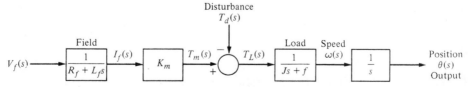

Fig. 2.15. Block diagram model of field controlled dc-motor.

Equations (2.58) and (2.59) represent the load torque so that

$$T_L(s) = Js^2\theta(s) + fs\theta(s) = T_m(s) - T_d(s). \tag{2.68}$$

The relations for the armature controlled dc-motor are shown schematically in Fig. 2.16. Using Eqs. (2.64), (2.67), and (2.68), or, alternatively, the block diagram, we obtain the transfer function

$$G(s) = \frac{\theta(s)}{V_a(s)} = \frac{K_m}{s[(R_a + L_a s)(Js + f) + K_b K_m]}$$

$$= \frac{K_m}{s(s^2 + 2\zeta\omega_n s + \omega_n^2)}. \tag{2.69}$$

However, for many dc-motors, the time constant of the armature, $\tau_a = L_a/R_a$, is negligible, and therefore

$$G(s) = \frac{\theta(s)}{V_a(s)} = \frac{K_m}{s[R_a(Js + f) + K_b K_m]} = \frac{[K_m/(R_a f + K_b K_m)]}{s(\tau_1 s + 1)}, \tag{2.70}$$

where the equivalent time constant $\tau_1 = R_a J/(R_a f + K_b K_m)$.

It is of interest to note that K_m is equal to K_b. This equality may be shown by considering the steady-state motor operation and the power balance when the rotor resistance is neglected. The power input to the rotor is $(K_b\omega)i_a$ and the power delivered to the shaft is $T\omega$. In the steady-state condition, the power input is equal to the power delivered to the shaft so that $(K_b\omega)i_a = T\omega$; and since $T = K_m i_a$ (Eq. 2.64), we find that $K_b = K_m$.

Electric motors are used for moving loads when a rapid response is not required and for relatively low power requirements. Actuators that operate as a result of hydraulic pressure are used for large loads. Figure 2.17 shows the usual ranges of use for electromechanical drives as contrasted to electrohydraulic drives. Typical applications are also shown on the figure.

Example 2.4. A useful actuator for the linear positioning of a mass is the hydraulic actuator shown in Table 2.6, Entry 9. The hydraulic actuator is capable of providing

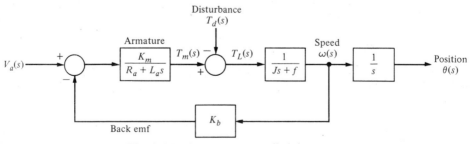

Fig. 2.16. Armature controlled dc-motor.

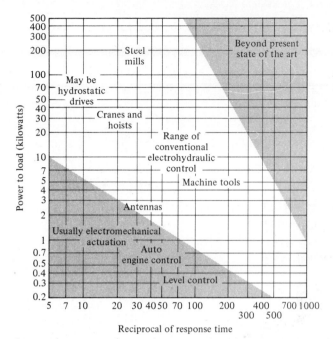

Fig. 2.17. Range of control response time and power to load for electromechanical and electrohydraulic devices.

a large power amplification. It will be assumed that the hydraulic fluid is available from a constant pressure source and that the compressibility of the fluid is negligible. A downward input displacement, x, moves the control valve, and thus fluid passes into the upper part of the cylinder and the piston is forced downward. A small, low-power displacement of $x(t)$ causes a larger, high-power displacement, $y(t)$. The volumetric fluid flow rate Q is related to the input displacement $x(t)$ and the differential pressure across the piston as $Q = g(x, P)$. Using the Taylor series linearization as in Eq. (2.11), we have

$$Q = \left(\frac{\partial g}{\partial x}\right)_{x_0 P_0} x + \left(\frac{\partial g}{\partial P}\right)_{P_0 x_0} P$$

$$= k_x x - k_P P, \tag{2.71}$$

where $g = g(x, P)$ and (x_0, P_0) is the operating point. The force developed by the actuator piston is equal to the area of the piston, A, multiplied by the pressure, P. This force is applied to the mass, and therefore we have

$$AP = M \frac{d^2 y}{dt^2} + f \frac{dy}{dt}. \tag{2.72}$$

Table 2.6 Transfer Functions of Dynamic Elements and Networks

Element or System	$G(s)$
1. Integrating circuit 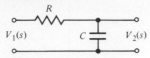	$\dfrac{V_2(s)}{V_1(s)} = \dfrac{1}{RCs + 1}$
2. Differentiating circuit 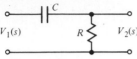	$\dfrac{V_2(s)}{V_1(s)} = \dfrac{RCs}{RCs + 1}$
3. Differentiating circuit 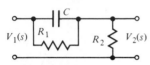	$\dfrac{V_2(s)}{V_1(s)} = \dfrac{s + 1/R_1 C}{s + (R_1 + R_2)/R_1 R_2 C}$
4. Lead-lag filter circuit 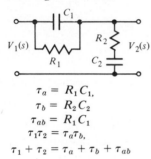 $\tau_a = R_1 C_1,$ $\tau_b = R_2 C_2$ $\tau_{ab} = R_1 C_1$ $\tau_1 \tau_2 = \tau_a \tau_b,$ $\tau_1 + \tau_2 = \tau_a + \tau_b + \tau_{ab}$	$\dfrac{V_2(s)}{V_1(s)} = \dfrac{(1 + s\tau_a)(1 + s\tau_b)}{\tau_a \tau_b s^2 + (\tau_a + \tau_b + \tau_{ab})s + 1}$ $= \dfrac{(1 + s\tau_a)(1 + s\tau_b)}{(1 + s\tau_1)(1 + s\tau_2)}$
5. dc-motor, field controlled 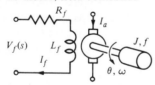	$\dfrac{\theta(s)}{V_f(s)} = \dfrac{K_m}{s(Js + f)(L_f s + R_f)}$
6. dc-motor, armature controlled 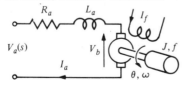	$\dfrac{\theta(s)}{V_a(s)} = \dfrac{K_m}{s[(R_a + L_a s)(Js + f) + K_b K_m]}$
7. ac-motor. two-phase control field 	$\dfrac{\theta(s)}{V_c(s)} = \dfrac{K_m}{s(\tau s + 1)}$ $\tau = J/(f - m)$ $m =$ slope of linearized torque-speed curve (normally negative)

46

Table 2.6   (Continued)

Element or System	$G(s)$
8. Amplidyne 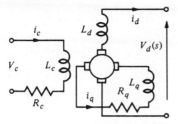	$$\dfrac{V_d(s)}{V_c(s)} = \dfrac{(K/R_c R_q)}{(s\tau_c + 1)(s\tau_q + 1)}$$ $$\tau_c = L_c/R_c, \quad \tau_q = L_q/R_q$$ For the unloaded case, $i_d \simeq 0, \tau_c \simeq \tau_q$, $0.05 \text{ sec} < \tau_c < 0.5 \text{ sec}$
9. Hydraulic actuator 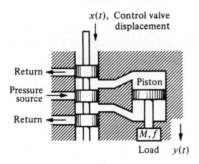	$$\dfrac{Y(s)}{X(s)} = \dfrac{K}{s(Ms + B)}$$ $$K = \dfrac{Ak_x}{k_P}, \quad B = \left(f + \dfrac{A^2}{k_P}\right)$$ $$k_x = \left.\dfrac{\partial g}{\partial x}\right|_{x0}, \quad k_P = \left.\dfrac{\partial g}{\partial P}\right|_{P0},$$ $$g = g(x, P) = \text{flow}$$ $$A = \text{area of piston}$$
10. Gear train 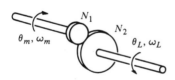	$$\text{Gear ratio} = n = \dfrac{N_1}{N_2}$$ $$N_2\theta_L = N_1\theta_m, \quad \theta_L = n\theta_m$$ $$\omega_L = n\omega_m$$
11. Potentiometer 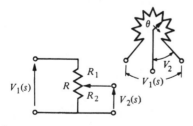	$$\dfrac{V_2(s)}{V_1(s)} = \dfrac{R_2}{R} = \dfrac{R_2}{R_1 + R_2}$$ $$\dfrac{R_2}{R} = \dfrac{\theta}{\theta_{max}}$$
12. Potentiometer error detector bridge 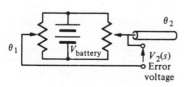	$$V_2(s) = k_s(\theta_1(s) - \theta_2(s))$$ $$V_2(s) = k_s\theta_{error}(s)$$ $$k_s = \dfrac{V_{battery}}{\theta_{max}}$$

Table 2.6 **(Continued)**

Element or System	$G(s)$

13. Tachometer

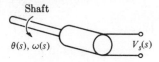

Shaft

$\theta(s), \omega(s)$ $V_s(s)$

$$V_2(s) = K_t\omega(s) = K_t s\theta(s);$$

$$K_t = \text{constant}$$

14. dc-amplifier

$+$ $+$
$V_1(s)$ $V_2(s)$

$$\frac{V_2(s)}{V_1(s)} = \frac{k_a}{s\tau + 1}$$

$R_0 = \text{output resistance}$

$C_0 = \text{output capacitance}$

$\tau = R_0C_0, \tau \ll 1$
    and is often negligible for servo-
    mechanism amplifier

15. Accelerometer

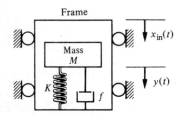

Frame

$x_{\text{in}}(t)$

Mass
M

K f $y(t)$

$$x_0(t) = y(t) - x_{\text{in}}(t),$$

$$\frac{X_0(s)}{X_{\text{in}}(s)} = \frac{-s^2}{s^2 + (f/M)s + K/M}$$

For low-frequency oscillations, where
$\omega < \omega_n,$

$$\frac{X_0(j\omega)}{X_{\text{in}}(j\omega)} \approx \frac{\omega^2}{K/M}$$

16. Thermal Heating System

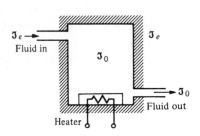

$\mathfrak{I}_e \rightarrow$ $\mathfrak{I}_e$
Fluid in

$\mathfrak{I}_0$

$\rightarrow \mathfrak{I}_0$
Fluid out

Heater

$$\frac{\mathfrak{I}(s)}{q(s)} = \frac{1}{C_t s + (QS + 1/R)}, \text{ where}$$

$\mathfrak{I} = \mathfrak{I}_0 - \mathfrak{I}_e = \text{temperature difference}$
                   due to thermal process

$C_t = \text{thermal capacitance}$

$Q = \text{fluid flow rate} = \text{constant}$

$S = \text{specific heat of water}$

$R_t = \text{thermal resistance of insulation}$

$q(s) = \text{rate of heat flow of heating element}$

Thus, substituting Eq. (2.71) into Eq. (2.72), we obtain

$$\frac{A}{k_P}(k_x x - Q) = M\frac{d^2 y}{dt^2} + f\frac{dy}{dt}. \tag{2.73}$$

Furthermore, the volumetric fluid flow is related to the piston movement as

$$Q = A\frac{dy}{dt}. \tag{2.74}$$

Then, substituting Eq. (2.74) into Eq. (2.73) and rearranging, we have

$$\frac{Ak_x}{k_P}x = M\frac{d^2 y}{dt^2} + \left(f + \frac{A^2}{k_P}\right)\frac{dy}{dt}. \tag{2.75}$$

Therefore, using the Laplace transformation, we have the transfer function

$$\frac{Y(s)}{X(s)} = \frac{K}{s(Ms + B)}, \tag{2.76}$$

where

$$K = \frac{Ak_x}{k_P} \quad \text{and} \quad B = \left(f + \frac{A^2}{k_P}\right).$$

Note that the transfer function of the hydraulic actuator is similar to that of the electric motor. Also, for an actuator operating at high pressure levels and requiring a rapid response of the load, the effect of the compressibility of the fluid must be accounted for [4, 5].

The SI units of the variables are given in Table B.1 in the Appendix. Also a complete set of conversion factors for the English system of units are given in Table B.2.

The transfer function concept and approach is very important since it provides the analyst and designer with a useful mathematical model of the system elements. We shall find the transfer function to be a continually valuable aid in the attempt to model dynamic systems. The approach is particularly useful since the s-plane poles and zeros of the transfer function represent the transient response of the system. The transfer functions of several dynamic elements are given in Table 2.6.

2.6 BLOCK DIAGRAM MODELS

The dynamic systems that comprise automatic control systems are represented mathematically by a set of simultaneous differential equations. As we have noted in the previous sections, the introduction of the Laplace transformation reduces the problem to the solution of a set of linear algebraic equations. Since control systems are concerned with the control of specific variables, the interrelationship of the controlled variables to the controlling variables is required. This relationship is typ-

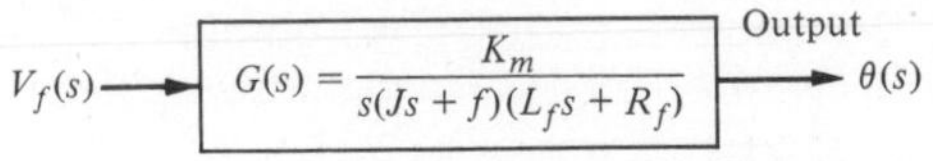

$$V_f(s) \longrightarrow \boxed{G(s) = \frac{K_m}{s(Js + f)(L_f s + R_f)}} \xrightarrow{\text{Output}} \theta(s)$$

Fig. 2.18. Block diagram of dc-motor.

ically represented by the transfer function of the subsystem relating the input and output variables. Therefore, one can assume, correctly, that the transfer function is an important relation for control engineering.

The importance of the cause and effect relationship of the transfer function is evidenced by the interest in representing the relationship of system variables by diagrammatic means. The *block diagram* representation of the systems relationships is prevalent in control system engineering. Block diagrams consist of *unidirectional,* operational blocks that represent the transfer function of the variables of interest. A block diagram of a field controlled dc-motor and load is shown in Fig. 2.18. The relationship between the displacement $\theta(s)$ and the input voltage $V_f(s)$ is clearly portrayed by the block diagram.

In order to represent a system with several variables under control, an interconnection of blocks is utilized. For example, the system shown in Fig. 2.19 has two input variables and two output variables. [38] Using transfer function relations, we can write the simultaneous equations for the output variables as

$$C_1(s) = G_{11}(s)\, R_1(s) + G_{12}(s)R_2(s), \tag{2.77}$$
$$C_2(s) = G_{21}(s)R_1(s) + G_{22}(s)R_2(s), \tag{2.78}$$

where $G_{ij}(s)$ is the transfer function relating the ith output variable to the jth input variable. The block diagram representing this set of equations is shown in Fig. 2.20. In general, for J inputs and I outputs, we write the simultaneous equation in matrix form as

$$
\begin{bmatrix} C_1(s) \\ C_2(s) \\ \vdots \\ C_I(s) \end{bmatrix} =
\begin{bmatrix} G_{11}(s) & \cdots & G_{1J}(s) \\ G_{21}(s) & \cdots & G_{2J}(s) \\ \vdots & & \vdots \\ G_{I1}(s) & \cdots & G_{IJ}(s) \end{bmatrix}
\begin{bmatrix} R_1(s) \\ R_2(s) \\ \vdots \\ R_J(s) \end{bmatrix}, \tag{2.79}
$$

or, simply,

$$\mathbf{C} = \mathbf{GR}. \tag{2.80}$$

Here the $\mathbf{C}$ and $\mathbf{R}$ matrices are column matrices containing the I output and the J input variables, respectively, and $\mathbf{G}$ is an I by J transfer function matrix. The matrix

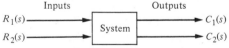

Fig. 2.19. General block representation of 2-input, 2-output system.

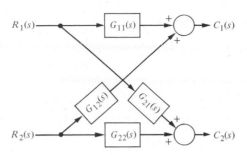

Fig. 2.20. Block diagram of interconnected system.

representation of the interrelationship of many variables is particularly valuable for complex multivariable control systems. An introduction to matrix algebra is provided in Appendix C for those unfamiliar with matrix algebra or who would find a review helpful [6].

The block diagram representation of a given system may often be reduced by block diagram reduction techniques to a simplified block diagram with fewer blocks than the original diagram. Since the transfer functions represent linear systems, the multiplication is commutative. Therefore, as in Table 2.7, Item 1, we have

$$X_3(s) = G_1(s)G_2(s)X_1(s) = G_2(s)G_1(s)X_1(s).$$

When two blocks are connected in cascade as in Item 1 of Table 2.7 we assume that

$$X_3(s) = G_2(s)G_1(s)X_1(s)$$

holds true. This assumes that when the first block is connected to the second block, loading of the first block is negligible. Loading and interaction between interconnected components or systems may occur. If loading of interconnected devices does occur, the engineer must account for this change in the transfer function and use the corrected transfer function in subsequent calculations.

Block diagram transformations and reduction techniques are derived by considering the algebra of the diagram variables. For example, consider the block diagram shown in Fig. 2.21. This negative feedback control system is described by the equation for the actuating signal

$$\begin{aligned} E_a(s) &= R(s) - B(s) \\ &= R(s) - H(s)C(s). \end{aligned} \tag{2.81}$$

Since the output is related to the actuating signal by $G(s)$, we have

$$C(s) = G(s)E_a(s), \tag{2.82}$$

and therefore

$$C(s) = G(s)(R(s) - H(s)C(s)). \tag{2.83}$$

Table 2.7 Block Diagram Transformations

Transformation	Original diagram	Equivalent diagram
1. Combining blocks in cascade	$X_1 \rightarrow \boxed{G_1(s)} \xrightarrow{X_2} \boxed{G_2(s)} \xrightarrow{X_3}$	$X_1 \rightarrow \boxed{G_1 G_2} \xrightarrow{X_3}$ or $X_1 \rightarrow \boxed{G_2 G_1} \xrightarrow{X_3}$
2. Moving a summing point behind a block		
3. Moving a pickoff point ahead of a block		
4. Moving a pickoff point behind a block		
5. Moving a summing point ahead of a block		
6. Eliminating a feedback loop		$X_1 \rightarrow \boxed{\dfrac{G}{1 \mp GH}} \xrightarrow{X_2}$

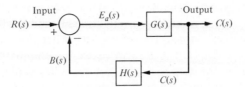

Fig. 2.21. Negative feedback control system.

Solving for $C(s)$, we obtain

$$C(s)(1 + G(s)H(s)) = G(s)R(s). \qquad (2.84)$$

Therefore the transfer function relating the output $C(s)$ to the input $R(s)$ is

$$\frac{C(s)}{R(s)} = \frac{G(s)}{1 + G(s)H(s)}. \qquad (2.85)$$

This *closed-loop transfer function* is particularly important since it represents many of the existing practical control systems.

The reduction of the block diagram shown in Fig. 2.21 to a single block representation is one example of several useful block diagram reductions. These diagram transformations are given in Table 2.7. All the transformations in Table 2.7 can be derived by simple algebraic manipulation of the equations representing the blocks. System analysis by the method of block diagram reduction has the advantage of affording a better understanding of the contribution of each component element than is possible to obtain by the manipulation of equations. The utility of the block diagram transformations will be illustrated by an example of a block diagram reduction.

Example 2.5. The block diagram of a multiple-loop feedback control system is shown in Fig. 2.22 [10]. It is interesting to note that the feedback signal $H_1(s)C(s)$

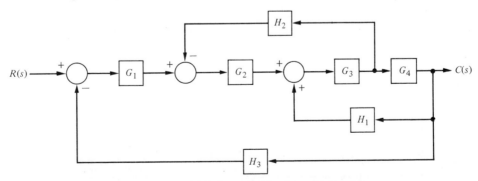

Fig. 2.22. Multiple loop feedback control system.

is a positive feedback signal and the loop $G_3(s)G_4(s)H_1(s)$ is called a *positive feed-back loop*. The block diagram reduction procedure is based on the utilization of rule 6 which eliminates feedback loops. Therefore, the other transformations are used in order to transform the diagram to a form ready for eliminating feedback loops. First, in order to eliminate the loop $G_3G_4H_1$, we move H_2 behind block G_4 by using rule 4, and therefore obtain Fig. 2.23(a). Eliminating the loop $G_3G_4H_1$ by using rule 6, we obtain Fig. 2.23(b). Then, eliminating the inner loop containing H_2/G_4, we obtain Fig. 2.23(c). Finally, by reducing the loop containing H_3 we obtain the closed-loop system transfer function as shown in Fig. 2.23(d). It is worthwhile to examine the form of the numerator and denominator of this closed-loop transfer function. We note that the numerator is composed of the cascade transfer function of the feedforward elements connecting the input $R(s)$ and the output $C(s)$. The denominator is comprised of 1 minus the sum of each loop transfer function. The sign of the loop $G_3G_4H_1$ is plus, since it is a positive feedback loop, while the loops

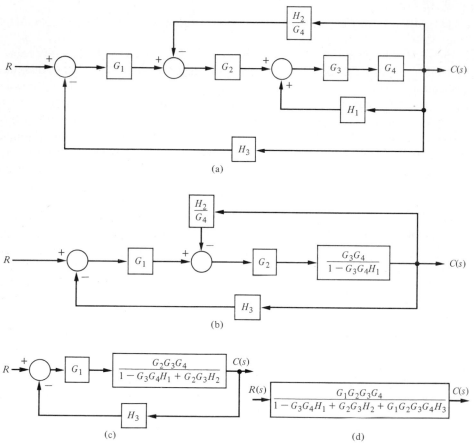

Fig. 2.23. Block diagram reduction of the system of Fig. 2.22.

$G_1G_2G_3G_4H_3$ and $G_2G_3H_2$ are negative feedback loops. The denominator can be rewritten as

$$q(s) = 1 - (+G_3G_4H_1 - G_2G_3H_2 - G_1G_2G_3G_4H_3) \qquad (2.86)$$

in order to illustrate this point. This form of the numerator and denominator is quite close to the general form for multiple-loop feedback systems as we shall find in the following section.

The block diagram representation of feedback control systems is a very valuable and widely used approach. The block diagram provides the analyst with a graphical representation of the interrelationships of controlled and input variables. Furthermore, the designer can readily visualize the possibilities for adding blocks to the existing system block diagram in order to alter and improve the system performance. The transition from the block diagram method to a method utilizing a line path representation instead of a block representation is readily accomplished and is presented in the following section.

2.7 SIGNAL FLOW GRAPH MODELS

Block diagrams are adequate for the representation of the interrelationships of controlled and input variables. However, for a system with reasonably complex interrelationships, the block diagram reduction procedure is cumbersome and often quite difficult to complete. An alternative method for determining the relationship between system variables has been developed by Mason and is based on a representation of the system by line segments [9]. The advantage of the line path method, called the signal flow graph method, is the availability of a flow graph gain formula which provides the relation between system variables without requiring any reduction procedure or manipulation of the flow graph.

The transition from a block diagram representation to a directed line segment representation is easy to accomplish by reconsidering the systems of the previous section. A *signal-flow graph* is a diagram consisting of nodes which are connected by several directed branches and is a graphical representation of a set of linear relations. Signal-flow graphs are particularly useful for feedback control systems because feedback theory is primarily concerned with the flow and processing of signals in systems. The basic element of a signal flow graph is a unidirectional path segment called a *branch* which relates the dependency of an input and an output variable in a manner equivalent to a block of a block diagram. Therefore, the branch relating the output of a dc-motor, $\theta(s)$, to the field voltage, $V_f(s)$, is similar to the block diagram of Fig. 2.18 and is shown in Fig. 2.24. The input and output points or junctions are called *nodes*. Similarly, the signal-flow graph representing Eqs.

$$V_f(s) \; O \xrightarrow{\quad G(s) \quad} O \; \theta(s)$$

Fig. 2.24. Signal flow graph of the dc-motor.

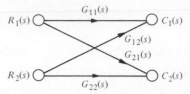

Fig. 2.25. Signal flow graph of interconnected system.

(2.77) and (2.78) and Fig. 2.20 is shown in Fig. 2.25 [38]. The relation between each variable is written next to the directional arrow. All branches leaving a node pass the nodal signal to the output node of each branch (unidirectionally). All branches entering a node summate as a total signal at the node. A *path* is a branch or a continuous sequence of branches which can be traversed from one signal (node) to another signal (node). A *loop* is a closed path which originates and terminates on the same node, and along the path no node is met twice. Therefore, reconsidering Fig. 2.25, we obtain

$$C_1(s) = G_{11}(s)R_1(s) + G_{12}(s)R_2(s), \tag{2.87}$$
$$C_2(s) = G_{21}(s)R_1(s) + G_{22}(s)R_2(s). \tag{2.88}$$

The flow graph is simply a pictorial method of writing a system of algebraic equations so as to indicate the interdependencies of the variables. As another example, consider the following set of simultaneous algebraic equations:

$$a_{11}x_1 + a_{12}x_2 + r_1 = x_1, \tag{2.89}$$
$$a_{21}x_1 + a_{22}x_2 + r_2 = x_2. \tag{2.90}$$

The two input variables are r_1 and r_2, and the output variables are x_1 and x_2. A signal flow graph representing Eqs. (2.89) and (2.90) is shown in Fig. 2.26. Equations (2.89) and (2.90) may be rewritten as

$$x_1(1 - a_{11}) + x_2(-a_{12}) = r_1, \tag{2.91}$$
$$x_1(-a_{21}) + x_2(1 - a_{22}) = r_2. \tag{2.92}$$

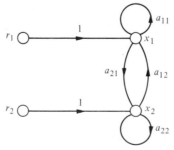

Fig. 2.26. Signal flow graph of two algebraic equations.

The simultaneous solution of Eqs. (2.91) and (2.92) using Cramer's rule results in the solutions

$$x_1 = \frac{(1 - a_{22})r_1 + a_{12}r_2}{(1 - a_{11})(1 - a_{22}) - a_{12}a_{21}} = \frac{(1 - a_{22})}{\Delta}r_1 + \frac{a_{12}}{\Delta}r_2, \tag{2.93}$$

$$x_2 = \frac{(1 - a_{11})r_2 + a_{21}r_1}{(1 - a_{11})(1 - a_{22}) - a_{12}a_{21}} = \frac{(1 - a_{11})}{\Delta}r_2 + \frac{a_{21}}{\Delta}r_1. \tag{2.94}$$

The denominator of the solution is the determinant Δ of the set of equations and is rewritten as

$$\begin{aligned} \Delta &= (1 - a_{11})(1 - a_{22}) - a_{12}a_{21} \\ &= 1 - a_{11} - a_{22} + a_{11}a_{22} - a_{12}a_{21}. \end{aligned} \tag{2.95}$$

In this case, the denominator is equal to 1 minus each self-loop a_{11}, a_{22}, and $a_{12}a_{21}$, plus the product of the two nontouching loops a_{11} and a_{22}.

The numerator for x_1 with the input r_1 is 1 times $(1 - a_{22})$ which is the value of Δ not touching the path 1 from r_1 to x_1. Therefore the numerator from r_2 to x_1 is simply a_{12} since the path through a_{12} touches all the loops. The numerator for x_2 is symmetrical to that of x_1.

In general, the linear dependence T_{ij} between the independent variable x_i (often called the input variable) and a dependent variable x_j is given by Mason's *loop rule* [8, 11]:

$$T_{ij} = \frac{\Sigma_k P_{ij_k} \Delta_{ij_k}}{\Delta}, \tag{2.96}$$

where

$$P_{ij_k} = k\text{th path from variable } x_i \text{ to variable } x_j,$$
$$\Delta = \text{determinant of the graph},$$
$$\Delta_{ij_k} = \text{cofactor of the path } P_{ij_k},$$

and the summation is taken over all possible k paths from x_i to x_j. The cofactor Δ_{ij_k} is the determinant with the loops touching the kth path removed. The determinant Δ is

$$\Delta = 1 - \sum_{n=1}^{N} L_n + \sum_{m=1,q=1}^{M,Q} L_m L_q - \sum L_r L_s L_t + \cdots, \tag{2.97}$$

where L_q equals the value of the qth loop transmittance. Therefore, the rule for evaluating Δ in terms of loops $L_1, L_2, L_3, \ldots, L_N$ is

$\Delta = 1 -$ (sum of all different loop gains)
$\quad +$ (sum of gain products of all combinations of 2 nontouching loops)
$\quad -$ (sum of the gain products of all combinations of 3 nontouching loops)
$\quad + \cdots$

Two loops are nontouching if they do not have any common nodes.

The gain formula is often used to relate the output variable $C(s)$ to the input variable $R(s)$ and is given in somewhat simplified form as

$$T = \frac{\Sigma_k P_k \, \Delta_k}{\Delta},$$ (2.98)

where $T(s) = C(s)/R(s)$. The path gain or transmittance P_k (or P_{ij_k}) is defined as the continuous succession of branches which are traversed in the direction of the arrows and with no node encountered more than once. A loop is defined as a closed path in which no node is encountered more than once per traversal.

Several examples will illustrate the utility and ease of this method. While the gain equation (2.96) appears to be formidable, one must remember that it represents a summation process, not a complicated solution process.

Example 2.6.   A 2-path signal flow graph is shown in Fig. 2.27. The paths connecting the input $R(s)$ and output $C(s)$ are

$$\text{path 1: } P_1 = G_1 G_2 G_3 G_4$$

and

$$\text{path 2: } P_2 = G_5 G_6 G_7 G_8.$$

There are four self-loops:

$$L_1 = G_2 H_2, \qquad L_2 = H_3 G_3, \qquad L_3 = G_6 H_6, \qquad L_4 = G_7 H_7.$$

Loops L_1 and L_2 do not touch L_3 and L_4. Therefore, the determinant is

$$\Delta = 1 - (L_1 + L_2 + L_3 + L_4)$$
$$+ (L_1 L_3 + L_1 L_4 + L_2 L_3 + L_2 L_4).$$ (2.99)

The cofactor of the determinant along path 1 is evaluated by removing the loops that touch path 1 from Δ. Therefore, we have

$$L_1 = L_2 = 0 \qquad \text{and} \qquad \Delta_1 = 1 - (L_3 + L_4).$$

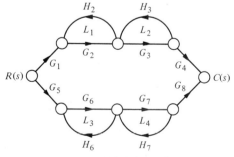

Fig. 2.27. Two-path interacting system.

Similarly, the cofactor for path 2 is

$$\Delta_2 = 1 - (L_1 + L_2).$$

Therefore, the transfer function of the system is

$$\frac{C(s)}{R(s)} = T(s) = \frac{P_1 \Delta_1 + P_2 \Delta_2}{\Delta} \tag{2.100}$$

$$= \frac{G_1 G_2 G_3 G_4 (1 - L_3 - L_4) + G_5 G_6 G_7 G_8 (1 - L_1 - L_2)}{1 - L_1 - L_2 - L_3 - L_4 + L_1 L_3 + L_1 L_4 + L_2 L_3 + L_2 L_4}. \tag{2.101}$$

The signal-flow graph gain formula provides a reasonably straightforward approach for the evaluation of complicated systems. In order to compare the method with block diagram reduction, which is really not much more difficult, let us reconsider the complex system of Example 2.4.

Example 2.7. A multiple-loop feedback system is shown in Fig. 2.22 in block diagram form. There is no reason to redraw the diagram in signal flow graph form so we shall proceed as usual by using the signal flow gain formula, Eq. (2.98). There is one forward path $P_1 = G_1 G_2 G_3 G_4$. The feedback loops are

$$L_1 = -G_2 G_3 H_2, \qquad L_2 = G_3 G_4 H_1, \qquad L_3 = -G_1 G_2 G_3 G_4 H_3.$$

All the loops have common nodes and therefore are all touching. Furthermore, the path P_1 touches all the loops and hence $\Delta_1 = 1$. Thus, the closed-loop transfer function is

$$T(s) = \frac{C(s)}{R(s)} = \frac{P_1 \Delta_1}{1 - L_1 - L_2 - L_3}$$

$$= \frac{G_1 G_2 G_3 G_4}{1 + G_2 G_3 H_2 - G_3 G_4 H_1 + G_1 G_2 G_3 G_4 H_3}. \tag{2.102}$$

Example 2.8. Finally, we shall consider a reasonably complex system which would be difficult to reduce by block diagram techniques. A system with several feedback loops and feedforward paths is shown in Fig. 2.28. The forward paths are

$$P_1 = G_1 G_2 G_3 G_4 G_5 G_6, \qquad P_2 = G_1 G_2 G_7 G_6, \qquad P_3 = G_1 G_2 G_3 G_4 G_8.$$

The feedback loops are

$$L_1 = G_2 G_3 G_4 G_5 H_2, \qquad L_2 = -G_5 G_6 H_1, \qquad L_3 = -G_8 H_1,$$
$$L_4 = -G_7 H_2 G_2, \qquad L_5 = -G_4 H_4, \qquad L_6 = -G_1 G_2 G_3 G_4 G_5 G_6 H_3,$$
$$L_7 = -G_1 G_2 G_7 G_6 H_3, \qquad L_8 = -G_1 G_2 G_3 G_4 G_8 H_3.$$

Loop L_5 does not touch loop L_4 and loop L_7; loop L_3 does not touch loop L_4; and all other loops touch. Therefore the determinant is

$$\Delta = 1 - (L_1 + L_2 + L_3 + L_4 + L_5 + L_6 + L_7 + L_8)$$
$$+ (L_5 L_7 + L_5 L_4 + L_3 L_4). \tag{2.103}$$

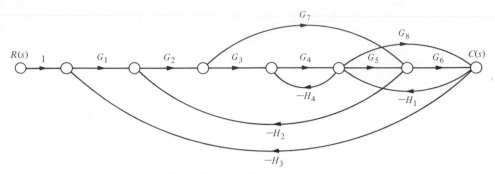

Fig. 2.28. Multiple-loop system.

The cofactors are

$$\Delta_1 = \Delta_3 = 1 \quad \text{and} \quad \Delta_2 = 1 - L_5 = 1 + G_4 H_4.$$

Finally, the transfer function is then

$$T(s) = \frac{C(s)}{R(s)} = \frac{P_1 + P_2 \Delta_2 + P_3}{\Delta}. \tag{2.104}$$

Signal-flow graphs and the signal-flow gain formula may be used profitably for the analysis of feedback control systems, analog computer diagrams, electronic amplifier circuits, statistical systems, mechanical systems, among many other examples.

2.8 SIMULATION OF CONTROL SYSTEMS

When a model is available for a component or system, a computer can be utilized to investigate the behavior of the system. A computer model of a system in a mathematical form suitable for demonstrating the system's behavior may be utilized to investigate various designs of a planned system without actually building the system itself. A *computer simulation* uses a model and the actual conditions of the system being modeled and actual input commands to which the system will be subjected.

A system may be simulated using analog or digital computers. An electronic analog computer is used to establish a model of a system, using the analogy between the voltage of the electronic amplifier and the variable of the system being modeled [1, 14, 15, 16, 17]. An electronic analog computer usually has available the mathematical functions of integration, multiplication by a constant, multiplication of two variables, and the summation of several variables, among others. These functions are often sufficient to develop a simulation model of a system. The analog simulation model of a second-order system is shown in Fig. 2.29 for the system with negative unity feedback and a plant transfer function

$$G(s) = \frac{C(s)}{E(s)} = \frac{K}{s(s + p)}. \tag{2.105}$$

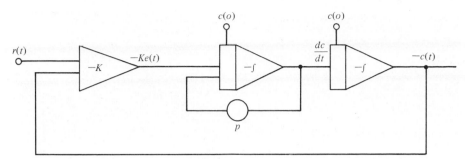

Fig. 2.29. An analog computer simulation model for a second-order system with negative unity feedback.

The differential equation necessary to yield the simulation of the plant is obtained by cross-multiplying in Eq. (2.105) to yield

$$s(s + p)C(s) = KE(s). \qquad (2.106)$$

Since $sC(s)$ is the derivative of $c(t)$ in the s-domain we have:

$$\frac{d^2c(t)}{dt^2} = -p\frac{dc(t)}{dt} + Ke(t). \qquad (2.107)$$

This equation is represented on the analog diagram by the integration in the center of the diagram with the output dc/dt. This analog simulation arrangement may be realized physically on an electronic analog computer in order to yield an output recording of the simulated response of the system. The parameters K and p may be varied in order to ascertain the effect of the parameter change.

A contemporary analog computer is shown in Fig. 2.30. This analog computer has a keyboard-directed system and incorporates digital logic for many laboratory projects.

Simulation models may also be utilized with digital computers. A computer simulation may be developed in common computer language such as FORTRAN or BASIC or in a language specifically developed for simulation [15, 42]. Three widely used languages for the simulation of systems operating in continuous time are CSMP, DYNAMO, and MIMIC. CSMP is an acronym for Continuous System Modeling Program and is an IBM product available for most IBM computers. A graphic feature is also available with CSMP. CSMP provides up to 42 functions such as integration and random number generation and including many nonlinear functions. A portion of a CSMP computer program for simulating the second-order control system of Eq. (2.105) is shown in Fig. 2.31. In this simulation, COUT = $c(t)$ and CDOT = dc/dt. The function REALPL simulates the function with one real pole, p. This simulation will yield an output printed at spacings of one-half second over a twenty-second interval.

A recent simulation of the world environment is an interesting and controversial example of the use of the simulation models [18, 19]. The model includes the vari-

Fig. 2.30. The EAI 1000 desk-top analog computer. (Courtesy of Electronic Associates, Inc., West Long Branch, New Jersey, U.S.A.)

ables population, pollution, resources, industrialization, and food supply. Some of the interconnections between a portion of the variables are shown in Fig. 2.32. One simulation run of the model is shown in Fig. 2.33, which assumes no major change in the physical, economic, or social relationships that have governed the development of the world system. The result of one simulation for the case is shown in Fig. 2.34: In order to avoid the food crisis, average land yield is assumed to double in 1975, the pollution generation per unit is reduced to one-fourth of its 1970 value,

```
DYNAMIC
        ERROR = RIN – COUT
        CONTL = GAIN*ERROR
        CDOT = REALPL (0.0, P, CONTL)
        COUT = INTGRL (0,0, CDOT)
PARAMETER P = 1.5, RIN = 1.0
PARAMETER GAIN = (1.0, 5.0, 10.0)
        TIMER FINTIM = 20.0, OUTDEL = 0.5
        PRINT COUT, CDOT
```

Fig. 2.31. A portion of a digital computer program written in the simulation language CSMP is shown for a second-order unity feedback system.

and nuclear power doubles the resource reserves. In the case of both Figs. 2.33 and 2.34, the world system is shown to collapse for this model as pollution grows rapidly and the population drops drastically. This simulation utilized the DYNAMO language and has been the subject of much discussion.

Assuming that a model and the simulation are reliably accurate, the advantages of computer simulation are [14]:

1. System performance can be observed under all conceivable conditions.

2. Results of field-system performance can be extrapolated with a simulation model for prediction purposes.

3. Decisions concerning future systems presently in a conceptual stage can be examined.

4. Trials of systems under test can be accomplished in a much-reduced period of time.

5. Simulation results can be obtained at lower cost than real experimentation.

6. Study of hypothetical situations can be achieved even when the hypothetical situation would be unrealizable in actual life at the present time.

7. Computer modeling and simulation is often the only feasible or safe technique to analyze and evaluate a system.

2.9 SUMMARY

In this chapter, we have been concerned with quantitative mathematical models of control components and systems. The differential equations describing the dynamic performance of physical systems were utilized to construct a mathematical model. The physical systems under consideration included mechanical, electrical, fluid, and thermodynamic systems. A linear approximation using a Taylor series expansion about the operating point was utilized to obtain a small-signal linear approximation for nonlinear control components. Then, with the approximation of a linear system, one may utilize the Laplace transformation and its related input-output

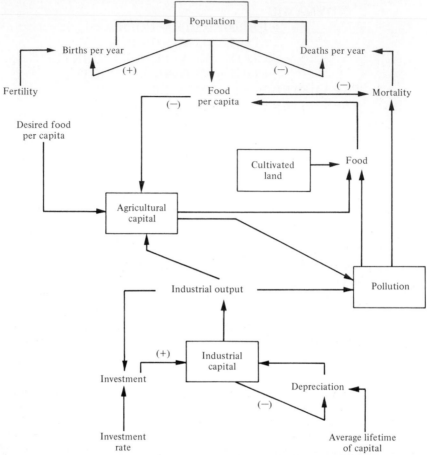

Fig. 2.32. The feedback loops of population, capital, agriculture, and pollution. Each arrow indicates a causal relationship which may include a time delay. (By permission from *The Limits to Growth,* by Donella H. Meadows, Dennis L. Meadows, Jørgen Randers, and William H. Behrens, 3rd. A Potomac Associates book, published by Universe Books, New York, 1972.)

relationship, the transfer function. The transfer function approach to linear systems allows the analyst to determine the response of the system to various input signals in terms of the location of the poles and zeros of the transfer function. Using transfer function notations, block diagram models of systems of interconnected components were developed. The block functions were obtained. Additionally, an alternative use of transfer function models in signal flow graph form was investigated. The signal flow graph gain formula was investigated. The signal flow graph gain formula was found to be useful for obtaining the relationship between system variables in a complex feedback system. The advantage of the signal flow graph method

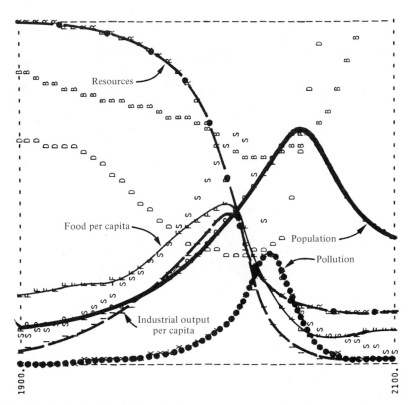

Fig. 2.33. The "standard" world model run assumes no major changes in the physical, economic, or social relationships that have historically governed the development of the world system. All variables plotted here follow historical values from 1900 to 1970. Food, industrial output, and population grow exponentially until the rapidly diminishing resource base forces a slowdown in industrial growth. Because of natural delays in the system, both population and pollution continue to increase for some time after the peak of industrialization. Population growth is finally halted by a rise in the death rate due to decreased food and medical services. (By permission from *The Limits to Growth,* by Donella H. Meadows, Dennis L. Meadows, Jørgen Randers, and William H. Behrens, 3rd. A Potomac Associates book, published by Universe Books, New York, 1972.)

was the availability of Mason's flow graph gain formula which provides the relation between system variables without requiring any reduction or manipulation of the flow graph. Thus, in Chapter 2, we have obtained a useful mathematical model for feedback control systems by developing the concept of a transfer function of a linear system and the relationship among system variables using block diagram and signal flow graph models. Finally, we considered the utility of the computer simulation of linear and nonlinear systems in order to determine the response of a system for several conditions of the system parameters and the environment.

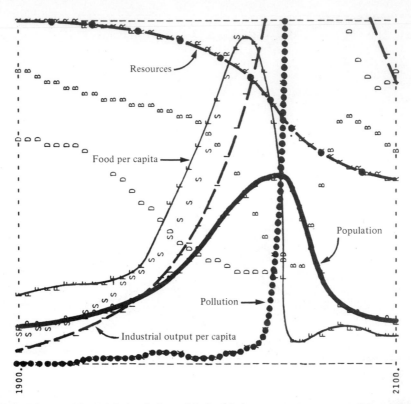

Fig. 2.34. The world model simulation with doubled resource reserves and the pollution per unit of industrial and agricultural output reduced to one-fourth of its 1970 value. Also, the average land yield is doubled in 1975. The combination of these three policies removes so many constraints to growth that population and industry reach very high levels. Although each unit of industrial production generates much less pollution, total production rises enough to create a pollution crisis that brings an end to growth. (By permission from *The Limits to Growth,* by Donella H. Meadows, Dennis L. Meadows, Jørgen Randers, and William H. Behrens, 3rd. A Potomac Associates book, published by Universe Books, New York, 1972.)

PROBLEMS

2.1. An electric circuit is shown in Fig. P2.1. Obtain a set of simultaneous integrodifferential equations representing the network.

2.2. A dynamic vibration absorber is shown in Fig. P2.2. This system is representative of many situations involving the vibration of machines containing unbalanced components. The parameters M_2 and k_{12} may be chosen so that the main mass M_1 does not vibrate when $F(t) = a \sin \omega_0 t$. (a) Obtain the differential equations describing the system. (b) Draw the analogous electrical circuit based on the force-current analogy.

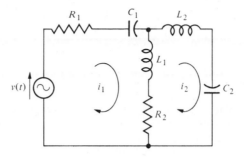

Figure P2.1

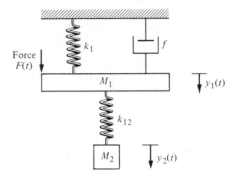

Figure P2.2

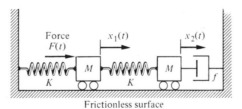

Frictionless surface

Figure P2.3

2.3. A coupled spring-mass system is shown in Fig. P2.3. The masses and springs are assumed to be equal. (a) Obtain the differential equations describing the system. (b) Draw an analogous electrical circuit based on the force-current analogy.

2.4. A nonlinear amplifier can be described by the following characteristic:

$$v_o(t) = \begin{cases} v_{in}^2 & v_{in} \geqslant 0 \\ -v_{in}^2 & v_{in} < 0 \end{cases}$$

The amplifier will be operated over a range for v_{in} of ± 0.5 volts at the operating point. Describe the amplifier by a linear approximation (a) when the operating point is $v_{in} = 0$; and

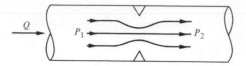

Figure P2.5

(b) when the operating point is $v_{in} = 1$ volt. Obtain a sketch of the nonlinear function and the approximation for each case.

2.5. Fluid flowing through an orifice can be represented by the nonlinear equation

$$Q = K(P_1 - P_2)^{1/2},$$

where the variables are shown in Fig. P2.5 and K is a constant. (a) Determine a linear approximation for the fluid flow equation. (b) What happens to the approximation obtained in (a) if the operating point is $P_1 - P_2 = 0$?

2.6. Using the Laplace transformation, obtain the current $I_2(s)$ of Problem 2.1. Assume that all the initial currents are zero, the initial voltage across capacitor C_1 is zero, $v(t)$ is zero, and the initial voltage across C_2 is 10 volts.

2.7. Obtain the transfer function of the differentiating circuit shown in Fig. P2.7.

2.8. A bridged-T network is often used in ac control systems as a filter network. The circuit of one bridged-T network is shown in Fig. P2.8. Show that the transfer function of the network is

$$\frac{V_{out}(s)}{V_{in}(s)} = \frac{1 + 2R_1Cs + R_1R_2C^2s^2}{1 + (2R_1 + R_2)Cs + R_1R_2C^2s^2}.$$

Draw the pole-zero diagram when $R_1 = 0.5$, $R_2 = 1$, and $C = 0.5$.

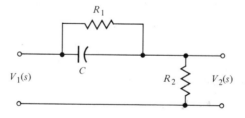

Figure P2.7

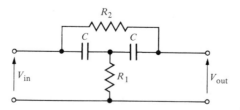

Figure P2.8

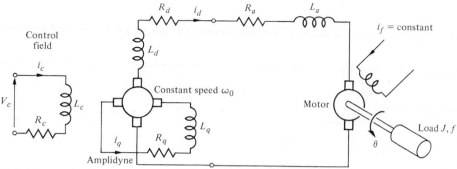

Figure P2.11

2.9. Determine the transfer function $X_1(s)/F(s)$ for the coupled spring-mass system of Problem 2.3. Draw the s-plane pole-zero diagram for low damping when $M = 1$, $f/K = 1$ and

$$\zeta = \frac{1}{2} \frac{f}{\sqrt{KM}} = 0.2.$$

2.10. Determine the transfer function $Y_1(s)/F(s)$ for the vibration absorber system of Problem 2.2. Determine the necessary parameters M_2 and k_{12} so that the mass M_1 does not vibrate when $F(t) = a \sin \omega_0 t$.

2.11. For electromechanical systems which require large power amplification, rotary amplifiers are often used. An amplidyne is a power amplifying rotary amplifier. An amplidyne and a servomotor are shown in Fig. P2.11. Obtain the transfer function $\theta(s)/V_c(s)$ and draw the block diagram of the system.

2.12. An electromechanical open-loop control system is shown in Fig. P2.12. The generator, driven at a constant speed, provides the field voltage for the motor. The motor has an inertia J_m and bearing friction f_m. Obtain the transfer function $\theta_L(s)/V_f(s)$ and draw a block diagram of the system. The generator voltage can be assumed to be proportional to the field current.

2.13. A fluid flow system is shown in Fig. P2.13, where an incompressible fluid is flowing into a open tank. Assuming that the flow is laminar, one may assume that the change in outflow ΔQ_2 is proportional to the change in head ΔH. At steady-state $Q_1 = Q_2$ and $Q_2 = kH^{1/2}$. Using a linear approximation, obtain the transfer function of the tank, $\Delta Q_2(s)/\Delta Q_1(s)$.

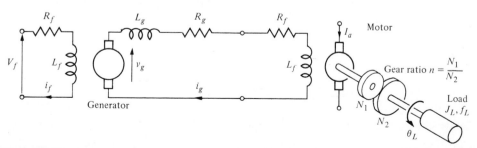

Figure P2.12

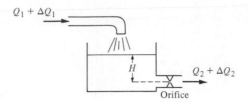

Figure P2.13

2.14. A rotating load is connected to a field-controlled dc electric motor through a gear system. The motor is assumed to be linear. A test results in the output load reaching a speed of 1 rad/sec within ½ sec when a constant 100 v is applied to the motor terminals. The output steady-state speed is 2 rad/sec. Determine the transfer function of the motor, $\theta(s)/V_f(s)$ in rad/v. The inductance of the field may be assumed to be negligible (see Fig. 2.15). Also, note that the application of 100 v to the motor terminals is a step input of 100 v in magnitude.

2.15. An algebraic set of equations may be rewritten in matrix form as

$$\begin{bmatrix} x_1 \\ x_2 \\ x_3 \end{bmatrix} = \begin{bmatrix} 3 & -1 & 0 \\ 0 & 1 & 4 \\ 2 & 0 & 1 \end{bmatrix} \begin{bmatrix} x_1 \\ x_2 \\ x_3 \end{bmatrix}.$$

Draw the signal-flow graph of the matrix equation. Compute the determinant of the matrix with the loop gain formula and with Cramer's algebraic formula.

2.16. Obtain a signal-flow graph to represent the following set of algebraic equations where x_1 and x_2 are to be considered the dependent variables and 8 and 13 are the inputs:

$$x_1 + 2x_2 = 8,$$
$$2x_1 + 3x_2 = 13.$$

Determine the value of each dependent variable by using the gain formula. After solving for x_1 by Mason's formula, verify the solution by using Cramer's rule.

2.17. A mechanical system is shown in Fig. P2.17, which is subjected to a known displacement $x_3(t)$ with respect to the reference. (a) Determine the two independent equations of motion. (b) Obtain the equations of motion in terms of the Laplace transform assuming that the initial conditions are zero. (c) Draw a signal-flow graph representing the system of equations. (d) Obtain the relationship between $X_1(s)$ and $X_3(s)$, $T_{13}(s)$, by using Mason's gain formula. Compare the work necessary to obtain $T_{13}(s)$ by matrix methods with Mason's gain formula approach.

2.18. An *LC* ladder network is shown in Fig. P2.18. One may write the equations describing the network as follows:

$$I_1 = (V_1 - V_a)Y_1, \qquad V_a = (I_1 - I_a)Z_2,$$
$$I_a = (V_a - V_2)Y_3, \qquad V_2 = I_a Z_4.$$

Construct a flow graph from the equations and determine the transfer function $V_2(s)/V_1(s)$.

2.19. The basic noninverting operational amplifier is shown in Fig. P2.19(a) and the signal flow representation of the equations of the circuit is shown in Fig. P2.19(b). (a) Write the

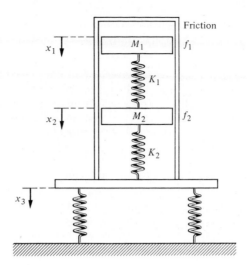

Figure P2.17

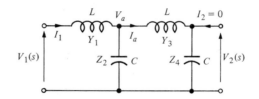

Figure P2.18

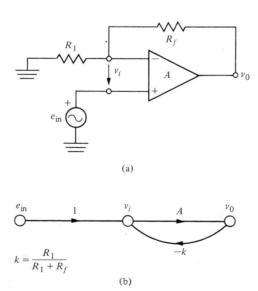

Figure P2.19

voltage equations and verify the representation in the flow graph. (b) Using the signal flow graph calculate the gain of the amplifier and verify that $T(s) = (R_1 + R_f)/R_1$ when $A \gg 10^3$.

2.20. The source follower amplifier provides lower output impedance and essentially unity gain. The circuit diagram is shown in Fig. P2.20(a) and the small signal model is shown in Fig. P2.20(b). This circuit uses an FET and provides a gain of approximately unity. Assume that $R_2 \gg R_1$ for biasing purposes and that $R_g \gg R_2$. (a) Solve for the amplifier gain. (b) Solve for the gain when $g_m = 2000 \ \mu mhos$ and $R_s = 10$ Kohms where $R_s = R_1 + R_2$. (c) Sketch a signal flow diagram that represents the circuit equations.

2.21. A hydraulic servomechanism with mechanical feedback is shown in Fig. P2.21 [12]. The power piston has an area equal to A. When the valve is moved a small amount Δz, then the oil will flow through to the cylinder at a rate $p \cdot \Delta z$, where p is the port coefficient. The input oil pressure will be assumed to be constant. (a) Determine the closed-loop signal-flow graph for this mechanical system. (b) Obtain the closed-loop transfer function $Y(s)/X(s)$.

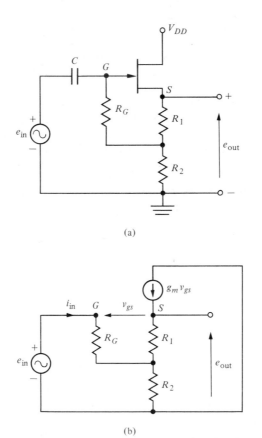

(a)

(b)

Fig. P2.20. The source follower or common drain amplifier using an FET.

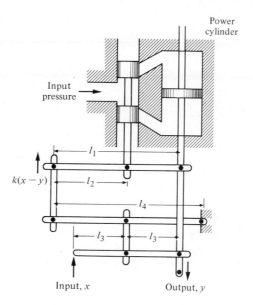

Figure P2.21

2.22. Figure P2.22 shows two pendulums suspended from frictionless pivots and connected at their midpoints by a spring [1]. Assume that each pendulum can be represented by a mass M at the end of a massless bar of length L. Also assume that the displacement is small and linear approximations can be used for $\sin \theta$ and $\cos \theta$. The spring located in the middle of the bars is unstretched when $\theta_1 = \theta_2$. The input force is represented by $f(t)$ which influences the left-hand bar only. (a) Obtain the equations of motion and draw a signal flow diagram for them. (b) Determine the transfer function $T(s) = \theta_1(s)/F(s)$. (c) Draw the location of the poles and zeros of $T(s)$ on the s-plane.

2.23. The small-signal equivalent circuit of a common-emitter transistor amplifier is shown

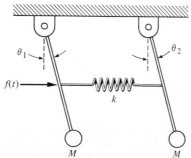

Fig. P2.22. The bars are each of length L and the spring is located at $L/2$.

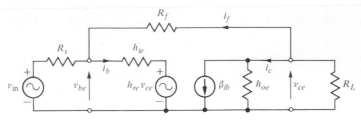

Figure P2.23

in Fig. P2.23. The transistor amplifier includes a feedback resistor R_f. Obtain a signal-flow graph model of the feedback amplifier and determine the input-output ratio v_{ce}/v_{in}.

2.24. A two-transistor series voltage feedback amplifier is shown in Fig. P2.24(a). This ac equivalent circuit neglects the bias resistors and the shunt capacitors. A signal flow graph representing the circuit is shown in Fig. P2.24(b). This flow graph neglects the effect of h_{re} which is usually an accurate approximation, and assumes that $(R_2 + R_L) \gg R_1$. (a) Determine the voltage gain, e_{out}/e_{in}. (b) Determine the current gain, i_{c_2}/i_{b_1}. (c) Determine the input impedance, e_{in}/i_{b_1}.

2.25. Often overlooked is the fact that H. S. Black, who is noted for developing a negative feedback amplifier in 1927, three years earlier had invented a circuit design technique known as feedforward correction [22, 23]. Recent experiments have shown that this technique offers the potential for yielding excellent amplifier stabilization. Black's amplifier is shown in Fig. P2.25(a) in the form recorded in 1924. The signal-flow graph is shown in Fig. P2.25(b) Determine the transfer function between the output $C(s)$ and the input $R(s)$ and between the output and the disturbance $D(s)$. $G(s)$ is used for the amplifier represented by μ in Fig. P2.25(a).

2.26. Natural gas pipeline systems are an important component of our national system of delivering energy to industry and residences. A two-chamber model of a natural gas pipeline system is shown in Fig. P.2.26. The linearized equations are [26, 27]:

$$\dot{p}_2 = \frac{-p_2}{\tau} + \frac{p_3}{\tau} + \frac{Q_1}{C},$$

$$\dot{p}_3 = \frac{p_2}{\tau} - \frac{p_3}{\tau} - \frac{Q_3}{C},$$

$$p_1 = p_2 + Q_1 R_1,$$

$$p_4 = p_3 - Q_3 R_2,$$

where $\tau = R_2 C$ and C = capacitance of each chamber of the line. Develop a signal-flow diagram of the system using an integrator $(1/s)$ between $\dot{p}$ and p. Then determine the pressure $P_4(s)$ as a function of $Q_1(s)$ and $Q_3(s)$. The purpose of a controller to be introduced later between $Q_3(s)$ and $Q_1(s)$ is to maintain a desired pressure response, $p_4(t)$, at the output pipe.

2.27. Ecology, the body of knowledge concerning the economy and function of nature, is of primary interest today. Ecology is concerned with the interrelationships of living organisms, plant or animal, and their environments [29, 30]. Environmental pollution is the unfavorable alteration of our surroundings as a by-product of man's actions. As an example, consider the

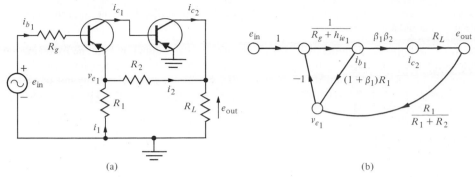

Figure P2.24

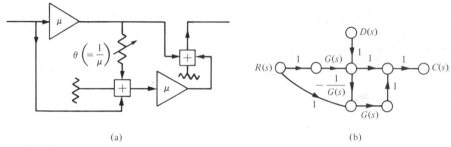

Figure P2.25

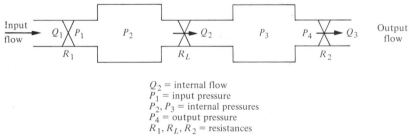

Q_2 = internal flow
P_1 = input pressure
P_2, P_3 = internal pressures
P_4 = output pressure
R_1, R_L, R_2 = resistances

Figure P2.26

changes that have taken place in Lake Erie during the last 50 years. The process of changing from low production of organic materials due to low amounts of nutrients to high production is referred to as *eutrophication*. Corresponding changes in plankton and fish also occur as eutrophication occurs. Lake Erie covers an area of 25,820 square kilometers, has a mean depth of 21 meters and a volume of 540 cubic kilometers. As nutrient enrichment proceeds the original plankton are replaced by blue-green algae which clog filters, release obnoxious odors and turn away the swimmers. In Lake Erie the eutrophication process has been accelerated by pollution from Ohio, Michigan, New York, and Canada. Devise a signal-flow model which includes the population along the lake and industrial development and other variables that you consider important.

2.28. A multiple loop model of an urban ecological system might include the following variables: number of people in the city (P), modernization (M), migration into the city (C), sanitation facilities (S), number of diseases (D), bacteria/area (B), and amount of garbage/area (G), where the symbol for the variable is given in the parentheses. The following causal loops are hypothesized:

1. $P \rightarrow G \rightarrow B \rightarrow D \rightarrow P$
2. $P \rightarrow M \rightarrow C \rightarrow P$
3. $P \rightarrow M \rightarrow S \rightarrow D \rightarrow P$
4. $P \rightarrow M \rightarrow S \rightarrow B \rightarrow D \rightarrow P$

Draw a signal-flow graph for these causal relationships, using appropriate gain symbols. Indicate whether you believe each gain transmission is positive or negative. For example, the causal link S to B is negative since improved sanitation facilities lead to reduced bacteria/area. Which of the four loops are positive feedback loops and which are negative feedback loops?

2.29. The living process associated with the cover of green chlorophyll during the lighted period in forests, lakes, and oceans is called photosynthesis. The source of the energy is the sun in the form of light. We will consider a simple nonlinear model which uses the variable p for production and the variable r for the return respiratory consumption. Together p and r function to provide a cycle of chemical elements. In summer p tends to exceed r and in the winter r exceeds p [31]. The input to the system is approximated by a square wave function of light intensity $I(t)$ with 12 hours of light and 12 hours of darkness. The governing equations are:

$$p(t) = an_2(t)I(t),$$
$$r(t) = bn_1(t),$$
$$\frac{dn_1}{dt} = p - r,$$
$$\frac{dn_2}{dt} = r - p.$$

The variable n_1 is the organic matter resulting from the photosynthesis process and n_2 is the material and CO_2 resulting from the respiratory process. Obtain the response of the system for several values of a and b, using an analog or digital simulation model..

2.30. The measurement or sensor element in a feedback system is important to the accuracy of the system [32]. The dynamic response of the sensor is often important. Most sensor elements possess a transfer function

$$H(s) = \frac{k}{\tau s + 1}.$$

Investigate several sensor elements available today and determine the accuracy available and the time constant of the sensor. Consider two of the following sensors: (1) linear position, (2) temperature with a thermistor, (3) strain measurement, (4) pressure.

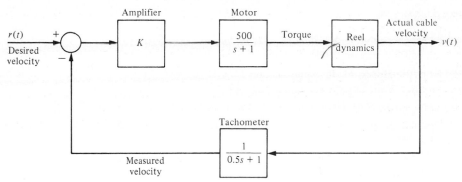

Figure P2.31

2.31. A cable reel control system uses a tachometer to measure the speed of the cable as it leaves the reel. The output of the tachometer is used to control the motor speed of the reel as the cable is unwound off the reel. The system is shown in Fig. P2.31. The radius of the reel, R, is 4 meters when full and 2 meters when empty. The moment of inertia of the reel is $I = 18.5R^4 - 221$. The rate of change of the radius is

$$\frac{dR}{dt} = \frac{-D^2\dot{\theta}}{2\pi W},$$

where W = width of the reel and D = diameter of the cable. The actual speed of the cable is $v(t) = R\dot{\theta}$. The desired cable speed is 50 m/sec. Develop a digital computer simulation of this system and obtain the response of the speed over 20 seconds for the three values of gain K = 0.5, 1.0, and 1.5. The reel angular velocity $\dot{\theta} = d\theta/dt$ is equal to $1/I$ times the integral of the torque. Note that the inertia changes with time as the reel is unwound. However, an equation for I within the simulation will account for this change.

2.32. An interacting control system with two inputs and two outputs is shown in Fig. P2.32. Solve for $C_1(s)/R_1(s)$ and $C_2(s)/R_1(s)$, when $R_2 = 0$.

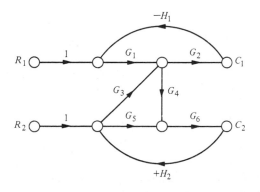

Figure P2.32

REFERENCES

1. C. M. Close and D. K. Frederick, *Modeling and Analysis of Dynamic Systems,* Houghton Mifflin, Boston, 1978.

2. R. E. Balzhiser, M. R. Samuels, and John D. Eliassen, *Engineering Thermodynamics,* Prentice-Hall, Englewood Cliffs, N.J., 1977.

3. T. N. Trick, *Introduction to Circuit Analysis,* Wiley, New York, 1977.

4. Doeblin, E. O., *System Dynamics: Modeling and Response,* Merrill Books, Columbus, Ohio, 1972.

5. J. L. Shearer, A. T. Murphy, and H. H. Richardson, *Dynamic Systems,* Addison-Wesley, Reading, Mass., 1967.

6. R. C. Dorf, *Matrix Algebra—A Programmed Introduction,* Wiley, New York, 1969.

7. R. J. Smith, *Circuits, Devices and Systems,* 3rd ed., Wiley, New York, 1976.

8. M. Athans et al, *Systems, Networks and Computation: Multivariable Methods,* McGraw-Hill, New York, 1974.

9. G. M. Swisher, *Introduction to Linear Systems Analysis,* Matrix Publishers, Champaign, Ill., 1976.

10. B. C. Kuo, *Automatic Control Systems,* 3rd ed., Prentice-Hall, Englewood Cliffs, N.J., 1975.

11. S. J. Mason, "Feedback Theory: Further Properties of Signal Flow Graphs," *Proceedings IRE,* **44,** 7, July 1956: pp. 920–926.

12. W. H. Hayt, Jr. and J. E. Kemmerly, *Engineering Circuit Analysis,* 3rd ed., McGraw-Hill, New York, 1978.

13. A. E. Fitzgerald, D. E. Higginbotham, and A. E. Grabel, *Basic Electrical Engineering,* McGraw-Hill, New York, 1975.

14. R. C. Dorf, *Introduction to Computers and Computer Science,* 2nd ed., Boyd and Fraser Publishing Co., San Francisco, 1977.

15. F. H. Speckhart and W. L. Green, *A Guide to Using CSMP—The Continuous System Modeling Program,* Prentice-Hall, Englewood Cliffs, N.J., 1976.

16. G. Gordon, *System Simulation,* 2nd ed., Prentice-Hall, Englewood Cliffs, N.J., 1978.

17. G. A. Korn and J. V. Wait, *Digital Continuous System Simulation,* Prentice-Hall, Englewood Cliffs, N.J., 1978.

18. D. H. Meadows et al. *The Limits to Growth,* Universe Books, New York, 1972.

19. H. S. D. Cole et al. *Models of Doom: A Critique of Limits to Growth,* Universe Books, New York, 1973.

20. K. Ogata, *System Dynamics,* Prentice-Hall, Englewood Cliffs, N.J., 1978.

21. J. F. Lindsay and S. Katz, *Dynamics of Physical Circuits and Systems,* Matrix Publishers, Champaign, Ill., 1978.

22. R. K. Jurgen, "Feedforward Correction: A Late-Blooming Design," *IEEE Spectrum,* April 1972; pp. 41–43.

23. H. S. Black, "Stabilized Feed-Back Amplifiers," *Electrical Engineering,* **53,** January

1934: pp. 114–120. Also in *Turning Points in American Electrical History,* J. E. Brittain, ed., IEEE Press, New York, 1977; pp. 359–361.

24. P. H. Garrett, *Analog Systems for Microprocessors and Minicomputers,* Reston Publishing Co., Reston, Va., 1978.

25. D. J. Comer, *Modern Electronic Circuit Design,* Addison-Wesley, Reading, Mass., 1976.

26. R. C. Dorf, *Energy, Resources, and Policy,* Addison-Wesley, Reading, Mass., 1978.

27. D. E. Whitney, "Comments on 'Optimization of Natural-Gas Pipeline Systems via Dynamic Programming,'" *IEEE Transactions on Automatic Control,* October 1971, pp. 518–520.

28. B. C. Patten, *System Analysis and Simulation in Ecology,* Academic Press, New York, 1976.

29. T. C. Foin, Jr., *Ecological Systems and the Environment,* Houghton Mifflin, New York, 1976.

30. P. R. Ehrlich, A. H. Ehrlich, and J. P. Holdren, *Ecoscience: Population, Resources, Environment,* W. H. Freeman and Co., San Francisco, 1977.

31. H. T. Odum, *Environment, Power and Society,* Wiley, New York, 1971.

32. G. Flynn, "The Heady Rise of Low Cost Sensors," *Product Engineering,* August 1978.

33. D. McLean, "Mathematical Models of Electrical Machines," *Measurement and Control,* **11,** June 1978; pp. 231–236.

34. J. L. Casti, *Dynamical Systems and Their Applications,* Academic Press, New York, 1977.

35. J. L. Hay, "Interactive Simulation on Minicomputers," *Simulation,* July 1978; pp. 1–5.

36. S. J. Bailey, "AC Motors Enrich Control Design Tradeoffs," *Control Engineering,* November 1978; pp. 34–38.

37. K. M. Abbott and J. D. Wheeler, "Simulation and Control of Thyristor Drives," *IEEE Transactions on Industrial Electronics and Control Instrumentation,* May 1978; pp. 130–137.

38. M. F. Witcher and T. J. McAvoy, "Interacting Control Systems," *ISA Transactions,* **16,** 3, 1977; pp. 35–41.

39. D. C. Karnopp and R. C. Rosenberg, *System Dynamics: A Unified Approach,* Wiley, New York, 1975.

40. N. H. Beachley and H. Harrison, *Introduction to Dynamic System Analysis,* Harper and Row, New York, 1979.

41. D. G. Luenberger, *Introduction to Dynamic Systems,* Wiley, New York, 1979.

42. W. J. Gajda and W. E. Biles, *Engineering: Modeling and Computation,* Houghton Mifflin, Boston, 1979.

3 / Feedback Control System Characteristics

3.1 OPEN- AND CLOSED-LOOP CONTROL SYSTEMS

Now that we are able to obtain mathematical models of the components of control systems, we shall examine the characteristics of control systems. A control system was defined in Section 1.1 as an interconnection of components forming a system configuration which will provide a desired system response. Since a desired system response is known, a signal proportional to the error between the desired and the actual response is generated. The utilization of this signal to control the process results in a closed-loop sequence of operations which is called a feedback system. This closed-loop sequence of operations is shown in Fig. 3.1. The introduction of feedback in order to improve the control system is often necessary. It is interesting that this is also the case for systems in nature, such as biological and physiological systems, and feedback is inherent in these systems. For example, the human heart-rate control system is a feedback control system.

In order to illustrate the characteristics and advantages of introducing feedback, we shall consider a simple, single-loop feedback system. While many control systems are not single-loop in character, a single-loop system is illustrative. A thorough comprehension of the benefits of feedback can best be obtained from the single-loop and then extended to multiloop systems.

An open-loop control system is shown in Fig. 3.2. For contrast, a closed-loop, negative feedback control system is shown in Fig. 3.3. The two control systems are shown in both block diagram and signal flow graph form, although signal flow graphs will be used predominantly for subsequent diagrams. The prime difference between the open- and closed-loop systems is the generation and utilization of the *error signal*. The closed-loop system, when operating correctly, operates so that the error will be reduced to a minimum value. The signal $E_a(s)$ is a measure of the error of the system and is equal to the error $E(s) = R(s) - C(s)$ when $H(s) = 1$.

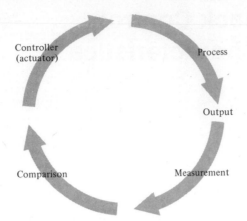

Fig. 3.1. Closed-loop system.

The output of the open-loop system is

$$C(s) = G(s)R(s). \tag{3.1}$$

The output of the closed-loop system is

$$C(s) = G(s)E_a(s) = G(s)(R(s) - H(s)C(s)),$$

and therefore

$$C(s) = \frac{G(s)}{1 + GH(s)} R(s). \tag{3.2}$$

The actuating error signal is

$$E_a(s) = \frac{1}{1 + GH(s)} R(s). \tag{3.3}$$

It is clear that in order to reduce the error, the magnitude of $1 + GH(s)$ must be greater than one over the range of s under consideration.

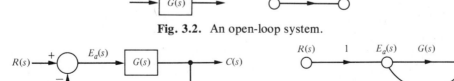

Fig. 3.2. An open-loop system.

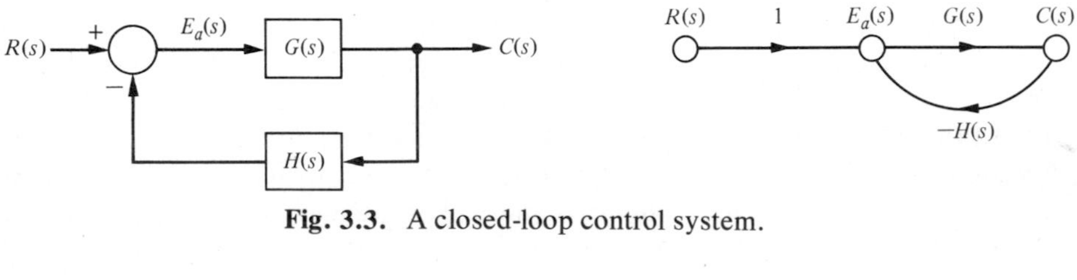

Fig. 3.3. A closed-loop control system.

3.2 SENSITIVITY OF CONTROL SYSTEMS TO PARAMETER VARIATIONS

A process, represented by the transfer function $G(s)$, whatever its nature, is subject to a changing environment, aging, ignorance of the exact values of the process parameters, and other natural factors which affect a control process. In the open-loop system, all these errors and changes result in a changing and inaccurate output. However, a closed-loop system senses the change in the output due to the process changes and attempts to correct the output. The *sensitivity* of a control system to parameter variations is of prime importance. A primary advantage of a closed-loop feedback control system is its ability to reduce the system's sensitivity [1, 2].

For the closed-loop case, if $GH(s) \gg 1$ for all complex frequencies of interest, then from Eq. (3.2) we obtain

$$C(s) \cong \frac{1}{H(s)} R(s). \tag{3.4}$$

That is, the output is affected only by $H(s)$ which may be a constant. If $H(s) = 1$, we have the desired result; that is, the output is equal to the input. However, before we use this approach for all control systems, we must note that the requirement that $G(s)H(s) \gg 1$ may cause the system response to be highly oscillatory and even unstable. But the fact that as we increase the magnitude of the loop transfer function $G(s)H(s)$, we reduce the effect of $G(s)$ on the output, is an exceedingly useful concept. Therefore, the *first advantage* of a feedback system is that the effect of the *variation of the parameters* of the process, $G(s)$, is reduced.

In order to illustrate the effect of parameter variations, let us consider a change in the process so that the new process is $G(s) + \Delta G(s)$. Then, in the open-loop case, the change in the transform of the output is

$$\Delta C(s) = \Delta G(s)R(s). \tag{3.5}$$

In the closed-loop system, we have

$$C(s) + \Delta C(s) = \frac{G(s) + \Delta G(s)}{1 + (G(s) + \Delta G(s))H(s)} R(s). \tag{3.6}$$

Then the change in the output is

$$\Delta C(s) = \frac{\Delta G(s)}{(1 + GH(s) + \Delta GH(s))(1 + GH(s))} R(s) \tag{3.7}$$

When $GH(s) \gg \Delta GH(s)$, as is often the case, we have

$$\Delta C(s) = \frac{\Delta G(s)}{(1 + GH(s))^2} R(s). \tag{3.8}$$

Examining Eq. (3.8), we note that the change in the output of the closed-loop system is reduced by the factor $1 + GH(s)$, which is usually much greater than one over the range of complex frequencies of interest. The factor $1 + GH(s)$ plays a very important role in the characteristics of feedback control systems.

The *system sensitivity* is defined as the ratio of the percentage change in the system transfer function to the percentage change of the process transfer function. The system transfer function is

$$T(s) = \frac{C(s)}{R(s)}, \tag{3.9}$$

and therefore the sensitivity is defined as

$$S = \frac{\Delta T(s)/T(s)}{\Delta G(s)/G(s)}. \tag{3.10}$$

In the limit, for small incremental changes, Eq. (3.10) becomes

$$S = \frac{\partial T/T}{\partial G/G} = \frac{\partial \ln T}{\partial \ln G}. \tag{3.11}$$

Clearly, from Eq. (3.5), the sensitivity of the open-loop system is equal to one. The sensitivity of the closed-loop is readily obtained by using Eq. (3.11). The system transfer function of the closed-loop system is

$$T(s) = \frac{G}{1 + GH}. \tag{3.12}$$

Therefore, the sensitivity of the feedback system is

$$S = \frac{\partial T}{\partial G} \cdot \frac{G}{T} = \frac{1}{(1 + GH)^2} \cdot \frac{G}{G/(1 + GH)} = \frac{1}{1 + GH(s)}. \tag{3.13}$$

Again we find that the sensitivity of the system may be reduced below that of the open-loop system by increasing $GH(s)$ over the frequency range of interest.

The sensitivity of the feedback system to changes in the feedback element $H(s)$ is

$$S_H^T = \frac{\partial T}{\partial H} \cdot \frac{H}{T} = \left(\frac{G}{1 + GH}\right)^2 \cdot \frac{-H}{G/(1 + GH)} = \frac{-GH}{1 + GH}. \tag{3.14}$$

When GH is large, the sensitivity approaches unity and the changes in $H(s)$ directly affect the output response. Therefore, it is important to use feedback components which will not vary with environmental changes or can be maintained constant.

Very often the transfer function of the system $T(s)$, is a fraction of the form: [1]

$$T(s, \alpha) = \frac{N(s,\alpha)}{D(s,\alpha)} \tag{3.15}$$

where α is a parameter that may be subject to variation due to the environment. Then we may obtain the sensitivity to α by rewriting Eq. (3.11) as:

$$S_\alpha^T = \frac{\partial \ln T}{\partial \ln \alpha} = \frac{\partial \ln N}{\partial \ln \alpha}\bigg|_{\alpha_0} - \frac{\partial \ln D}{\partial \ln \alpha}\bigg|_{\alpha_0} \tag{3.16}$$

$$= S_\alpha^N - S_\alpha^D,$$

where α_0 is the normal value of the parameter.

The ability to reduce the effect of the variation of parameters of a control system by adding a feedback loop is an important advantage of feedback control systems. To obtain highly accurate open-loop systems, the components of the open-loop $G(s)$ must be selected carefully in order to meet the exact specifications. However, a closed-loop system allows $G(s)$ to be less accurately specified since the sensitivity to changes or errors in $G(s)$ is reduced by the loop gain $1 + GH(s)$. This benefit of closed-loop systems is a profound advantage for the electronic amplifiers of the communication industry. A simple example will illustrate the value of feedback for reducing sensitivity.

Example 3.1.   The integrated circuit operational amplifier can be fabricated on a single chip of silicon and sold for less than a dollar. As a result, IC operational amplifiers (op amps) are widely used. The model symbol of an op amp is shown in Fig. 3.4(a). We can assume that the gain A is at least 10^4 [14]. The basic inverting amplifier circuit is shown in Fig. 3.4(b). Because of the high input impedance of the op amp, the amplifier input current is negligibly small. At node n we may write the current equation as

$$\frac{e_{\text{in}} - v_n}{R_1} + \frac{v_o - v_n}{R_f} = 0. \tag{3.17}$$

Since the gain of the amplifier is A, $v_o = Av_n$ and therefore

$$v_n = \frac{v_o}{A} \tag{3.18}$$

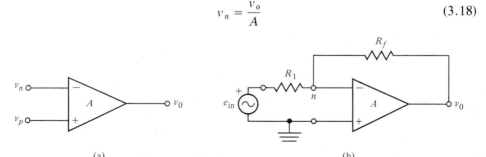

(a)   (b)

Fig. 3.4.   (a) An operational amplifier model symbol. (b) Inverting amplifier circuit.

we may substitute Eq. (3.18) into Eq. (3.17), obtaining

$$\frac{e_{\text{in}}}{R_1} - \frac{v_o}{AR_1} + \frac{v_o}{R_f} - \frac{v_o}{AR_f} = 0. \tag{3.19}$$

Solving for the output voltage, we have

$$v_o = \frac{A(R_f/R_1)e_{\text{in}}}{\dfrac{R_f}{R_1} - A}. \tag{3.20}$$

Alternatively we may rewrite Eq. (3.20) as follows:

$$\frac{v_o}{e_{\text{in}}} = \frac{A}{1 - A(R_1/R_f)} = \frac{A}{1 - Ak}, \tag{3.21}$$

where $k = R_1/R_f$. The signal flow graph representation of the inverting amplifier is shown in Fig. 3.5. Note that when $A \gg 1$ we have

$$\frac{v_o}{e_{\text{in}}} = -\frac{R_f}{R_1}. \tag{3.22}$$

The feedback factor in the diagram is $H(s) = k$ and the open loop transfer function is $G(s) = A$.

The op amp is subject to variations in the amplification A. The sensitivity of the open loop is unity. The sensitivity of the closed-loop amplifier is

$$S_A^T = \frac{\partial T/T}{\partial A/A} = \frac{1}{1 - GH} = \frac{1}{1 - AK}. \tag{3.23}$$

If $A = 10^4$ and $k = 0.1$, we have

$$S_A^T = \frac{1}{1 - 10^3}, \tag{3.24}$$

or the magnitude of the sensitivity is approximately equal to 0.001, which is one-thousandth of the magnitude of the open loop sensitivity. The sensitivity due to changes in the feedback resistance R_1 (or the factor k) is

$$S_k^T = \frac{GH}{1 - GH} = \frac{Ak}{1 - Ak}, \tag{3.25}$$

and the sensitivity to k is approximately equal to 1.

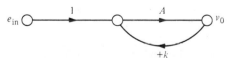

Fig. 3.5. Signal flow graph of inverting amplifier.

We shall return to the concept of sensitivity in subsequent chapters to emphasize the importance of sensitivity in the design and analysis of control systems.

3.3 CONTROL OF THE TRANSIENT RESPONSE OF CONTROL SYSTEMS

One of the most important characteristics of control systems is their transient response. Since the purpose of control systems is to provide a desired response, the transient response of control systems often must be adjusted until it is satisfactory. If an open-loop control system does not provide a satisfactory response, then the process, $G(s)$, must be replaced with a suitable process. By contrast, a closed-loop system can often be adjusted to yield the desired response by adjusting the feedback loop parameters. It should be noted that it is often possible to alter the response of an open-loop system by inserting a suitable cascade filter, $G_1(s)$, preceding the process, $G(s)$, as shown in Fig. 3.6. Then it is necessary to design the cascade transfer function $G_1(s)G(s)$ so that the resulting transfer function provides the desired transient response.

In order to make this concept more readily comprehensible, let us consider a specific control system which may be operated in an open- or closed-loop manner. A speed control system, which is shown in Fig. 3.7, is often used in industrial processes to move materials and products. Several very important speed control systems are used in steel mills for rolling the steel sheets and moving the steel through the mill. The transfer function of the open-loop system was obtained in Eq. (2.70) and for $\omega(s)/V_a(s)$ we have

$$\frac{\omega(s)}{V_a(s)} = G(s) = \frac{K_1}{(\tau_1 s + 1)}, \qquad (3.26)$$

where

$$K_1 = \frac{K_m}{(R_a f + K_b K_m)} \quad \text{and} \quad \tau_1 = \frac{R_a J}{(R_a f + K_b K_m)}.$$

In the case of a steel mill, the inertia of the rolls is quite large and a large armature controlled motor is required. If the steel rolls are subjected to a step command for a speed change of

$$V_A(s) = \frac{k_2 E}{s}, \qquad (3.27)$$

the output response is

$$\omega(s) = G(s)V_A(s). \qquad (3.28)$$

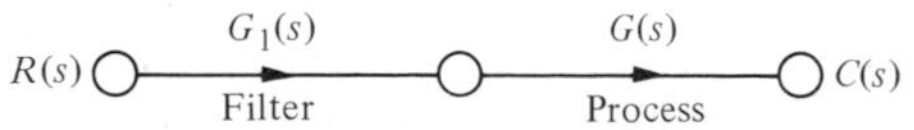

$$R(s) \circ\!\!-\!\!\xrightarrow[\text{Filter}]{G_1(s)}\!\!-\!\!\circ\!\!-\!\!\xrightarrow[\text{Process}]{G(s)}\!\!-\!\!\circ\, C(s)$$

Fig. 3.6. Cascade filter open-loop system.

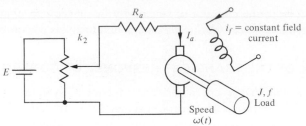

Fig. 3.7. Open-loop speed control system.

The transient speed change is then

$$\omega(t) = K_1(k_2E)(1 - e^{-t/\tau_1}). \tag{3.29}$$

If this overdamped transient response is too slow, there is little choice but to choose another motor with a different time constant τ_1, if possible. However, since τ_1 is dominated by the load inertia, little hope for much alteration of the transient response remains.

A closed-loop speed control system is easily obtained by utilizing a tachometer to generate a voltage proportional to the speed as shown in Fig. 3.8. This voltage is subtracted from the potentiometer voltage and amplified as shown in Fig. 3.8. A practical transistor amplifier circuit for accomplishing this feedback in low power applications is shown in Fig. 3.9 [3]. The closed-loop transfer function is

$$\frac{\omega(s)}{R(s)} = \frac{K_aG(s)}{1 + K_aK_tG(s)}$$
$$= \frac{K_aK_1}{\tau_1s + 1 + K_aK_tK_1}$$
$$= \frac{K_aK_1/\tau_1}{s + [(1 + K_aK_tK_1)/\tau_1]} \tag{3.30}$$

The amplifier gain K_a may be adjusted to meet the required transient response specifications. Also, the tachometer gain constant K_t may be varied, if necessary.

The transient response to a step change in the input command is then

$$\omega(t) = \frac{K_aK_1}{(1 + K_aK_tK_1)} (k_2E)(1 - e^{-pt}), \tag{3.31}$$

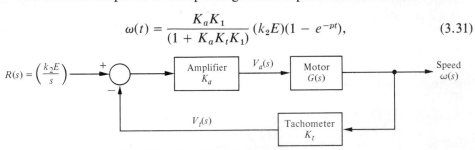

Fig. 3.8. Closed-loop speed control system.

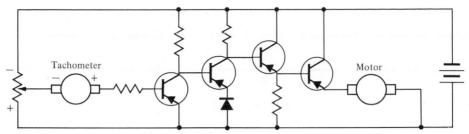

Fig. 3.9. Transistorized speed control system.

where $p = (1 + K_a K_t K_1)/\tau_1$. Since the load inertia is assumed to be very large, we alter the response by increasing K_a, and we have the approximate response

$$\omega(t) \simeq \frac{1}{K_t}(k_2 E)\left(1 - e^{\frac{-(K_a K_t K_1)t}{\tau_1}}\right). \tag{3.32}$$

For a typical application the open-loop pole might be $1/\tau_1 = 0.10$, while the closed-loop pole could be at least $(K_a K_t K_1)/\tau_1 = 10$, a factor of 100 in the improvement of the speed of response. It should be noted that in order to attain the gain $K_a K_t K_1$, the amplifier gain K_a must be reasonably large, and the armature voltage signal to the motor and its associated torque signal will be larger for the closed-loop than for the open-loop operation. Therefore, a larger motor will be required in order to avoid saturation of the motor.

Also, while we are considering this speed control system, it will be worthwhile to determine the sensitivity of the open- and closed-loop systems. As before, the sensitivity of the open-loop system to a variation in the motor constant or the potentiometer constant k_2 is unity. The sensitivity of the closed-loop system to a variation in K_m is

$$\begin{aligned}
S^T_{K_m} &= \frac{1}{1 + GH(s)} \\
&= \frac{1}{1 + K_a K_t G(s)} \\
&= \frac{[s + (1/\tau_1)]}{[s + (K_a K_t K_1 + 1)/\tau_1]}.
\end{aligned} \tag{3.33}$$

Using the typical values given in the previous paragraph, we have

$$S^T_{K_m} = \frac{(s + 0.10)}{(s + 10)}. \tag{3.34}$$

We find that the sensitivity is a function of s and must be evaluated for various values of frequency. This type of frequency analysis is straightforward but will be

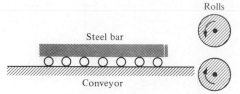

Fig. 3.10.   Steel rolling mill.

deferred until a later chapter. However, it is clearly seen that at a specific frequency, for example, $s = j\omega = j1$, the magnitude of the sensitivity is approximately $\left| S_{K_m}^T \right| \cong 0.1$.

3.4 DISTURBANCE SIGNALS IN A FEEDBACK CONTROL SYSTEM

The third most important effect of feedback in a control system is the control and partial elimination of the effect of disturbance signals. Many control systems are subject to extraneous disturbance signals which cause the system to provide an inaccurate output. Electronic amplifiers have inherent noise generated within the integrated circuits or transistors; radar antennas are subjected to wind gusts, and many systems generate unwanted distortion signals due to nonlinear elements. Feedback systems have the beneficial aspect that the effect of distortion, noise, and unwanted disturbances can be effectively reduced.

As a specific example of a system with an unwanted disturbance, let us reconsider the speed control system for a steel rolling mill [12]. Rolls passing steel through are subject to large load changes or disturbances. As a steel bar approaches the rolls (see Fig. 3.10), the rolls turn unloaded. However, when the bar engages in the rolls, the load on the rolls increases immediately to a large value. This loading effect can be approximated by a step change of disturbance torque as shown in Fig. 3.11. Alternatively, we might examine the speed-torque curves of a typical motor as shown in Fig. 3.12.

The transfer function model of an armature controlled dc-motor with a load torque disturbance was determined in Example 2.3 and is shown in Fig. 3.11, where

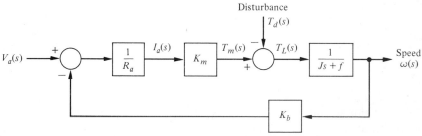

Fig. 3.11.   Open-loop speed control system.

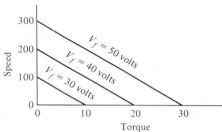

Fig. 3.12. Motor speed-torque curves.

it is assumed that L_a is negligible. Change in speed due to load disturbance is

$$\omega(s) = \left(\frac{-1}{Js + f + (K_m K_b/R_a)}\right) T_d(s). \tag{3.35}$$

The steady-state error in speed due to the load torque $T_d(s) = D/s$ is found by using the final-value theorem. Therefore, for the open-loop system, we have

$$\lim_{t \to \infty} \omega(t) = \lim_{s \to \infty} s\omega(s) = \lim_{s \to 0} s\left(\frac{-1}{Js + f + (K_m K_b/R_a)}\right)\left(\frac{D}{s}\right)$$

$$= \frac{-D}{f + (K_m K_b/R_a)}. \tag{3.36}$$

The closed-loop speed control system is shown in block diagram form in Fig. 3.13. The closed-loop system is shown in the more general signal flow graph form in Fig. 3.14. The output, $\omega(s)$, of the closed-loop system of Fig. 3.14 can be obtained by utilizing the signal-flow gain formula and is

$$\omega(s) = \frac{-G_2(s)}{1 + G_1(s)G_2(s)H(s)} T_d(s). \tag{3.37}$$

Then, if $G_1 G_2 H(s)$ is much greater than one over the range of s, we obtain the approximate result

$$\omega(s) \simeq \frac{-1}{G_1(s)H(s)} T_d(s). \tag{3.38}$$

Therefore, if $G_1(s)$ is made sufficiently large, the effect of the disturbance can be decreased by closed-loop feedback.

The output for the speed control system of Fig. 3.13 due to the load disturbance, when the input $R(s) = 0$, may be obtained by using Mason's formula as

$$\omega(s) = \frac{-[1/(Js + f)]}{1 + (K_t K_a K_m/R_a)[1/(Js + f)] + (K_m K_b/R_a)[1/(Js + f)]} T_d(s).$$

$$= \frac{-1}{Js + f + (K_m/R_a)(K_t K_a + K_b)} T_d(s). \tag{3.39}$$

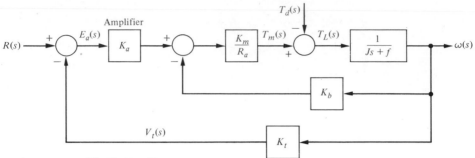

Fig. 3.13. Closed-loop speed tachometer control system.

Again, the steady-state output is obtained by utilizing the final-value theorem, and we have

$$\lim_{t\to\infty} \omega(t) = \lim_{s\to 0} (s\omega(s))$$

$$= \frac{-1}{f + (K_m/R_a)(K_t K_a + K_b)} D; \tag{3.40}$$

when the amplifier gain is sufficiently high, we have

$$\omega(\infty) \simeq \frac{-R_a}{K_a K_m K_t} D. \tag{3.41}$$

The ratio of closed-loop to open-loop steady-state speed output due to an undesirable disturbance is

$$\frac{\omega_c(\infty)}{\omega_0(\infty)} = \frac{R_a f + K_m K_b}{K_a K_m K_t} \tag{3.42}$$

and is usually less than 0.02.

This advantage of a feedback speed control system can also be illustrated by considering the speed-torque curves for the closed-loop system. The closed-loop system speed-torque curves are shown in Fig. 3.15. Clearly, the improvement of the feedback system is evidenced by the almost horizontal curves which indicate that the speed is almost independent of the load torque.

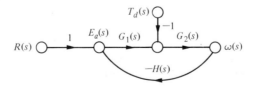

Fig. 3.14. Signal-flow graph of closed-loop system.

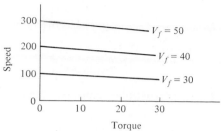

Fig. 3.15. The closed-loop system speed-torque curves.

In general, a primary reason for introducing feedback is the ability to alleviate the effects of disturbances and noise signals occurring within the feedback loop. A noise signal which is prevalent in many systems is the noise generated by the measurement sensor. This disturbance or noise, $N(s) = T_d(s)$, can be represented as shown in Fig. 3.16. The effect of the noise on the output is

$$C(s) = \frac{-G_1G_2H_2(s)}{1 + G_1G_2H_1H_2(s)} N(s), \qquad (3.43)$$

which is approximately

$$C(s) \cong - \frac{1}{H_1(s)} N(s). \qquad (3.44)$$

Clearly, the designer must obtain a maximum value of $H_1(s)$, which is equivalent to maximizing the signal-to-noise ratio of the measurement sensor. This necessity is equivalent to requiring that the feedback elements $H(s)$ be well-designed and operated with minimum noise, drift, and parameter variation. This is equivalent to the requirement determined from the sensitivity function, Eq. (3.14), which showed that $S_H^T \cong 1$. Therefore, one must be aware of the necessity of assuring the quality and constancy of the feedback sensors and elements. This is usually possible since the feedback elements operate at low power levels and can be well designed at reasonable cost.

The equivalency of sensitivity S_G^T and the response of the closed-loop system to

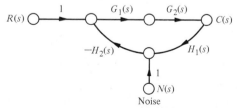

Fig. 3.16. Closed-loop control system with measurement noise.

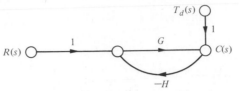

Fig. 3.17. Closed-loop control system with output noise.

a disturbance input can be illustrated by considering Fig. 3.14. The sensitivity of the system to G_2 is

$$S_{G_2}^T = \frac{1}{1 + G_1 G_2 H(s)} \cong \frac{1}{G_1 G_2 H(s)}. \tag{3.45}$$

The effect of the disturbance on the output is

$$\frac{C(s)}{T_d(s)} = \frac{-G_2(s)}{1 + G_1 G_2 H(s)} \cong \frac{-1}{G_1 H(s)}. \tag{3.46}$$

In both cases, we found that the undesired effects could be alleviated by increasing $G_1(s) = K_a$, the amplifier gain. The utilization of feedback in control systems is primarily to reduce the sensitivity of the system parameter variations and the effect of disturbance inputs. It is noteworthy that the effort to reduce the effects of parameter variations or disturbances is equivalent and we have the fortunate circumstance that they reduce simultaneously. As a final illustration of this fact, we note that for the system shown in Fig. 3.17, the effect of the noise or disturbance on the output is

$$\frac{C(s)}{T_d(s)} = \frac{1}{1 + GH(s)}, \tag{3.47}$$

which is identically equal to the sensitivity S_G^T.

Quite often, noise is present at the input to the control system. For example, the signal at the input to the system might be $r(t) + n(t)$, where $r(t)$ is the desired system response and $n(t)$ is the noise signal. The feedback control system, in this case, will simply process the noise as well as the input signal $r(t)$ and will not be able to improve the signal-noise ratio which is present at the input to the system. However, if the frequency spectrums of the noise and input signals are of a different character, the output signal-noise ratio can be maximized, often by simply designing a closed-loop system transfer function which has a low-pass frequency response.

3.5 STEADY-STATE ERROR

A feedback control system is valuable since it provides the engineer with the ability to adjust the transient response. In addition, as we have seen, the sensitivity of the

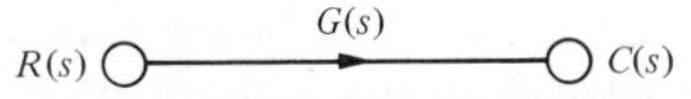

$G(s)$

$R(s) \bigcirc \xrightarrow{\hspace{3cm}} \bigcirc C(s)$

Fig. 3.18. Open-loop control system.

system and the effect of disturbances can be reduced significantly. However, as a further requirement, one must examine and compare the final steady-state error for an open-loop and a closed-loop system.

The error of the open-loop system shown in Fig. 3.18 is

$$
\begin{aligned}
E_0(s) &= R(s) - C(s) \\
&= (1 - G(s))R(s).
\end{aligned}
\tag{3.48}
$$

The error of the closed-loop system shown in Fig. 3.19, when $H(s) = 1$, is

$$
E_c(s) = \frac{1}{1 + G(s)} R(s).
\tag{3.49}
$$

In order to calculate the steady-state error, we utilize the final-value theorem which is

$$
\lim_{t \to \infty} e(t) = \lim_{s \to 0} sE(s).
\tag{3.50}
$$

Therefore, using a unit step input as a comparable input, we obtain for the open-loop system

$$
\begin{aligned}
e_0(\infty) &= \lim_{s \to 0} s(1 - G(s)) \left(\frac{1}{s} \right) \\
&= \lim_{s \to 0} (1 - G(s)) \\
&= 1 - G(0).
\end{aligned}
\tag{3.51}
$$

For the closed-loop system, when $H(s) = 1$, we have

$$
\begin{aligned}
e_c(\infty) &= \lim_{s \to 0} s \left(\frac{1}{1 + G(s)} \right) \left(\frac{1}{s} \right) \\
&= \frac{1}{1 + G(0)}.
\end{aligned}
\tag{3.52}
$$

The value of $G(s)$ when $s = 0$ is often called the dc-gain and is normally greater than one. Therefore, the open-loop system will usually have a steady-state error of

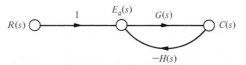

Fig. 3.19. Closed-loop control system.

significant magnitude. By contrast, the closed-loop system with a reasonably large dc-loop gain $GH(0)$ will have a small steady-state error.

Upon examination of Eq. (3.51), one notes that the open-loop control system can possess a zero steady-state error by simply adjusting and calibrating the dc-gain, $G(0)$, of the system, so that $G(0) = 1$. Therefore, one may logically ask, what is the advantage of the closed-loop system in this case? Again, we return to the concept of the sensitivity of the system to parameter changes as our answer to this question. In the open-loop system, one may calibrate the system so that $G(0) = 1$, but during the operation of the system it is inevitable that the parameters of $G(s)$ will change due to environmental changes and the dc-gain of the system will no longer be equal to one. However, since it is an open-loop system the steady-state error will remain other than zero until the system is maintained and recalibrated. By contrast, the closed-loop feedback system continually monitors the steady-state error and provides an actuating signal in order to reduce the steady-state error. Thus, we find that it is the sensitivity of the system to parameter drift, environmental effects, and calibration errors that encourages the introduction of negative feedback.

The advantage of the closed-loop system in reducing the steady-state error of the system resulting from parameter changes and calibration errors may be illustrated by an example. Let us consider a system with a process transfer function

$$G(s) = \frac{K}{\tau s + 1},$$
(3.53)

which would represent a thermal control process, a voltage regulator, or a water-level control process. For a specific setting of the desired input variable, which may be represented by the normalized unit step input function, we have $R(s) = 1/s$. Then, the steady-state error of the open-loop system is, as in Eq. (3.51),

$$e_0(\infty) = 1 - G(0) = 1 - K$$
(3.54)

when a consistent set of dimensional units are utilized for $R(s)$ and K.

The steady-state error for the closed-loop system, with unity-feedback, is

$$e_c(\infty) = \frac{1}{1 + G(0)} = \frac{1}{1 + K}.$$
(3.55)

For the open-loop system, one would calibrate the system so that $K = 1$ and the steady-state error is zero. For the closed-loop system, one would set a large gain K, for example, $K = 100$. Then the closed-loop system steady-state error is $e_c(\infty) = 1/101$.

If the calibration of the gain setting drifts or changes in some way by $\Delta K/K = 0.1$, a 10% change, the open-loop steady-state error is $\Delta e_0(\infty) = 0.1$ and the percent change from the calibrated setting is

$$\frac{\Delta e_0(\infty)}{|r(t)|} = \frac{0.10}{1}$$
(3.56)

or 10%. By contrast, the steady-state error of the closed-loop system, with $\Delta K/K = 0.1$, is $e_c(\infty) = \frac{1}{91}$, if the gain decreases. Thus, the change in the

$$\Delta e_c(\infty) = \frac{1}{101} - \frac{1}{91}, \qquad (3.57)$$

and the relative change is

$$\frac{\Delta e_c(\infty)}{|r(t)|} = 0.0011 \qquad (3.58)$$

or 0.11%. This is indeed a significant improvement.

3.6 THE COST OF FEEDBACK

The addition of feedback to a control system results in the advantages outlined in the previous sections. However, it is natural that these advantages have an attendant cost. The cost of feedback is first manifested in the increased number of *components* and the *complexity* of the system. In order to add the feedback, it is necessary to consider several feedback components, of which the measurement component (sensor) is the key component. The sensor is often the most expensive component in a control system. Furthermore, the sensor introduces noise and inaccuracies into the system.

The second cost of feedback is the *loss of gain*. For example, in a single-loop system, the open-loop gain is $G(s)$ and is reduced to $G(s)/(1 + G(s))$ in a unity negative feedback system. The reduction in closed-loop gain is $1/(1 + G(s))$, which is exactly the factor that reduces the sensitivity of the system to parameter variations and disturbances. Usually, we have open-loop gain to spare, and we are more than willing to trade it for increased control of the system response.

However, we should note that it is the gain of the input-output transmittance that is reduced. The control system does possess a substantial power gain which is fully utilized in the closed-loop system.

Finally, a cost of feedback is the introduction of the possibility of *instability*. While the open-loop system is stable, the closed-loop system may not be always stable. The question of the stability of a closed-loop system is deferred until Chapter 5, where it can be treated more completely.

The addition of feedback to dynamic systems results in several additional problems for the designer. However, for most cases, the advantages far outweigh the disadvantages, and a feedback system is utilized. Therefore, it is necessary to consider the additional complexity and the problem of stability when designing a control system. One complex control system is shown in Fig. 3.20.

It has become clear that it is desired that the output of the system $C(s)$ equal the input $R(s)$. However, upon reflection, one might ask, "Why not simply set the transfer function $G(s) = C(s)/R(s)$ equal to 1?" (See Fig. 3.2.) The answer to this question becomes apparent once we recall that the process (or plant) $G(s)$ was

Fig. 3.20. The Viking Lander is the most complex robot (remote space vehicle) ever constructed and proved remarkably dependable. Here Viking II demonstrates its long boom arm for collecting soil samples. The Lander houses numerous experiments, all conducted under automatic control. Viking II landed on Mars on September 3, 1976. (Courtesy of Martin Marietta Corp.)

necessary in order to provide the desired output; that is, the transfer function $G(s)$ represents a real process and possesses dynamics which cannot be neglected. If we set $G(s)$ equal to 1, we imply that the output is directly connected to the input. However, one must recall that a specific output, such as temperature, shaft rotation, or engine speed, is desired while the input might be a potentiometer setting or a voltage. The process $G(s)$ is necessary in order to provide the physical process between $R(s)$ and $C(s)$. Therefore, a transfer function $G(s) = 1$ is unrealizable, and we must settle for a practical transfer function.

3.7 SUMMARY

The fundamental reasons for using feedback, despite its cost and additional complexity, are as follows:

1. the decrease in the sensitivity of the system to variations in the parameters of the process $G(s)$;
2. the ease of control and adjustment of the transient response of the system;
3. the improvement in the rejection of the disturbance and noise signals within the system;
4. the improvement in the reduction of the steady-state error of the system.

Feedback control systems possess many beneficial characteristics, and it is not surprising that one finds a multitude of feedback control systems in industry, government, and nature.

PROBLEMS

3.1. The open-loop transfer function of a fluid-flow system obtained in Problem 2.13 can be written as

$$G(s) = \frac{\Delta Q_2(s)}{\Delta Q_1(s)} = \frac{1}{\tau s + 1},$$

where $\tau = RC$, R is a constant equivalent to the resistance offered by the orifice so that $1/R = \frac{1}{2} k H_0^{-1/2}$, and C = the cross-sectional area of the tank. Since $\Delta H = R \, \Delta Q_2$, we have for the transfer function relating the head to the input change:

$$G_1(s) = \frac{\Delta H(s)}{\Delta Q_1(s)} = \frac{R}{RCs + 1}.$$

For a closed-loop feedback system, a float level sensor and valve may be used as shown in Fig. P3.1.[4] Assuming the float is a negligible mass, the valve is controlled so that a reduction in the flow rate, ΔQ_1, is proportional to an increase in head, ΔH, or $\Delta Q_1 = -K \, \Delta H$. Draw a closed-loop flow graph or block diagram. Determine and compare the open-loop and

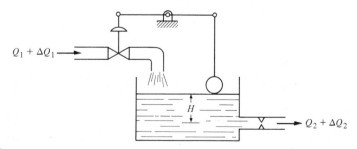

$Q_1 + \Delta Q_1 \longrightarrow$

H

$Q_2 + \Delta Q_2$

Figure P3.1

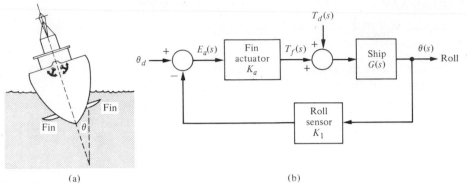

Figure P3.2

(a)   (b)

closed-loop system for (a) sensitivity to changes in the equivalent coefficient R and the feedback coefficient K; (b) ability to reduce the effects of a disturbance in the level $\delta H(s)$; (c) steady-state error of the level (head) for a step change of the input $\Delta Q_1(s)$.

3.2. It is important to ensure passenger comfort on ships by stabilizing the ship's oscillations due to waves [4, 5]. Most ship stabilization systems use fins or hydrofoils projecting into the water in order to generate a stabilization torque on the ship. A simple diagram of a ship stabilization system is shown in Fig. P3.2. The rolling motion of a ship can be regarded as an oscillating pendulum with a deviation from the vertical of θ degrees and a typical period of 3 sec. The transfer function of a typical ship is

$$G(s) = \frac{\omega_n^2}{s^2 + 2\zeta\omega_n s + \omega_n^2},$$

where $\omega_n = 2\pi/T = 2$, $T = 3.14$ sec, and $\zeta = 0.10$. With this low damping factor ζ, the oscillations continue for several cycles and the rolling amplitude can reach 18° for the expected amplitude of waves in a normal sea. Determine and compare the open-loop and closed-loop system for (a) sensitivity to changes in the actuator constant K_a and the roll sensor K_1; and (b) the ability to reduce the effects of the disturbance of the waves. Note that the desired roll $\theta_d(s)$ is zero degrees.

3.3. One of the most important variables that must be controlled in industrial and chemical systems is temperature. A simple representation of a thermal control system is shown in Fig. P3.3 [9, 10]. The temperature $\mathfrak{I}$ of the process is controlled by the heater with a resistance R. An approximate representation of the dynamics linearly relates the heat loss from the process to the temperature difference ($\mathfrak{I} - \mathfrak{I}_e$). This relation holds if the temperature difference is relatively small and the energy storage of the heater and the vessel walls is negligible. Also, it is assumed that the voltage connected to the heater e_h, is proportional to $e_{desired}$ or $e_h = kE_b = k_a E_b e(t)$, where k_a is the constant of the actuator. Then the linearized open-loop response of the system is

$$\mathfrak{I}(s) = \frac{(k_1 k_a E_b)}{\tau s + 1} E(s) + \frac{\mathfrak{I}_e(s)}{\tau s + 1},$$

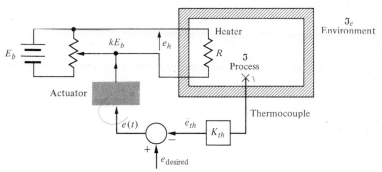

Figure P3.3

where

$$\tau = MC/\rho A,$$
$$M = \text{mass in tank},$$
$$A = \text{surface area of tank},$$
$$\rho = \text{heat transfer constant},$$
$$C = \text{specific heat constant},$$
$$k_1 = \text{a dimensionality constant [4]}.$$
$$e_{th} = \text{output voltage of thermocouple}.$$

Determine and compare the open-loop and closed-loop systems for (a) sensitivity to changes in the constant $K = k_1 k_a E_b$; (b) ability to reduce the effects of a step disturbance in the environmental temperature $\Delta \Im_e(s)$; and (c) the steady-state error of the temperature controller for a step change in the input, e_{desired}.

3.4. A control system has two forward signal paths as shown in Fig. P3.4. (a) Determine the overall transfer function $T(s) = C(s)/R(s)$. (b) Calculate the sensitivity S_G^T using Eq. (3.16). (c) Does the sensitivity depend on $U(s)$ or $M(s)$?

3.5. Large microwave antennas have become increasingly important for missile tracking, radio astronomy, and satellite tracking. A large antenna, for example, with a diameter of 60 ft, is subject to large wind gust torques. A proposed antenna is required to have an error of less than 0.20° in a 35 mph wind [6]. Experiments show that this wind force exerts a maximum disturbance at the antenna of 200,000 lb-ft at 35 mph, or equivalent to 20 volts at the input, $T_d(s)$, to the amplidyne. Also, one problem of driving large antennas is the form of the system transfer function with possesses a structural resonance. The antenna servosystem is

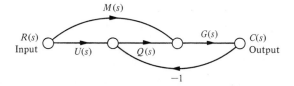

Figure P3.4

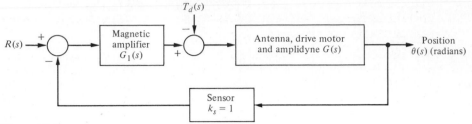

Figure P3.5

shown in Fig. P3.5. The transfer function of the antenna, drive motor, and amplidyne is approximated by

$$G(s) = \frac{\omega_n^2}{s^2 + 2\zeta\omega_n s + \omega_n^2},$$

where $\zeta = 0.4$ and $\omega_n = 10$. The transfer function of the magnetic amplifier is approximately

$$G_1(s) = \frac{k_a}{\tau s + 1},$$

where $\tau = 0.20$ sec. (a) Determine the sensitivity of the system to a change of the parameter k_a. (b) The system is subjected to a disturbance $T_d(s) = 15/s$. Determine the required magnitude of k_a in order to maintain the steady-state error of the system less than $0.20°$ when the input $R(s)$ is zero. (c) Determine the error of the system when subjected to a disturbance $T_d(s) = 15/s$ when the system is operating as an open-loop system ($k_s = 0$).

3.6. An automatic speed control system will be necessary for passenger cars traveling on the automatic highways of the future [7, 8]. A model of a feedback speed control system for a standard vehicle is shown in Fig. P3.6. The load disturbance due to a percent grade, $\Delta D(s)$, is shown also. The engine gain K_e varies within the range of 10 to 1,000 for various models of automobiles. The engine time constant, τ_e, is 20 sec. (a) Determine the sensitivity of the system to changes in the engine gain K_e. (b) Determine the effect of the load torque on the speed. (c) Determine the constant percent grade $\Delta D(s) = \Delta d/s$ for which the engine stalls in

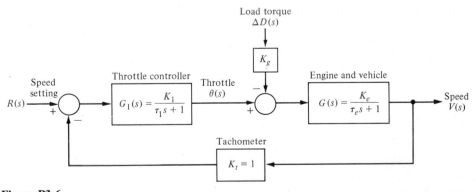

Figure P3.6

terms of the gain factors. Note that since the grade is constant, the steady-state solution is sufficient. Assume that $R(s) = 30/s$ km/hr and that $K_e K_1 \gg 1$. When $(K_g/K_1) = 2$, what percent grade Δd would cause the automobile to stall?

3.7. The dynamic behavior of a complex business organization may be considered to be a feedback control system. A simple, single-loop model of a management control system is shown in Fig. P3.7. The inertia of the organization opposes the act of accomplishing the prescribed task and possesses a delay τ. Management senses the error $E_a(s)$, and assesses and sums up the situation which, in effect, is approximated by $G_1(s)$. The gains, K_1 and K_2, represent the per-unit level of effort of management and engineering, respectively. The feedback signal, $B(s)$, is proportional to the output plus the rate of change of the output. (a) Determine the sensitivity of the system to changes in the management gain, K_1. (b) Determine the effect of the disturbance on the output by obtaining the transfer function $C(s)/D(s)$. (c) Determine the steady-state error for the system with a disturbance $D(s) = d/s$ and the input $R(s) = 0$. (d) Determine the steady-state error when $R(s) = 1/s$ and $D(s) = 0$.

3.8. A simplified model of the human speech system is shown in Fig. P3.8 [11]. Speech control is a feedback process occurring naturally in humans, and when the system becomes unstable, a person stutters. The model shown does not duplicate the central nervous system, but it does allow the analyst to determine the effects of various elements in the system. The vocal tract is a resonant cavity and is approximated by $G_2(s)$, where $\omega_n = 2\pi(500)$, $\zeta = 0.032$, and, normally, $K_2 = 0.01$. The central nervous system is approximated by the synaptic delay

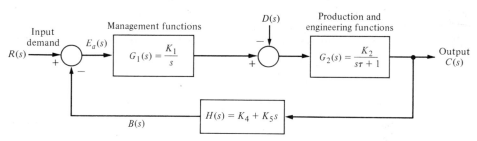

Figure P3.7

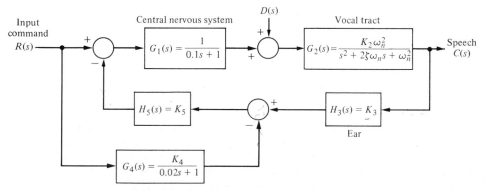

Figure P3.8

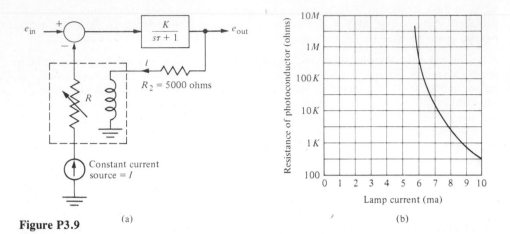

Figure P3.9   (a)   (b)

of the nerves. (a) Determine the sensitivity of the system to the parameters K_2 and K_4 in order to investigate their effect in the case of stuttering. (b) Determine the effect of the disturbance on the output. The disturbance represents an extraneous noise signal generated within the central nervous system.

3.9. A useful unidirectional sensing device is the photoemitter sensor [9]. A light source is sensitive to the emitter current flowing and alters the resistance of the photosensor. Both the light source and photoconductor are packaged in a single 4-terminal device. This device provides a large gain and total isolation. A feedback circuit utilizing this device is shown in Fig. P3.9(a), and the nonlinear resistance-current characteristic is shown in Fig. 3.9(b) for the Raytheon CK1116. The resistance curve can be represented by the equation [9]

$$\log_{10} R = \frac{0.175}{(i - 0.005)^{1/2}}.$$

The normal operating point is obtained when $e_{out} = 35$ v, and $e_{in} = 2.0$ v. (a) Determine the closed-loop transfer function of the system. (b) Determine the sensitivity of the system to changes in the gain K.

3.10. For a paper processing plant, it is important to maintain a constant tension on the continuous sheet of paper between the windoff and windup rolls. The tension varies as the widths of the rolls change, and an adjustment in the take-up motor speed is necessary, as shown in Fig. P3.10. If the windup motor speed is uncontrolled, as the paper transfers from the windoff roll to the windup roll, the velocity v_0 decreases and the tension of the paper drops [10, 11]. The three-roller and spring combination provides a measure of the tension of the paper. The spring force is equal to $k_1 y$ and linear differential transformer, rectifier, and amplifier may be represented by $e_0 = -k_2 y$. Therefore, the measure of the tension is described by the relation $2T(s) = k_1 y$, where y is the deviation from the equilibrium condition and $T(s)$ is the vertical component of the deviation in tension from the equilibrium condition. The time constant of the motor is $\tau = L_a/R_a$ and the linear velocity of the windup roll is twice the angular velocity of the motor; that is, $v_0(t) = 2\omega_0(t)$. The equation of the motor is then

$$E_0(s) = \frac{1}{K_m} [\tau s \omega_0(s) + \omega_0(s)] + k_3 \Delta T(s),$$

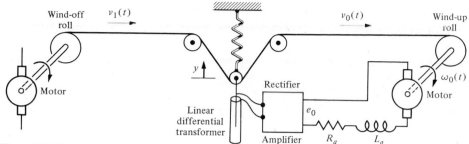

Figure P3.10

where ΔT = a tension disturbance. (a) Draw the closed-loop block diagram for the system, including the disturbance $\Delta T(s)$. (b) Add the effect of a disturbance in the windoff roll velocity $\Delta V_1(s)$ to the block diagram. (c) Determine the sensitivity of the system to the motor constant K_m. (d) Determine the steady-state error in the tension when a step disturbance in the input velocity $\Delta V_1(s) = A/s$ occurs.

3.11. One important objective of the paper-making process is to maintain uniform consistency of the stock output as it progresses to drying and rolling. A diagram of the thick stock consistency dilution control system is shown in Fig. P3.11(a). The amount of water added determines the consistency. The signal flow diagram of the system is shown in Fig. P3.11(b). Let $H(s) = 1$ and

$$G_c(s) = \frac{K}{(10s + 1)},$$

$$G(s) = \frac{1}{(2s + 1)}.$$

Determine:

(a) the closed loop transfer function $T(s) = C(s)/R(s)$,

(b) the sensitivity S_K^T,

(c) the steady state error for a step change in the desired consistency $R(s) = A/s$.

(d) Calculate the value of K required for an allowable steady-state error of one-percent.

3.12. Two feedback systems are shown in signal-flow diagram form in Fig. P3.12(a) and (b). (a) Evaluate the closed-loop transfer functions T_1 and T_2 for each system. (b) Show that $T_1 = T_2 = 100$ when $K_1 = K_2 = 100$. (c) Compare the sensitivities of the two systems with respect to the parameter K_1 for the nominal values of $K_1 = K_2 = 100$.

3.13. One form of closed loop transfer function is

$$T(s) = \frac{G_1(s) + kG_2(s)}{G_3(s) + kG_4(s)}.$$

(a) Use Eq. (3.16) to show that [1]

$$S_k^T = \frac{k(G_2G_3 - G_1G_4)}{(G_3 + kG_4)(G_1 + kG_2)}.$$

(b) Determine the sensitivity of the system shown in Fig. P3.13, using the equation verified in part (a).

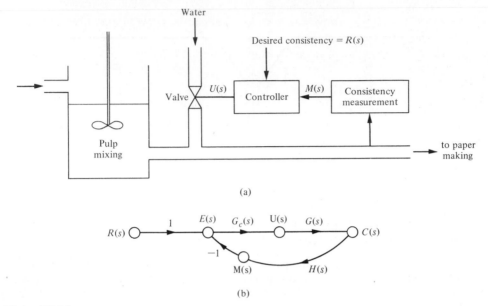

(a)

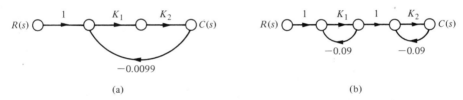

(b)

Figure P3.11

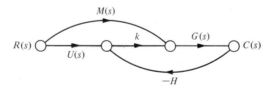

(a)   (b)

Figure P3.12

Figure P3.13

REFERENCES

1. P. M. Frank, *Introduction to System Sensitivity Theory,* Academic Press, New York, 1978.

2. J. B. Cruz, Jr., *Feedback Systems,* McGraw-Hill, New York, 1972, Ch. 2.

3. *Motomatic Speed Control,* Electro-Craft Corp., Hopkins, Minn, 1978.

4. J. Hall, "Guide to Level Monitoring," *Instruments and Control Systems,* October 1978; pp. 25–33.

5. P. LaFrance, "Ship Hydrodynamics," *Physics Today,* June 1978; pp. 34–42.

6. D. V. Stallard, "Servo Problems and Techniques in Large Antennas," *Proceedings of the JACC,* 1965; pp. 313–325.

7. P. Walker, "Getting Serious About EV Motors," *Machine Design,* May 11, 1978; pp. 108–112.

8. C. P. Gilmore, "Cruise Control," *Popular Science,* February 1978; pp. 50–54.

9. C. D. Johnson, *Process Control Instrumentation Technology,* Wiley, New York, 1977.

10. J. Martin, Jr., et al, "Comparison of Tuning Methods for Temperature Control of a Chemical Reactor," *ISA Transactions,* **16,** 4, 1977; pp. 53–57.

11. B. R. Butler, Jr., "The Stuttering Problem Considered From an Automatic Control Point of View," *Proceedings San Diego Symposium for Biomedical Engineering,* 1964; pp. 47–62.

12. P. C. Sen and M. L. MacDonald, "Thyristorized DC Drives With Regenerative Braking and Speed Reversal," *IEEE Transactions on Industrial Electronics and Control Instrumentation,* November 1978; pp. 347–354.

13. M. Fjeld, "Application of Modern Control Concepts on a Kraft Paper Machine," *Automatica,* **14,** 1978; pp. 107–117.

14. G. C. Temes and J. W. LaPatra, *Circuit Synthesis and Design,* McGraw-Hill, New York, 1977, Ch. 7.

4 / The Performance of Feedback Control Systems

4.1 INTRODUCTION

The ability to adjust the transient and steady-state performance is a distinct advantage of feedback control systems. In order to analyze and design control systems, we must define and measure the performance of a system. Then, based on the desired performance of a control system, the parameters of the system may be adjusted in order to provide the desired response. Since control systems are inherently dynamic systems, the performance is usually specified in terms of both the time response for a specific input signal and the resulting steady-state error.

The design *specifications* for control systems normally include several time-response indices for a specified input command as well as a desired steady-state accuracy. However, often in the course of any design, the specifications are revised in order to effect a compromise. Therefore, specifications are seldom a rigid set of requirements, but rather a first attempt at listing a desired performance. The effective compromise and adjustment of specifications can be graphically illustrated by examining Fig. 4.1. Clearly, the parameter p may minimize the performance measure M_2 by selecting p as a very small value. However, this results in large measure M_1, an undesirable situation. Obviously, if the performance measures are equally important, the crossover point at $p_{\min}$ provides the best compromise. This type of compromise is normally encountered in control system design. It is clear that if the original specifications called for both M_1 and M_2 to be zero, the specifications could not be simultaneously met and the specifications would have to be altered to allow for the compromise resulting with $p_{\min}$.

The specifications stated in terms of the measures of performance indicate to the designer the quality of the system. In other words, the performance measures are an answer to the question: How well does the system perform the task it was designed for?

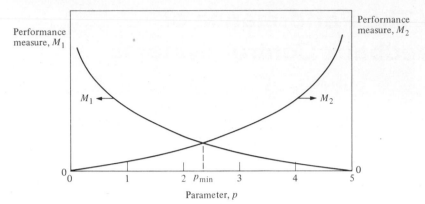

Fig. 4.1. Two performance measures vs. parameter p.

4.2 TIME-DOMAIN PERFORMANCE SPECIFICATIONS

The time-domain performance specifications are important indices since control systems are inherently time-domain systems. That is, the system transient or time performance is the response of prime interest for control systems. It is necessary to determine initially if the system is stable by utilizing the techniques of ensuing chapters. If the system is stable, then the response to a specific input signal will provide several measures of the performance. However, since the actual input signal of the system is usually unknown, a standard *test input signal* is normally chosen. This approach is quite useful since there is a reasonable correlation between the response of a system to a standard test input and the system's ability to perform under normal operating conditions. Furthermore, using a standard input allows the designer to compare several competing designs. Also, many control systems experience input signals very similar to the standard test signals.

The standard test input signals commonly used are (1) the step input, (2) the ramp input, and (3) the parabolic input. These inputs are shown in Fig. 4.2. The equations representing these test signals are given in Table 4.1, where the Laplace transform can be obtained by using Table 2.5. The ramp signal is the integral of the step input, and the parabola is simply the integral of the ramp input. A *unit impulse*

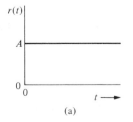
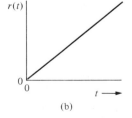
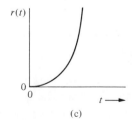

Fig. 4.2. Test input signals.

Table 4.1

Test signal	$r(t)$	$R(s)$
Step	$r(t) = A, t > 0$ $= 0, t < 0$	$R(s) = A/s$
Ramp	$r(t) = At, t > 0$ $= 0, t < 0$	$R(s) = A/s^2$
Parabolic	$r(t) = At^2, t > 0$ $= 0, t < 0$	$R(s) = 2A/s^3$

function is also useful for test signal purposes. The unit impulse is based on a rectangular function $f_\epsilon(t)$ such that

$$f_\epsilon(t) = \begin{cases} 1/\epsilon, & 0 \leqslant t \leqslant \epsilon, \\ 0, & t > \epsilon, \end{cases}$$

where $\epsilon > 0$. As ϵ approaches zero, the function $f_\epsilon(t)$ approaches the impulse function $\delta(t)$, which has the following properties:

$$\int_0^\infty \delta(t)\, dt = 1,$$

$$\int_0^\infty \delta(t - a)g(t) = g(a). \tag{4.1}$$

The impulse input is useful when one considers the convolution integral for an output $c(t)$ in terms of an input $r(t)$, which is written as

$$c(t) = \int_0^t g(t - \tau)r(\tau)\, d\tau$$
$$= \mathcal{L}^{-1}\{G(s)R(s)\}. \tag{4.2}$$

This relationship is shown in block diagram form in Fig. 4.3. Clearly, if the input is an impulse function of unit amplitude, we have

$$c(t) = \int_0^t g(t - \tau)\, \delta(\tau)\, d\tau. \tag{4.3}$$

The integral only has a value at $\tau = 0$, and therefore

$$c(t) = g(t),$$

the impulse response of the system $G(s)$. The impulse response test signal can often

Fig. 4.3. Open-loop control system.

be used for a dynamic system by subjecting the system to a large amplitude, narrow width pulse of area A.

The standard test signals are of the general form

$$r(t) = t^n, \tag{4.4}$$

and the Laplace transform is

$$R(s) = \frac{n!}{s^{n+1}}. \tag{4.5}$$

Clearly, one may relate the response to one test signal to the response of another test signal of the form of Eq. (4.4). The step input signal is the easiest to generate and evaluate and is usually chosen for performance tests.

Initially, let us consider a single-loop second-order system and determine its response to a unit step input. A closed-loop feedback control system is shown in Fig. 4.4. The closed-loop output is

$$
\begin{aligned}
C(s) &= \frac{G(s)}{1 + G(s)} R(s) \\
&= \frac{K}{s^2 + ps + K} R(s).
\end{aligned}
\tag{4.6}
$$

Utilizing the generalized notation of Section 2.4, we may rewrite Eq. (4.6) as

$$C(s) = \frac{\omega_n^2}{s^2 + 2\zeta\omega_n s + \omega_n^2} R(s). \tag{4.7}$$

With a unit step input, we obtain

$$C(s) = \frac{\omega_n^2}{s(s^2 + 2\zeta\omega_n s + \omega_n^2)}, \tag{4.8}$$

for which the transient output, as obtained from the Laplace transform table in Appendix A, is

$$c(t) = 1 - \frac{1}{\beta} e^{-\zeta\omega_n t} \sin(\omega_n \beta t + \theta), \tag{4.9}$$

where $\beta = \sqrt{1 - \zeta^2}$ and $\theta = \tan^{-1}\beta/\zeta$. The transient response of this second-order system for various values of the damping ratio ζ is shown in Fig. 4.5. As ζ

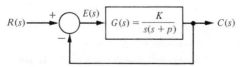

Fig. 4.4. Closed-loop control system.

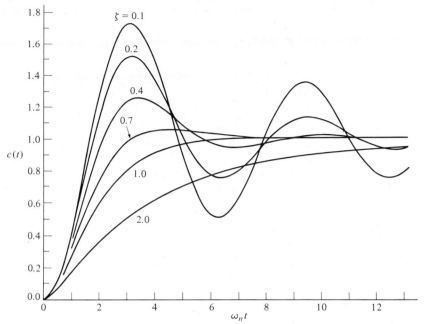

Fig. 4.5. Transient response of a second-order system (Eq. 4.9) for a step input.

decreases, the closed-loop roots approach the imaginary axis and the response becomes increasingly oscillatory.

The Laplace transform of the unit impulse is $R(s) = 1$, and therefore the output for an impulse is

$$C(s) = \frac{\omega_n^2}{s^2 + 2\zeta\omega_n s + \omega_n^2}, \tag{4.10}$$

which is $T(s) = C(s)/R(s)$, the transfer function of the closed-loop system. The transient response for an impulse function input is then

$$c(t) = \frac{\omega_n}{\beta} e^{-\zeta\omega_n t} \sin \omega_n \beta t, \tag{4.11}$$

which is simply the derivative of the response to a step input. The impulse response of the second-order system is shown in Fig. 4.6 for several values of the damping ratio, ζ. Clearly, one is able to select several alternative performance measures from the transient response of the system for either a step or impulse input.

Standard performance measures are usually defined in terms of the step response of a system as shown in Fig. 4.7. The swiftness of the response is measured by the rise time T_r and the peak time. For underdamped systems with an overshoot, the 0–100% rise time is a useful index. If the system is overdamped, then the peak time is not defined and the 10–90% rise time, T_{r_1}, is normally used.

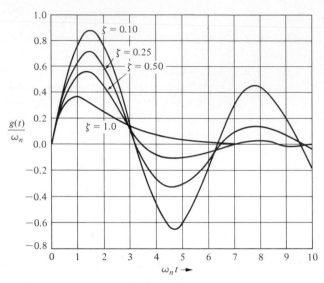

Fig. 4.6. Response of a second-order system for an impulse function input.

The similarity with which the actual response matches the step input is measured by the percent overshoot and settling time T_s. The *percent overshoot,* P.O., is defined as

$$\text{P.O.} = \frac{M_{p_t} - 1}{1} \times 100\% \qquad (4.12)$$

for a unit step input, where M_{p_t} is the peak value of the time response. The *settling time, T_s,* is defined as the time required for the system to settle within a certain

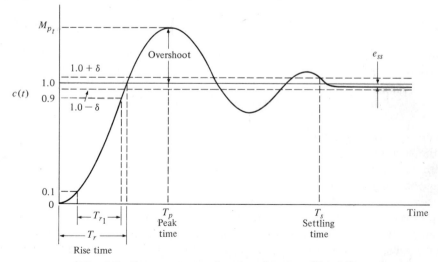

Fig. 4.7. Step response of a control system (Eq. 4.9).

percentage δ of the input amplitude. This band of $\pm\delta$ is shown in Fig. 4.7. For the second-order system with closed-loop damping constant $\zeta\omega_n$, the response remains within 2% after four time constants, or

$$T_s = 4\tau = \frac{4}{\zeta\omega_n}. \tag{4.13}$$

Therefore, we will define the settling time as four time constants of the dominant response. Finally, the steady-state error of the system may be measured on the step response of the system as shown in Fig. 4.7.

Therefore, the transient response of the system may be described in terms of

(1) the swiftness of response, T_r and T_p;

(2) the closeness of the response to the desired M_{p_t} and T_s.

As nature would have it, these are contradictory requirements and a compromise must be obtained. In order to obtain an explicit relation for M_{p_t} and T_p as a function of ζ, one can differentiate Eq. (4.9) and set it equal to zero. Alternatively, one may utilize the differentiation property of the Laplace transform which may be written as

$$\mathcal{L}\left\{\frac{dc(t)}{dt}\right\} = sC(s)$$

when the initial value of $c(t)$ is zero. Therefore, we may acquire the derivative of $c(t)$ by multiplying Eq. (4.8) by s and thus obtaining the right side of Eq. (4.10). Taking the inverse transform of the right side of Eq. (4.10) we obtain Eq. (4.11), which is equal to zero when $\omega_n\beta t = \pi$. Therefore we find that the peak time relationship for this second-order system is

$$T_p = \frac{\pi}{\omega_n\sqrt{1 - \zeta^2}}, \tag{4.14}$$

and the peak response as

$$M_{p_t} = 1 + e^{-\zeta\pi/\sqrt{1-\zeta^2}}. \tag{4.15}$$

Therefore, the percent overshoot is

$$\text{P.O.} = 100e^{-\zeta\pi/\sqrt{1-\zeta^2}}. \tag{4.16}$$

The percent overshoot vs. the damping ratio ζ is shown in Fig. 4.8. Also, the normalized peak time, $\omega_n T_p$, is shown vs. the damping ratio ζ in Fig. 4.8. Again, we are confronted with a necessary compromise between the swiftness of response and the allowable overshoot.

The curves presented in Fig. 4.8 are only exact for the second-order system of Eq. (4.8). However, they provide a remarkably good source of data, since many systems possess a dominant pair of roots and the step response can be estimated by utilizing Fig. 4.8. This approach, while an approximation, avoids the evaluation

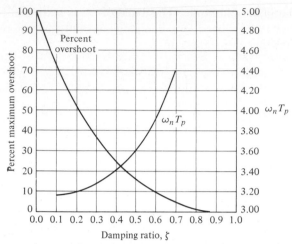

Fig. 4.8. Percent overshoot and peak time versus damping ratio ζ for a second-order system (Eq. 4.8).

of the inverse Laplace transformation in order to determine the percent overshoot and other performance measures. For example, for a third-order system with a closed-loop transfer function

$$T(s) = \frac{1}{(s^2 + 2\zeta s + 1)(\gamma s + 1)},$$
(4.17)

the s-plane diagram is shown in Fig. 4.9. This third-order system is normalized with $\omega_n = 1$. It was ascertained experimentally that the performance as indicated by the percent overshoot, M_{p_t}, and the settling time, T_s, was represented by the second-order system curves when [4]

$$|1/\gamma| \geq 10|\zeta\omega_n|.$$

In other words, the response of a third-order system can be approximated by the

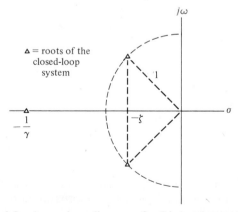

Fig. 4.9. An s-plane diagram of a third-order system.

dominant roots of the second-order system as long as the real part of the dominant roots is less than $\frac{1}{10}$ of the real part of the third root.

Using a computer simulation, when $\zeta = .45$ one can determine the response of a system to a unit step input. When $\gamma = 2.25$ we find that the response is overdamped since the real part of the complex poles is $-.45$, while the real pole is equal to $-.444$. The settling time is found via the simulation to be 12.8 seconds. If $\gamma = .90$ or $1/\gamma = 1.11$ is compared to $\zeta\omega_n = .45$ of the complex poles we find that the overshoot is 12% and the settling time is 6.4 seconds. If the complex roots were entirely dominant we would expect the overshoot to be 20% and the settling time to be $4/\zeta\omega_n = 4.4$ seconds.

Also, we must note that the performance measures of Fig. 4.8 are only correct for a transfer function without finite zeros. If the transfer function of a system possesses finite zeros and they are located relatively near the dominant poles, then the zeros will materially affect the transient response of the system [5].

The transient response of a system with one zero and two poles may be affected by the location of the zero [5]. The percent overshoot for a step input as a function of $a/\zeta\omega_n$ is given in Fig. 4.10 for the system transfer function

$$T(s) = \frac{(\omega_n^2/a)(s + a)}{s^2 + 2\zeta\omega_n s + \omega_n^2}$$

The correlation of the time-domain response of a system with the s-plane location of the poles of the closed-loop transfer function is very useful for selecting the specifications of a system. In order to clearly illustrate the utility of the s-plane, let us consider a simple example.

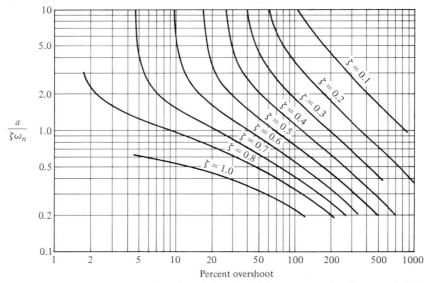

Fig. 4.10. Percent overshoot as a function of ζ and ω_n when a second-order transfer function contains a zero. (From *Introduction to Automatic Control Systems*, by R. N. Clark. New York, Wiley, 1962, redrawn with permission.)

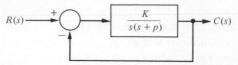

Fig. 4.11. Single-loop feedback control system.

Example 4.1.   A single-loop feedback control system is shown in Fig. 4.11. It is desired to select the gain K and the parameter p so that the time-domain specifications will be satisfied. The transient response to a step should be as fast in responding as reasonable and with an overshoot of less than 5%. Furthermore, the settling time should be less than four seconds. The minimum damping ratio ζ for an overshoot of 4.3% is 0.707. This damping ratio is shown graphically in Fig. 4.12. Since the settling time is

$$T_s = \frac{4}{\zeta\omega_n} \leqslant 4 \text{ sec}, \tag{4.18}$$

we require that the real part of the complex poles of $T(s)$ is

$$\zeta\omega_n \geqslant 1.$$

This region is also shown in Fig. 4.12. The region that will satisfy both time-domain requirements is shown cross-hatched on the s-plane of Fig. 4.12. If the closed-loop roots are chosen as the limiting point, in order to provide the fastest response, as r_1 and $\hat{r}_1$, then $r_1 = -1 + j1$ and $\hat{r}_1 = -1 - j1$. Therefore, $\zeta = 1/\sqrt{2}$ and $\omega_n = 1/\zeta = \sqrt{2}$. The closed-loop transfer function is

$$T(s) = \frac{G(s)}{1 + G(s)} = \frac{K}{s^2 + ps + K}$$

$$= \frac{\omega_n^2}{s^2 + 2\zeta\omega_n s + \omega_n^2}. \tag{4.19}$$

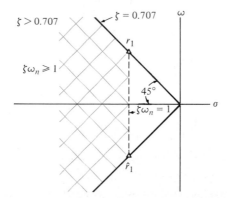

Fig. 4.12. Specifications and root locations on the s-plane.

Therefore, we require that $K = \omega_n^2 = 2$ and $p = 2\zeta\omega_n = 2$. A full comprehension of the correlation between the closed-loop root location and the system transient response is important to the system analyst and designer. Therefore, we shall consider the matter more fully in the following section.

4.3 THE S-PLANE ROOT LOCATION AND THE TRANSIENT RESPONSE

The transient response of a closed-loop feedback control system can be described in terms of the location of the poles of the transfer function. The closed-loop transfer function is written in general as

$$T(s) = \frac{C(s)}{R(s)} = \frac{\Sigma P_i(s)\, \Delta_i(s)}{\Delta(s)}, \tag{4.20}$$

where $\Delta(s) = 0$ is the characteristic equation of the system. For the single-loop system of Fig. 4.11, the characteristic equation reduces to $1 + G(s) = 0$. It is the poles and zeros of $T(s)$ that determine the transient response. However, for a closed-loop system, the poles of $T(s)$ are the roots of the characteristic $\Delta(s) = 0$ and the poles of $\Sigma P_i(s)\, \Delta_i(s)$. The output of a system without repeated roots and a unit step input can be formulated as a partial fraction expansion as

$$C(s) = \frac{1}{s} + \sum_{i=1}^{M} \frac{A_i}{s + \sigma_i} + \sum_{k=1}^{N} \frac{B_k}{s^2 + 2\alpha_k s + (\alpha_k^2 + \omega_k^2)}, \tag{4.21}$$

where the A_i and B_k are the residues. The roots of the system must be either $s = -\sigma_i$ or complex conjugate pairs as $s = -\alpha_k \pm j\omega_k$. Then the inverse transform results in the transient response as a sum of terms as follows:

$$c(t) = 1 + \sum_{i=1}^{M} A_i e^{-\sigma_i t} + \sum_{k=1}^{N} B_k \left(\frac{1}{\omega_k}\right) e^{-\alpha_k t} \sin \omega_k t. \tag{4.22}$$

The transient response is composed of the steady-state output, exponential terms, and damped sinusoidal terms. Obviously, in order for the response to be stable, that is, bounded for a step input, one must require that the real part of the roots, σ_i or α_k, be in the left-hand portion of the s-plane. The impulse response for various root locations is shown in Fig. 4.13. The information imparted by the location of the roots is graphic indeed and usually well worth the effort of determining the location of the roots in the s-plane.

4.4 THE STEADY-STATE ERROR OF FEEDBACK CONTROL SYSTEMS

One of the fundamental reasons for using feedback, despite its cost and increased complexity, is the attendant improvement in the reduction of the steady-state error of the system. As was illustrated in Section 3.5, the steady-state error of a stable closed-loop system is usually several orders of magnitude smaller than the error of

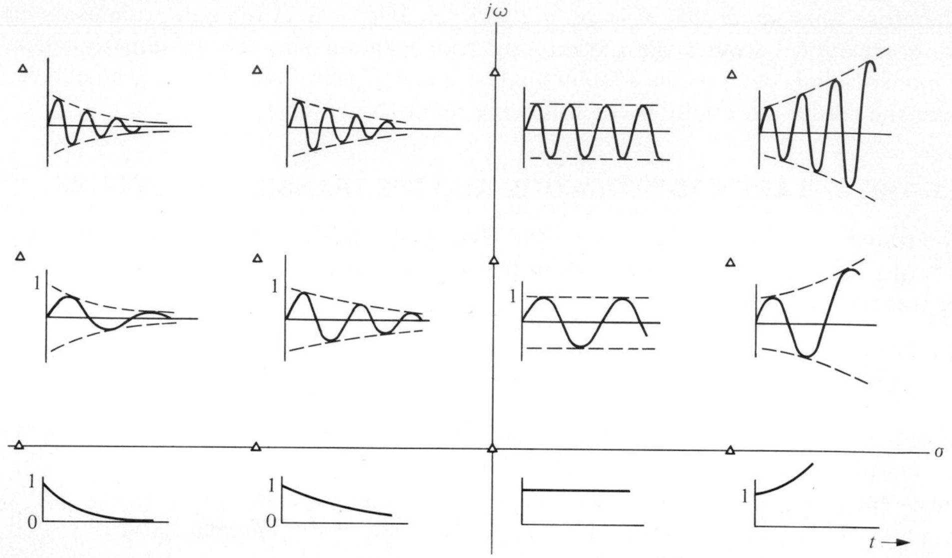

Fig. 4.13. Impulse response for various root locations in the s-plane. (The conjugate root is not shown.)

the open-loop system. The system actuating signal, which is a measure of the system error, is denoted as $E_a(s)$. However, the actual system error is $E(s) = R(s) - C(s)$. Considering the closed-loop feedback system of Fig. 4.14, we have

$$E(s) = R(s) - \frac{G(s)}{1 + GH(s)} R(s) = \frac{[1 + GH(s) - G(s)]}{1 + GH(s)} R(s). \qquad (4.23)$$

The system error is equal to the actuating signal when $H(s) = 1$, which is a common situation, and then

$$E(s) = \frac{1}{1 + G(s)} R(s).$$

The steady-state error, when $H(s) = 1$, is then

$$\lim_{t\to\infty} e(t) = e_{ss} = \lim_{s\to 0} \frac{sR(s)}{1 + G(s)}. \qquad (4.24)$$

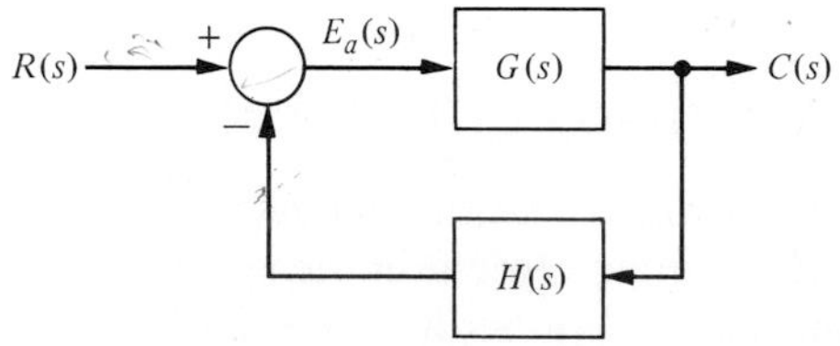

Fig. 4.14. Closed-loop control system.

It is useful to determine the steady-state error of the system for the three standard test inputs for a unity feedback system ($H(s) = 1.$).

Step Input

The steady-state error for a step input is therefore

$$e_{ss} = \lim_{s \to 0} \frac{s(A/s)}{1 + G(s)} = \frac{A}{1 + G(0)}. \tag{4.25}$$

Clearly, it is the form of the loop transfer function $GH(s)$ that determines the steady-state error. The loop transfer function is written in general form as

$$G(s) = \frac{K\prod_{i=1}^{M} (s + z_i)}{s^N \prod_{k=1}^{Q} (s + p_k)}, \tag{4.26}$$

where $\prod$ denotes the product of the factors. Therefore, the loop transfer function as s approaches zero depends upon the number of integrations N. If N is greater than zero, then $G(0)$ approaches infinity and the steady-state error approaches zero. The number of integrations is often indicated by labeling a system with a *type number* which is simply equal to N.

Therefore, for a type zero system, $N = 0$, the steady-state error is

$$e_{ss} = \frac{A}{1 + G(0)}$$

$$= \frac{A}{1 + (K\prod_{i=1}^{M} z_i / \prod_{k=1}^{Q} p_k)}. \tag{4.27}$$

The constant $G(0)$ is denoted by K_p, the position error constant, so that

$$e_{ss} = \frac{A}{1 + K_p}. \tag{4.28}$$

Clearly, the steady-state error for a unit step input with one integration or more, $N \geqslant 1$, is zero since

$$e_{ss} = \lim_{s \to 0} \frac{A}{1 + (K\prod z_i / s^N \prod p_k)}$$

$$= \lim_{s \to 0} \frac{As^N}{s^N + (K\prod z_i / \prod p_k)} = 0. \tag{4.29}$$

Ramp Input

The steady-state error for a ramp (velocity) input is

$$e_{ss} = \lim_{s \to 0} \frac{s(A/s^2)}{1 + G(s)} = \lim_{s \to 0} \frac{A}{s + sG(s)} = \lim_{s \to 0} \frac{A}{sG(s)}. \tag{4.30}$$

Again, the steady-state error depends upon the number of integrations N. For a type zero system, $N = 0$, the steady-state error is infinite. For a type one system, $N = 1$, the error is

$$e_{ss} = \lim_{s \to 0} \frac{A}{s\{[K\prod(s + z_i)]/[s\prod(s + p_k)]\}}$$

$$= \frac{A}{(K\prod z_i/\prod p_k)} = \frac{A}{K_v}, \tag{4.31}$$

where K_v is designated the velocity error constant. When the transfer function possesses two or more integrations, $N \geq 2$, we obtain a steady-state error of zero.

Acceleration Input

When the system input is $r(t) = At^2/2$, the steady-state error is then

$$e_{ss} = \lim_{s \to 0} \frac{s(A/s^3)}{1 + G(s)}$$

$$= \lim_{s \to 0} \frac{A}{s^2 G(s)}. \tag{4.32}$$

The steady-state error is infinite for one integration; and for two integrations, $N = 2$, we obtain

$$e_{ss} = \frac{A}{K\prod z_i/\prod p_k} = \frac{A}{K_a}, \tag{4.33}$$

where K_a is designated the acceleration constant. When the number of integrations equals or exceeds three, then the steady-state error of the system is zero.

Table 4.2 Summary of Steady-State Errors

Number of integrations in $G(s)$, type number	Step, $r(t) = A$, $R(s) = A/s$	Ramp, At, A/s^2	Parabola, $At^2/2$, A/s^3
0	$e_{ss} = \dfrac{A}{1 + K_p}$	Infinite	Infinite
1	$e_{ss} = 0$	$\dfrac{A}{K_v}$	Infinite
2	$e_{ss} = 0$	0	$\dfrac{A}{K_a}$

Control systems are often described in terms of their type number and the error constants, K_p, K_r, and K_a. Definitions for the error constants and the steady-state error for the three inputs are summarized in Table 4.2. The usefulness of the error constants can be illustrated by considering a simple example.

Example 4.2. An automatic speed control system for an automobile was outlined in Problem 3.6. This system is commonly called cruise control. The block diagram of a specific speed control system is shown in Fig. 4.15. The throttle controller, $G_1(s)$, is

$$G_1(s) = K_1 + K_2/s. \tag{4.34}$$

The steady-state error of the system for a step input when $K_2 = 0$ and $G(s) = K_1$ is therefore

$$e_{ss} = \frac{A}{1 + K_p}, \tag{4.35}$$

where $K_p = K_e K_1$. When K_2 is greater than zero, we have a type one system,

$$G_1(s) = \frac{K_1 s + K_2}{s},$$

and the steady-state error is zero for a step input.

If the speed command was a ramp input, the steady-state error is then

$$e_{ss} = \frac{A}{K_v}, \tag{4.36}$$

where

$$K_v = \lim_{s \to 0} s G_1(s) G(s) = K_2 K_e.$$

The transient response of the automobile to a triangular wave input when $G_1(s) = (K_1 s + K_2)/s$ is shown in Fig. 4.16. The transient response clearly shows the effect of the steady-state error, which may not be objectionable if K_v is sufficiently large.

The error constants, K_p, K_v, and K_a, of a control system describe the ability of a system to reduce or eliminate the steady-state error. Therefore, they are utilized as numerical measures of the steady-state performance. The designer determines the error constants for a given system and attempts to determine methods of

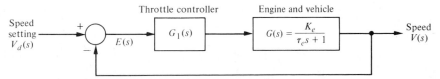

Fig. 4.15. An automobile speed control system.

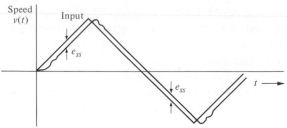

Fig. 4.16. Triangular wave response.

increasing the error constants while maintaining an acceptable transient response. In the case of the automobile speed control system, it is desirable to increase the gain factor $K_e K_2$ in order to increase K_v and reduce the steady-state error. However, an increase in $K_e K_2$ results in an attendant decrease in the damping ratio, ζ, of the system and therefore a more oscillatory response to a step input. Again, a compromise would be determined which would provide the largest K_v based on the smallest ζ allowable.

4.5 PERFORMANCE INDICES

An increased amount of emphasis on the mathematical formulation and measurement of control system performance can be found in the recent literature on automatic control. A *performance index* is a quantitative measure of the performance of a system and is chosen so that emphasis is given to the important system specifications. Modern control theory assumes that the systems engineer can specify quantitatively the required system performance. Then a performance index can be calculated or measured and used to evaluate the system's performance. A quantitative measure of the performance of a system is necessary for the operation of modern adaptive control systems, for automatic parameter optimization of a control system, and for the design of optimum systems.

Whether the aim is to improve the design of a system or to design an adaptive control system, a performance index must be chosen and measured. Then the system is considered an *optimum control system* when the system parameters are adjusted so that the index reaches an extremum value, commonly a minimum value. A performance index, in order to be useful, must be a number that is always positive or zero. Then the best system is defined as the system which minimizes this index.

A suitable performance index is the integral of the square of the error, ISE, which is defined as

$$I_1 = \int_0^T e^2(t)\, dt. \tag{4.37}$$

The upper limit T is a finite time chosen somewhat arbitrarily so that the integral approaches a steady-state value. It is usually convenient to choose T as the settling time, T_s. The step response for a specific feedback control system is shown in Fig. 4.17(b); and the error, in Fig. 4.17(c). The error squared is shown in Fig. 4.17(d); and the integral of the error squared, in Fig. 4.17(e). This criterion will discriminate between excessively overdamped systems and excessively underdamped systems. The minimum value of the integral occurs for a compromise value of the damping. The performance index of Eq. (4.37) is easily adapted for practical measurements, since a squaring circuit is readily obtained. Furthermore, the squared error is mathematically convenient for analytical and computational purposes.

Another readily instrumented performance criterion is the integral of the absolute magnitude of the error, IAE, which is written as

$$I_2 = \int_0^T |e(t)| \, dt. \tag{4.38}$$

This index is particularly useful for analog computer simulation studies. In order to

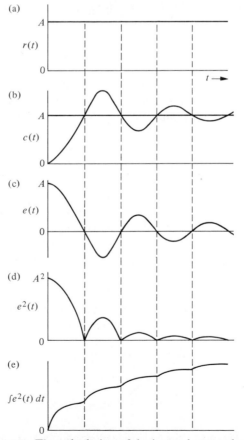

Fig. 4.17. The calculation of the integral squared error.

reduce the contribution of the large initial error to the value of the performance integral, as well as to place an emphasis on errors occurring later in the response, the following index has been proposed [6]:

$$I_3 = \int_0^T t \, | \, e(t) | \, dt. \tag{4.39}$$

This performance index is designated the integral of time multiplied by absolute error, ITAE. Another similar index is the integral of time multiplied by the squared error, ITSE, which is

$$I_4 = \int_0^T t e^2(t) \, dt. \tag{4.40}$$

The performance index I_3, ITAE, provides the best selectivity of the performance indices; that is, the minimum value of the integral is readily discernible as the system parameters are varied. The general form of the performance integral is

$$I = \int_0^T f(e(t), r(t), c(t), t) \, dt, \tag{4.41}$$

where f is a function of the error, input, output, and time. Clearly, one can obtain numerous indices based on various combinations of the system variables and time. It is worth noting that the minimization of IAE or ISE is often of practical significance. For example, the minimization of a performance index can be directly related to the minimization of fuel consumption for aircraft and space vehicles.

Performance indices are useful for the analysis and design of control systems. Two examples will illustrate the utility of this approach.

Example 4.3. A single-loop feedback control system is shown in Fig. 4.18, where the natural frequency is the normalized value, $\omega_n = 1$. The closed-loop transfer function is then

$$T(s) = \frac{1}{s^2 + 2\zeta s + 1}. \tag{4.42}$$

Three performance indices—ISE, ITSE, and ITAE—calculated for various values of the damping ratio ζ and for a step input are shown in Fig. 4.19. These curves

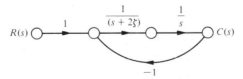

Fig. 4.18. Single-loop feedback control system.

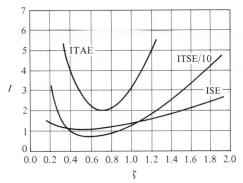

Fig. 4.19. Three performance criteria for a second-order system. (Courtesy of Professor R. C. H. Wheeler, U.S. Naval Postgraduate School.)

show the selectivity of the ITAE index in comparison with the ISE index. The value of the damping ratio ζ selected on the basis of ITAE is 0.7, which, for a second-order system, results in a swift response to a step with a 5% overshoot.

Example 4.4. The signal-flow graph of a space vehicle attitude control system is shown in Fig. 4.20 [7]. It is desired to select the magnitude of the gain K_3 in order to minimize the effect of the disturbance $U(s)$. The disturbance in this case is equivalent to an initial attitude error. The closed-loop transfer function for the disturbance is obtained by using the signal-flow gain formula as

$$\frac{C(s)}{U(s)} = \frac{P_1(s)\,\Delta_1(s)}{\Delta(s)}$$

$$= \frac{1 \cdot (1 + K_1 K_3 s^{-1})}{1 + K_1 K_3 s^{-1} + K_1 K_2 K_p s^{-2}}$$

$$= \frac{s(s + K_1 K_3)}{s^2 + K_1 K_3 s + K_1 K_2 K_p}. \tag{4.43}$$

Typical values for the constants are $K_1 = 0.5$ and $K_1 K_2 K_p = 2.5$. Then the natural frequency of the vehicle is $f_n = \sqrt{2.5}/2\pi = 0.25$ cycles/sec. For a unit step disturbance, the minimum ISE can be analytically calculated. The attitude $c(t)$ is

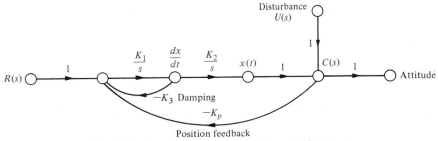

Fig. 4.20. A space vehicle attitude control system.

$$c(t) = \frac{\sqrt{10}}{\beta}\left[e^{-0.25K_3 t}\sin\left(\frac{\beta}{2}t + \psi\right)\right],\tag{4.44}$$

where $\beta = K_3\sqrt{(K_3^2/8) - 5}$. Squaring $c(t)$ and integrating the result, we have

$$\begin{aligned}I &= \int_0^\infty \frac{10}{\beta^2}e^{-0.5K_3 t}\sin^2\left(\frac{\beta}{2}t + \psi\right)dt\\&= \int_0^\infty \frac{10}{\beta^2}e^{-0.5K_3 t}\left(\frac{1}{2} - \frac{1}{2}\cos(\beta t + 2\psi)\right)dt\\&= \left(\frac{1}{K_3} + 0.1K_3\right)\end{aligned}\tag{4.45}$$

Differentiating I and equating the result to zero, we obtain

$$\frac{dI}{dK_3} = -K_3^{-2} + 0.1 = 0.\tag{4.46}$$

Therefore, the minimum ISE is obtained when $K_3 = \sqrt{10} = 3.2$. This value of K_3 corresponds to a damping ratio ζ of 0.50. The values of ISE and IAE for this system are plotted in Fig. 4.21. The minimum for the IAE performance index is obtained when $K_3 = 4.2$ and $\zeta = 0.665$. While the ISE criterion is not as selective as the IAE criterion, it is clear that it is possible to solve analytically for the minimum value of ISE. The minimum of IAE is obtained by measuring the actual value of IAE for several values of the parameter of interest.

A control system is optimum when the selected performance index is minimized. However, the optimum value of the parameters depends directly upon the definition of optimum, that is, the performance index. Therefore, in the two examples, we found that the optimum setting varied for different performance indices.

The coefficients that will minimize the ITAE performance criterion for a step input have been determined for the general closed-loop transfer function [6]:

$$T(s) = \frac{C(s)}{R(s)} = \frac{b_0}{s^n + b_{n-1}s^{n-1} + \cdots + b_1 s + b_0}.\tag{4.47}$$

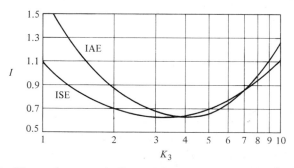

Fig. 4.21. The performance indices of the attitude control system vs. K_3.

Table 4.3 The Optimum Coefficients of $T(s)$ Based on the ITAE Criterion for a Step Input

$$s + \omega_n$$
$$s^2 + 1.4\omega_n s + \omega_n^2$$
$$s^3 + 1.75\omega_n s^2 + 2.15\omega_n^2 s + \omega_n^3$$
$$s^4 + 2.1\omega_n s^3 + 3.4\omega_n^2 s^2 + 2.7\omega_n^3 s + \omega_n^4$$
$$s^5 + 2.8\omega_n s^4 + 5.0\omega_n^2 s^3 + 5.5\omega_n^3 s^2 + 3.4\omega_n^4 s^4 + \omega_n^5$$
$$s^6 + 3.25\omega_n s^5 + 6.60\omega_n^2 s^4 + 8.60\omega_n^3 s^3 + 7.45\omega_n^4 s^2 + 3.95\omega_n^5 s + \omega_n^6$$

This transfer function has a steady-state error equal to zero for a step input. The optimum coefficients for the ITAE criterion are given in Table 4.3. The responses using optimum coefficients for a step input are given in Fig. 4.22 for ISE, IAE, and ITAE. Other standard forms based on different performance indices are available and can be useful in aiding the designer to determine the range of coefficients for a specific problem. A final example will illustrate the utility of the standard forms for ITAE.

Example 4.5. A very accurate and rapidly responding control system is required for a system which allows live actors to seemingly perform inside of complex miniature sets [9]. The two-camera system is shown in Fig. 4.23(a), where one camera is trained on the actor and the other on the miniature set. The challenge is to obtain

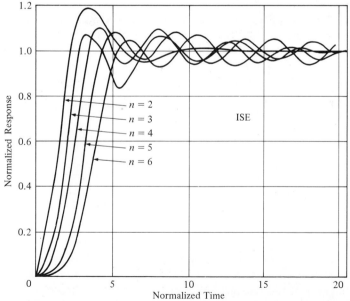

Fig. 4.22. (a) Step responses of a normalized transfer function using optimum coefficients for ISE. Next page: (b) IAE and (c) ITAE.

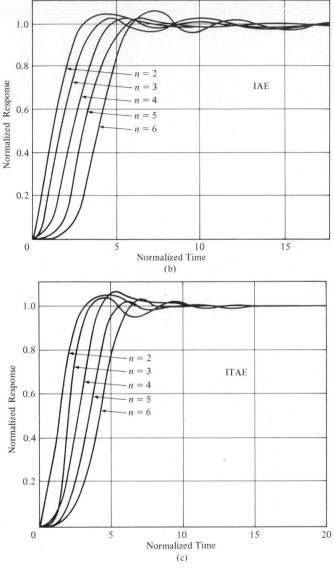

Fig. 4.22 (b) and (c) .

rapid and accurate coordination of the two cameras by using sensor information from the foreground camera to control the movement of the background camera. The block diagram of the background camera system is shown in Fig. 4.23(b) for one axis of movement of the background camera. Closed-loop transfer function is

$$T(s) = \frac{K_a K_m \omega_0^2}{s^3 + 2\zeta\omega_0 s^2 + \omega_0^2 s + K_a K_m \omega_0^2}. \tag{4.48}$$

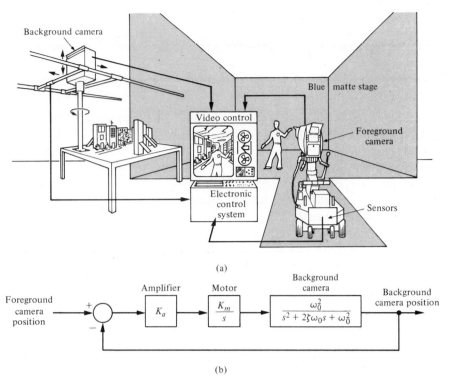

(a)

(b)

Fig. 4.23 The foreground camera, which may be either a film or video camera, is trained on the blue cyclorama stage. The electronic servo control installation permits the slaving, by means of electronic servo devices, of the two video film cameras. The background camera reaches into the miniature set with a periscope lens, and instantaneously reproduces all movements of the foreground camera in the scale of the miniature. The video control installation allows the composite image to be monitored and recorded live. Part (a) reprinted with permission from Electronic Design 24, (11, May 24, 1976. Copyright © Hayden Publishing Co., Inc., 1976.)

The standard form for a third-order system given in Table 4.3 requires that

$$2\zeta\omega_0 = 1.75\omega_n, \qquad \omega_0^2 = 2.15\omega_n^2, \qquad K_a K_m \omega_0^2 = \omega_n^3.$$

Since a rapid response is required, a large ω_n will be selected so that the settling time will be less than one second. Thus, ω_n will be set equal to 50 rad/sec. Then, for an ITAE system, it is necessary that the parameters of the camera dynamics be

$$\omega_0 = 73 \text{ rad/sec}$$

and

$$\zeta = 0.60.$$

The amplifier and motor gain are required to be

$$K_a K_m = \frac{\omega_n^3}{\omega_0^2} = \frac{\omega_n^3}{2.15\omega_n^2} = \frac{\omega_n}{2.15} = 23.2.$$

Then, the closed-loop transfer function is

$$T(s) = \frac{125,000}{s^3 + 87.5s + 5375s + 125,000}$$

$$= \frac{125,000}{(s + 35.5)(s + 26 + j53.4)(s + 26 - j53.4)}.$$
(4.49)

The locations of the closed-loop roots dictated by the ITAE system are shown in Fig. 4.24. The damping ratio of the dominant complex roots is $\zeta = 0.44$, and using Fig. 4.8 one may estimate the overshoot to be approximately 20%. The settling time is approximately $4(\frac{1}{26}) = 0.15$ seconds. This is only an approximation since the complex conjugate roots are not dominant; however it does indicate the magnitude of the performance measures. The actual response to a step input using a computer simulation showed the overshoot to be only 2% and the settling time equal to 0.08 seconds.

For a ramp input, the coefficients have been determined that minimize the ITAE criterion for the general closed-loop transfer function [6]:

$$T(s) = \frac{b_1s + b_0}{s^n + b_{n-1}s^{n-1} + \cdots + b_1s + b_0}$$
(4.50)

This transfer function has a steady-state error equal to zero for a ramp input. The optimum coefficients for this transfer function are given in Table 4.4. The transfer function, Eq. (4.50), implies that the plant $G(s)$ has two or more pure integrations as required to provide zero steady-state error.

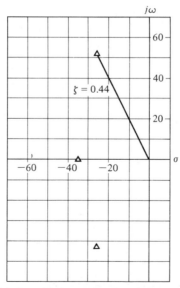

Fig. 4.24. The closed-loop roots of a minimum ITAE system.

Table 4.4 The Optimum Coefficients of $T(s)$ Based on the ITAE Criterion for a Ramp Input

$$s^2 + 3.2\omega_n s + \omega_n^2$$
$$s^3 + 1.75\omega_n s^2 + 3.25\omega_n^2 s + \omega_n^3$$
$$s^4 + 2.41\omega_n s^3 + 4.93\omega_n^2 s^2 + 5.14\omega_n^3 s + \omega_n^4$$
$$s^5 + 2.19\omega_n s^4 + 6.50\omega_n^2 s^3 + 6.30\omega_n^3 s^2 + 5.24\omega_n^4 s + \omega_n^5$$

4.6 THE SIMPLIFICATION OF LINEAR SYSTEMS

It is quite useful to study complex systems with high-order transfer functions by using lower-order approximate models. Thus, for example, a fourth-order system could be approximated by a second-order system leading to a use of the indices in Fig. 4.8. There are several methods now available for reducing the order of a systems transfer function [10, 11].

We will let the high-order system be described by the transfer function

$$H(s) = K \frac{a_m s^m + a_{m-1} s^{m-1} + \cdots + a_1 s + 1}{b_n s^n + b_{n-1} s^{n-1} + \cdots + b_1 s + 1}, \tag{4.51}$$

in which the poles are in the left-hand s plane and $m \leqslant n$. The lower-order approximate transfer function is

$$L(s) = K \frac{c_p s^p + \cdots + c_1 s + 1}{d_g s^g + \cdots + d_1 s + 1} \tag{4.52}$$

where $p \leqslant g < n$. Notice that the gain constant K is the same for the original and approximate system in order to insure the same steady-state response. The method outlined in the following paragraph is based on selecting c_i and d_i in such a way that $L(s)$ has a frequency response (see Chapter 7) very close to that of $H(s)$ [10, 11]. This is equivalent to stating that $H(j\omega)/L(j\omega)$ is required to deviate the least amount from unity for various frequencies. The c and d coefficients are obtained by utilizing the following equation:

$$M^{(k)}(s) = \frac{d^k}{ds^k} M(s) \tag{4.53}$$

and

$$\Delta^{(k)}(s) = \frac{d^k}{ds^k} \Delta(s), \tag{4.54}$$

where $M(s)$ and $\Delta(s)$ are the numerator and denominator polynomials of $H(s)/L(s)$ respectively. We also define

$$M_{2q} = \sum_{k=0}^{2q} \frac{(-1)^{k+q} M^{(k)}(0) M^{(2q-k)}(0)}{k!(2q-k)!}, \qquad q = 0, 1, 2 \ldots \tag{4.55}$$

and a completely identical equation for Δ_{2q}. The solutions for the c and d coefficients are obtained by equating

$$M_{2q} = \Delta_{2q} \tag{4.56}$$

for $q = 1, 2, \ldots$ up to the number required to solve for the unknown coefficients [10].

Let us consider an example in order to clarify the use of these equations.

Example 4.6   Consider the third-order system

$$H(s) = \frac{6}{s^3 + 6s^2 + 11s + 6} = \frac{1}{1 + \frac{11}{6}s + s^2 + \frac{1}{6}s^3}, \tag{4.57}$$

and we wish to use a second-order model

$$L(s) = \frac{1}{1 + d_1 s + d_2 s^2}. \tag{4.58}$$

Now, $M(s) = 1 + d_1 s + d_2 s^2$ and $\Delta(s) = 1 + \frac{11}{6}s + s^2 + \frac{1}{6}s^3$. Then

$$M^0(s) = 1 + d_1 s + d_2 s^2 \tag{4.59}$$

and $M^0(0) = 1$. Similarly,

$$M^1 = \frac{d}{ds}(1 + d_1 s + d_2 s^2) = d_1 + 2\,d_2 s. \tag{4.60}$$

Therefore, $M^1(0) = d_1$. Continuing this process, we find that

$$\begin{array}{ll}
M^0(0) = 1 & \Delta^0(0) = 1 \\
M^1(0) = d_1 & \Delta^1(0) = \frac{11}{6} \\
M^2(0) = 2d_2 & \Delta^2(0) = 2 \\
M^3(0) = 0 & \Delta^3(0) = 1
\end{array} \tag{4.61}$$

We now equate $M_{2q} = \Delta_{2q}$ for $q = 1$ and 2. We find that for $q = 1$,

$$\begin{aligned}
M_2 &= (-1)\frac{M^0(0)M^2(0)}{2} + \frac{M^1(0)M^1(0)}{1} + (-1)\frac{M^2(0)M^0(0)}{2} \\
&= -d_2 + d_1^2 - d_2 = -2d_2 + d_1^2.
\end{aligned} \tag{4.62}$$

Then, since the equation for Δ_2 is identical, we have

$$\begin{aligned}
\Delta_2 &= -\frac{\Delta^0(0)\,\Delta^2(0)}{2} + \frac{\Delta^1(0)\,\Delta^1(0)}{1} + (-1)\frac{\Delta^2(0)\,\Delta^0(0)}{2} \\
&= -1 + \frac{121}{36} - 1 = \frac{49}{36}.
\end{aligned} \tag{4.63}$$

Therefore, since $M_2 = \Delta_2$, we have

$$-2d_2 + d_1^2 = \frac{49}{36}. \tag{4.64}$$

Completing the process for $M_4 = \Delta_4$, when $q = 2$, we obtain

$$d_2^2 = \frac{7}{18}. \tag{4.65}$$

Then the solution for $L(s)$ is $d_1 = 1.615$ and $d_2 = 0.625$. (The other sets of solutions are rejected since they lead to unstable poles.) It is interesting to see that the poles of $H(s)$ are at $s = -1, -2, -3$ whereas the poles of $L(s)$ are at $s = -1.029$ and -1.555. The lower-order system transfer function is

$$\begin{aligned} L(s) &= \frac{1}{1 + 1.615s + 0.625s^2} \\ &= \frac{1.60}{s^2 + 2.584s + 1.60}. \end{aligned} \tag{4.66}$$

Since the lower-order model has two poles we can estimate that we would obtain a slightly overdamped response with a settling time of approximately 4 seconds.

It is sometimes desirable to retain the dominant poles of the original system, $H(s)$, in the low-order model. This can be accomplished by specifying the denominator of $L(s)$ to be the dominant poles of $H(s)$ and allow the numerator of $L(s)$ to be subject to approximation.

4.7 SUMMARY

In this chapter we have considered the definition and measurement of the performance of a feedback control system. The concept of a performance measure or index was discussed and the usefulness of standard test signals was outlined. Then several performance measures for a standard step input test signal were delineated. For example, the overshoot, peak time, and settling time of the response of the system under test for a step input signal were considered. The fact that often the specifications on the desired response are contradictory was noted and the concept of a design compromise was proposed. The relationship between the location of the s-plane root of the system transfer function and the system response was discussed. A most important measure of system performance is the steady-state error for specific test input signals. Thus, the relationship of the steady-state error of a system in terms of the system parameters was developed by utilizing the final-value theorem. Finally, the utility of an integral performance index was outlined and several examples of design which minimized a system's performance index were completed. Thus, we have been concerned with the definition and usefulness of quantitative measures of the performance of feedback control systems.

PROBLEMS

4.1. An important problem for television camera systems is the jumping or wobbling of the picture due to the movement of the camera. This effect occurs when the camera is mounted in a moving truck or airplane. A system called the Dynalens system has been designed which

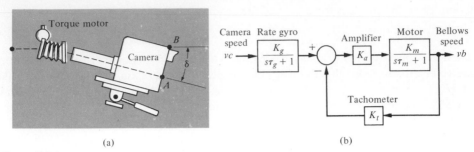

Figure P4.1

reduces the effect of rapid scanning motion and is shown in Fig. P4.1 [12]. A maximum scanning motion of 25°/sec is expected. Let $K_g = K_t = 1$ and assume that τ_g is negligible. (a) Determine the error of the system $E(s)$. (b) Determine the necessary loop gain, $K_a K_m K_t$, when a 1°/sec steady-state error is allowable. (c) The motor time constant is 0.40 sec. Determine the necessary loop gain so that the settling time of v_b is less than or equal to 0.04 sec.

4.2. Determine the error constants for the fluid level control system of Problem 3.1. Determine the necessary loop gain, KR, in order to maintain the steady-state error of the head less than 5% of the magnitude of a step change in the input ΔQ_1.

4.3. A specific closed-loop control system is to be designed for an underdamped response to a step input. The specifications for the system are

$$30\% > \text{percent overshoot} > 10\%,$$
$$\text{Settling time} < 0.4 \text{ sec.}$$

(a) Identify the desired area for the dominant roots of the system. (b) Determine the smallest value of a third root, r_3, if the complex conjugate roots are to represent the dominant response. (c) The closed-loop system transfer function $T(s)$ is third-order and the feedback has a unity gain. Determine the forward transfer function $G(s) = C(s)/E(s)$ when the settling time is 0.4 sec and the percent overshoot is 30%.

4.4. The open-loop transfer function of a unity negative feedback system is

$$G(s) = \frac{K}{s(s + 2)}.$$

A system response to a step input is specified as follows:

$$\text{peak time } T_p = 1 \text{ sec,} \qquad \text{percent overshoot} = 5\%.$$

(a) Determine whether both specifications can be met simultaneously. (b) If the specifications cannot be met simultaneously, determine a compromise value for K so that the peak time and percent overshoot specifications are relaxed the same percentage.

4.5. A major problem confronting engineers today is the optimum use of energy resources. Many engineers are now working on solar energy systems for heating homes [13]. One system uses solar collectors and thermal storage as shown in Fig. P4.5(a). The block diagram of the system is shown in Fig. P4.5(b). The collectors, thermal storage, and the dynamics of the house are represented by $G(s)$. Assume that $\tau_1 = 1$ sec and $\tau_2 = 0$ (an approximation). (a)

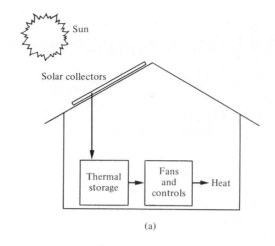

Solar collectors

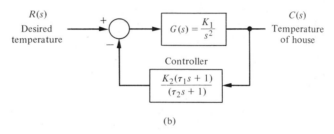

(b)

Fig. P4.5. A solar heated house with a thermal storage system.

Determine the gain $K = K_1K_2$ required so that the response to a step command is as rapid as reasonable with an overshoot of less than 5%. (b) Determine the steady state error of the system for a step and ramp input. (c) Determine the value of K_1K_2 for an ITAE optimal system for (1) a step input and (2) a ramp input.

4.6. A simple model of a management control system is shown in Fig. P4.6. The production lags the production command with a time constant τ. The rate of growth of inventory is assumed to grow proportionately to the production x_1 minus the sales $u_2(t)$. It is desired to maintain the steady-state inventory at a level of A units. Measurement of the inventory and the sales is available without any delay in time. (a) Construct a feedback structure so that the inventory is held at the level A, independent of the sales level. (b) Determine the management gain K which will provide a rapid response to sales changes and yet maintain the inventory

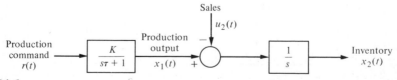

Figure P4.6

within 20% of the desired level. Assume that the time constant $\tau = 1$ week. (c) Determine a desirable set of parameters K and τ so that the response to a step production command results in a minimum ITAE performance index. (d) Determine the error constants for the system designed on the basis of minimum ITAE.

4.7. Space missions require astronauts to leave their spacecraft and operate in space. In order to allow the astronaut to work with his hands in the vehicle and the space environment, it is necessary to provide a maneuvering control system that does not require hand or foot control. Therefore a voice controller has been proposed which is shown in a simplified diagram in Fig. P4.7 [15]. The gas jet controller operates on voice commands and can be approximately represented by a proportional gain K_2. The inertia of the man and equipment with his arms at his sides is 25 Kg-m². (a) Determine the necessary gain K_3 to maintain a steady-state error equal to 1-cm. when the input is a ramp $r(t) = t$ (meters). (b) With this gain K_3, determine the necessary gain $K_1 K_2$ in order to restrict the percent overshoot to 10%. (c) Determine, analytically, the gain $K_1 K_2$ in order to minimize the ISE performance index.

4.8. The dynamic behavior of a management control system was discussed in Problem 3.7. Typical values of the parameters of a business firm are

$$K_1 K_2 = 0.50, \qquad \tau = 10 \text{ months}, \qquad K_4 = 1, \qquad K_5 = 7.6.$$

(a) Determine the response of the business firm to a step input demand by estimating the peak time, the percent overshoot, and the settling time in months. (b) The response of the business firm is judged to be too slow and it is decided to increase the effort in the management and engineering functions so that both K_1 and K_2 are doubled. Estimate the response of this system to a step input command. (c) The business firm during a period of large growth is experiencing a continuously increasing demand so that $r(t) = t$. Determine the ability of the system to satisfy this demand by determining the K_v and steady-state error of the system for the gain $K_1 K_2$ of both (a) and (b).

4.9. The antenna that receives and transmits signals to the Telstar communication satellite is the largest horn antenna ever built. The microwave antenna is 177 ft long, weighs 340 tons, and rolls on a circular track. A photo of the antenna is shown in Fig. P4.9. The Telstar satellite is 34 inches in diameter and moves about 16,000 mph at an altitude of 2,500 miles. The antenna must be positioned accurately to ¹⁄₂₀ of a degree, since the microwave beam is 0.2° wide and highly attenuated by the large distance. If the antenna is following the moving satellite, determine the K_v necessary for the system.

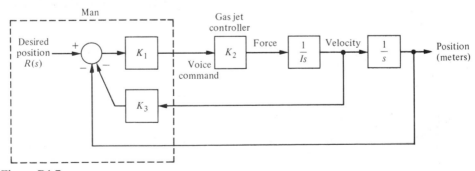

Figure P4.7

Fig. P4.9. A model of the antenna for the Telstar System at Andover, Maine. (Photo courtesy of Bell Telephone Laboratories, Inc.)

4.10. Significant efforts have been made to introduce recycling of aluminum beverage cans to many town and cities. A simplified block diagram of the process of recycling in a town is shown in Fig. P4.10. The emphasis on a collection campaign is represented by the gain K. The disturbance $D(s)$ represents loss of cans that have been discarded in garbage collection or kept rather than recycled. Assume that $\tau_1 = 1$ month and $\tau_2 = 0.5$ month. (a) Estimate the response of the collection system to a step response in the number of cans in use. Calculate

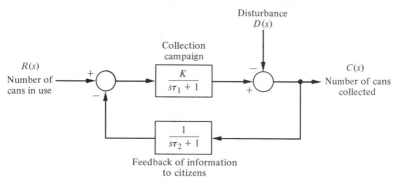

Figure P4.10

the gain necessary for a system with critical damping. (b) Calculate the steady-state error for a step input for the gain determined in (a). (c) A step disturbance $D(s) = D/s$ occurs. Calculate the steady-state effect on the number of cans collected.

4.11. A simple unity feedback control system has a process transfer function

$$\frac{C(s)}{E(s)} = G(s) = \frac{K}{s}.$$

The system input is a step function with an amplitude A. The initial condition of the system at time t_0 is $c(t_0) = Q$, where $c(t)$ is the output of the system. The performance index is defined as

$$I = \int_0^\infty e^2(t) \, dt.$$

(a) Show that $I = (A - Q)^2/2K$. (b) Determine the gain K that will minimize the performance index I. Is this gain a practical value? (c) Select a practical value of gain and determine the resulting value of the performance index.

4.12. Train travel between cities will increase as trains are developed which travel at high speeds, making the train travel time from city center to city center equivalent to airline travel time. The Japanese National Railway has a train called the Bullet Express (Fig. P4.12) that travels between Tokyo and Osaka on the Tokaido line. This train travels the 320 miles in three hours and ten minutes, an average speed of 101 miles per hour [16]. This speed will be increased as new systems are used, such as magnetically levitated systems to float vehicles above an aluminum guideway. In order to maintain a desired speed, a speed control system

Fig. P4.12. Fifty minutes out of Tokyo, the Bullet Express whizzes past Mt. Fuji. Photo courtesy of the Japan National Tourist Organization.

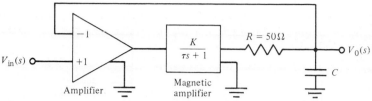

Figure P4.15

is proposed which yields a zero steady-state error to a ramp input. A third-order system is sufficient. Determine the optimum system for an ITAE performance criterion and estimate the system response when $\omega_n = 4$.

4.13. It is desired to approximate a fourth-order system by a lower-order model. The transfer function of the original system is

$$H(s) = \frac{s^3 + 7s^2 + 24s + 24}{s^4 + 10s^3 + 35s^2 + 50s + 24} = \frac{s^3 + 7s^2 + 24s + 24}{(s + 1)(s + 2)(s + 3)(s + 4)}.$$

Show that if we obtain a second order model by the method of Section 4.6, and we do not specify the poles and the zero of $L(s)$, we have

$$L_1(s) = \frac{0.2917s + 1}{0.399s^2 + 1.375s + 1} = \frac{.731(s + 3.428)}{(s + 1.043)(s + 2.4)}.$$

4.14. For the original system of Problem 4.13, it is desired to find the lower-order model when the poles of the second-order model are specified as -1 and -2 and the model has one unspecified zero. Show that this low-order model is

$$L(s) = \frac{0.986s + 2}{s^2 + 3s + 2} = \frac{0.986(s + 2.028)}{(s + 1)(s + 2)}.$$

4.15. A magnetic amplifier with a low output impedance is shown in Fig. P4.15 in cascade with a low pass filter and a pre-amplifier. The amplifier has a high input impedance and a gain of one and is used for adding the signals as shown. Select a value for the capacitance C so that the transfer function $V_0(s)/V_{in}(s)$ has a damping ratio of 0.7. The time constant of the magnetic amplifier is equal to one second and the gain is $K = 20$. Calculate the settling time of the resulting system.

4.16. Electronic pacemakers for a human heart are currently available. These devices regulate the speed of the heart pump. A closed-loop system which includes a pacemaker and the measurement of the heart rate is proposed as shown in Fig. P4.16. The transfer function of the heart pump and the pacemaker is found to be

$$G(s) = \frac{K}{s(s/10 + 1)}.$$

It is desired to design the amplifier gain to yield a tightly controlled system with a settling time to a step disturbance of less than one second. The overshoot to a step in desired heart rate should be less than 15%.

(a) Find a suitable range of K.

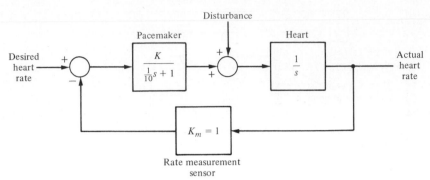

Figure P4.16

(b) If the nominal value of K is $K = 10$, find the sensitivity of the system to small changes in K.

(c) Evaluate the sensitivity at DC (set $s = 0$).

(d) Evaluate the magnitude of the sensitivity at the normal heart rate of 60 beats/minute.

4.17. Consider the original third-order system given in Example 4.6. Determine a first-order model with one pole unspecified and no zeros which will represent the third-order system.

REFERENCES

1. C. M. Clare and D. K. Frederick, *Modeling and Analysis of Dynamic Systems,* Houghton Mifflin, Boston, 1978.

2. F. H. Raven, *Automatic Control Engineering,* 3rd ed., McGraw-Hill, New York, 1978.

3. J. Capanale, "Freely Adjust Damping and Period in this Active 2nd-Order Filter," *Electronic Design News,* April 5, 1978; pp. 89–91.

4. P. R. Clement, "A Note on Third-Order Linear Systems," *IRE Transactions on Automatic Control,* June 1960; p. 151.

5. R. N. Clark, *Introduction to Automatic Control Systems,* Wiley, New York, 1962; pp. 115–124.

6. D. Graham and R. C. Lathrop, "The Synthesis of Optimum Response: Criteria and Standard Forms, Part 2," *Transactions of the AIEE,* **72,** November 1953; pp. 273–288.

7. R. A. Bruns, "Computer Investigation of Two Important Criteria for Adaptive Control Systems," *Jet Propulsion Laboratories Technical Report No. 32-191,* November 1961.

8. J. J. D'Azzo and C. H. Houpis, *Linear Control System Analysis and Design,* McGraw-Hill, New York, 1975.

9. "Analog Servo System Puts Live Actors in Mini Sets," *Electronic Design,* **11,** May 24, 1976.

10. T. C. Hsia, "On the Simplification of Linear Systems," *IEEE Transactions on Automatic Control,* June 1972; pp. 372–374.

11. E. J. Davison, "A Method for Simplifying Linear Dynamic Systems," *IEEE Transactions on Automatic Control,* January 1966; pp. 93–101.

12. J. DeLaCierva, "Rate Servo Keeps TV Picture Clear," *Control Engineering,* May 1965; p. 112.

13. H. E. Harpster and A. H. Eltimsahy, "A Heat Pump Driven Optimal System," *IEEE Transactions on Industry Applications,* August 1978; pp. 357–363.

14. R. N. Ogbudinkpa, "Optimal Recycling of Aluminum Beverage Cans," *J. of Environmental Systems,* **7,** 4, 1978; pp. 343–353.

15. W. E. Drissel, et al., "Study of an Attitude Control System for the Astronaut Maneuvering Unit," *NASA Contract Report CR-198,* March 1965.

16. D. Bartruff, "The Best Trains in the World," *Saturday Review,* January 15, 1973; p. 19.

5 / The Stability of
Linear Feedback Systems

5.1 THE CONCEPT OF STABILITY

The transient response of a feedback control system is of primary interest and must be investigated. A very important characteristic of the transient performance of a system is the *stability* of the system. A *stable system* is defined as a system with a bounded system response. That is, if the system is subjected to a bounded input or disturbance and the response is bounded in magnitude, the system is said to be stable.

The concept of stability can be illustrated by considering a right circular cone placed on a plane horizontal surface [1]. If the cone is resting on its base and is tipped slightly, it returns to its original equilibrium position. This position and response is said to be stable. If the cone rests on its side and is displaced slightly, it rolls with no tendency to leave the position on its side. This position is designated as the neutral stability. On the other hand, if the cone is placed on its tip and released, it falls onto its side. This position is said to be unstable. These three positions are illustrated in Fig. 5.1.

The stability of a dynamic system is defined in a similar manner. The response to a displacement, or initial condition, will result in either a decreasing, neutral, or increasing response. Specifically, it follows from the definition of stability that a linear system is stable if and only if the absolute value of its impulse response, $g(t)$, integrated over an infinite range is finite. That is, in terms of the convolution integral Eq. (4.1) for a bounded input, one requires that $\int_0^\infty |g(t)| \, dt$ be finite. The location in the s-plane of the poles of a system indicate the resulting transient response. The poles in the left-hand portion of the s-plane result in a decreasing response for disturbance inputs. Similarly, poles on the $j\omega$-axis and in the right-hand plane result in a neutral and an increasing response, respectively, for a disturbance input. This division of the s-plane is shown in Fig. 5.2. Clearly the poles of desirable dynamic systems must lie in the left-hand portion of the s-plane.

An example of an unstable system is shown in Fig. 5.3. The first bridge across

145

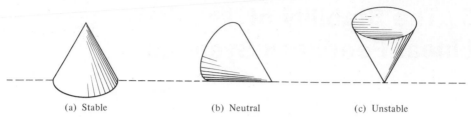

(a) Stable   (b) Neutral   (c) Unstable

Fig. 5.1. The stability of a cone.

the Tacoma Narrows at Puget Sound, Washington, was opened to traffic on July 1, 1940. The bridge was found to oscillate whenever the wind blew. After four months, on November 7, 1940, a wind produced an oscillation which grew in amplitude until the bridge broke apart. Figure 5.3(a) shows the condition of beginning oscillation while Fig. 5.3(b) shows the catastrophic failure [11].

In terms of linear systems, we recognize that the stability requirement may be defined in terms of the location of the poles of the closed-loop transfer function. The closed-loop system transfer function is written as

$$T(s) = \frac{p(s)}{q(s)} = \frac{K\prod_{i=1}^{M}(s + z_i)}{s^N\prod_{k=1}^{Q}(s + \sigma_k)\prod_{m=1}^{R}(s^2 + 2\alpha_m s + (\alpha_m^2 + \omega_m^2))}, \qquad (5.1)$$

where $q(s) = \Delta(s)$ is the characteristic equation whose roots are the poles of the closed-loop system. The output response for an impulse function input is then

$$c(t) = \sum_{k=1}^{Q} A_k e^{-\sigma_k t} + \sum_{m=1}^{R} B_m \left(\frac{1}{\omega_m}\right) e^{-\alpha_m t} \sin \omega_m t \qquad (5.2)$$

when $N = 0$. Clearly, in order to obtain a bounded response, the poles of the closed-loop system must be in the left-hand portion of the s-plane. Thus, *a necessary and sufficient condition that a feedback system be stable is that all the poles of the system transfer function have negative real parts.*

In order to ascertain the stability of a feedback control system, one could determine the roots of the characteristic equation $q(s)$. However, we are first interested in determining the answer to the question: "Is the system stable?" If we calculate

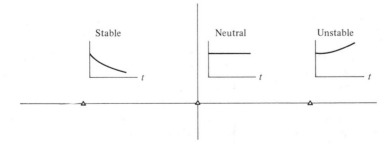

Fig. 5.2. Stability in the s-plane.

Figure 5.3(a)

Figure 5.3(b) (Photos courtesy of Mrs. F. B. Farquharson.)

the roots of the characteristic equation in order to answer this question, we have determined much more information than is necessary. Therefore, several methods have been developed which provide the required "yes" or "no" answer to the stability question. The three approaches to the question of stability are: (1) The s-plane approach, (2) the frequency plane ($j\omega$) approach, and (3) the time-domain approach. The real frequency ($j\omega$) approach is outlined in Chapter 8, and the discussion of the time-domain approach is deferred until Chapter 9.

5.2 THE ROUTH-HURWITZ STABILITY CRITERION

The discussion and determination of stability has occupied the interest of many engineers. Maxwell and Vishnegradsky first considered the question of stability of dynamic systems. In the late 1800's, A. Hurwitz and E. J. Routh published independently a method of investigating the stability of a linear system [2, 3]. The Routh-Hurwitz stability method provides an answer to the question of stability by considering the characteristic equation of the system. The characteristic equation in the Laplace variable is written as

$$\Delta(s) = q(s) = a_n s^n + a_{n-1} s^{n-1} + \cdots + a_1 s + a_0 = 0. \tag{5.3}$$

In order to ascertain the stability of the system, it is necessary to determine if any of the roots of $q(s)$ lie in the right-half of the s-plane. If Eq. (5.3) is written in factored form, we have

$$a_n(s - r_1)(s - r_2) \cdots (s - r_n) = 0, \tag{5.4}$$

where $r_i = i$th root of the characteristic equation. Multiplying the factors together, we find that

$$\begin{aligned}
q(s) = a_n s^n &- a_n(r_1 + r_2 + \cdots + r_n)s^{n-1} \\
&+ a_n(r_1 r_2 + r_2 r_3 + r_1 r_3 + \cdots)s^{n-2} \\
&- a_n(r_1 r_2 r_3 + r_1 r_2 r_4 \cdots)s^{n-3} + \cdots \\
&+ a_n(-1)^n r_1 r_2 r_3 \cdots r_n = 0.
\end{aligned} \tag{5.5}$$

In other words, for an nth-degree equation, we obtain

$$\begin{aligned}
q(s) = a_n s^n &- a_n(\text{sum of all the roots})s^{n-1} \\
&+ a_n(\text{sum of the products of the roots taken 2 at a time})s^{n-2} \\
&- a_n(\text{sum of the products of the roots taken 3 at a time})s^{n-3} \\
&+ \cdots + a_n(-1)^n(\text{product of all } n \text{ roots}) = 0.
\end{aligned} \tag{5.6}$$

Examining Eq. (5.5), we note that all the coefficients of the polynomial must have the same sign if all the roots are in the left-hand plane. Also, it is necessary, for a stable system, that all the coefficients be nonzero. However, while these requirements are necessary, they are not sufficient; that is, if they are not satisfied, we immediately know the system is unstable. However, if they are satisfied, we must procced to ascertain the stability of the system. For example, when the characteristic equation is

$$q(s) = (s + 2)(s^2 - s + 4) = (s^3 + s^2 + 2s + 8), \tag{5.7}$$

the system is unstable and yet the polynomial possesses all positive coefficients.

The Routh-Hurwitz criterion is a necessary and sufficient criterion for the stability of linear systems. The method was originally developed in terms of determinants, but we shall utilize the more convenient array formulation.

The Routh-Hurwitz criterion is based on ordering the coefficients of the characteristic equation

$$a_n s^n + a_{n-1} s^{n-1} + a_{n-2} s^{n-2} + \cdots + a_1 s + a_0 = 0 \tag{5.8}$$

into an array or schedule as follows [4]:

$$
\begin{array}{c|cccc}
s^n & a_n & a_{n-2} & a_{n-4} & \cdots \\
s^{n-1} & a_{n-1} & a_{n-3} & a_{n-5} & \cdots
\end{array}
$$

Then, further rows of the schedule are completed as follows:

$$
\begin{array}{c|ccc}
s^n & a_n & a_{n-2} & a_{n-4} \\
s^{n-1} & a_{n-1} & a_{n-3} & a_{n-5} \\
s^{n-2} & b_{n-1} & b_{n-3} & b_{n-5} \\
s^{n-3} & c_{n-1} & c_{n-3} & c_{n-5} \\
\vdots & \vdots & \vdots & \vdots \\
s^0 & h_{n-1} & &
\end{array}
$$

where

$$b_{n-1} = \frac{(a_{n-1})(a_{n-2}) - a_n(a_{n-3})}{a_{n-1}} = \frac{-1}{a_{n-1}} \begin{vmatrix} a_n & a_{n-2} \\ a_{n-1} & a_{n-3} \end{vmatrix},$$

$$b_{n-3} = -\frac{1}{a_{n-1}} \begin{vmatrix} a_n & a_{n-4} \\ a_{n-1} & a_{n-5} \end{vmatrix},$$

and

$$c_{n-1} = \frac{-1}{b_{n-1}} \begin{vmatrix} a_{n-1} & a_{n-3} \\ b_{n-1} & b_{n-3} \end{vmatrix},$$

and so on. The algorithm for calculating the entries in the array can be followed on a determinant basis or by using the form of the equation for b_{n-1}.

The Routh-Hurwitz criterion states that the number of roots of $q(s)$ with positive real parts is equal to the number of changes in sign of the first column of the array. This criterion requires that there be no changes in sign in the first column for a stable system. This requirement is both necessary and sufficient.

There are three distinct cases which must be treated separately, requiring suitable modifications of the array calculation procedure. The three cases are: (1) No element in the first column is zero; (2) there is a zero in the first column but some other elements of the row containing the zero in the first column are nonzero; and (3) there is a zero in the first column and the other elements of the row containing the zero are also zero.

In order to clearly illustrate this method, several examples will be presented for each case.

CASE 1. *No element in the first column is zero.*

Example 5.1. The characteristic equation of a second-order system is

$$q(s) = a_2 s^2 + a_1 s + a_0.$$

The array is written as

$$
\begin{array}{c|cc}
s^2 & a_2 & a_0 \\
s & a_1 & 0 \\
s^0 & b_1 & 0
\end{array}
$$

where

$$b_1 = \frac{a_1 a_0 - (0)a_2}{a_1} = \frac{-1}{a_1}\begin{vmatrix} a_2 & a_0 \\ a_1 & 0 \end{vmatrix} = a_0.$$

Therefore, the requirement for a stable second-order system is simply that all the coefficients be positive.

Example 5.2. The characteristic equation of a third-order system is

$$q(s) = a_3 s^3 + a_2 s^2 + a_1 s + a_0.$$

The array is

$$
\begin{array}{c|cc}
s^3 & a_3 & a_1 \\
s^2 & a_2 & a_0 \\
s^1 & b_1 & 0 \\
s^0 & c_1 & 0
\end{array}
$$

where

$$b_1 = \frac{a_2 a_1 - a_0 a_3}{a_2} \quad \text{and} \quad c_1 = \frac{b_1 a_0}{b_1} = a_0.$$

For the third-order system to be stable, it is necessary and sufficient that the coefficients be positive and $a_2 a_1 \geqslant a_0 a_3$. The condition when $a_2 a_1 = a_0 a_3$ results in a borderline stability case, and one pair of roots lies on the imaginary axis in the s-plane. This borderline case is recognized as Case 3 since there is a zero in the first column when $a_2 a_1 = a_0 a_3$, and it will be discussed under Case 3.

As a final example of characteristic equations which result in no zero elements in the first row, let us consider a polynomial

$$q(s) = (s - 1 + j\sqrt{7})(s - 1 - j\sqrt{7})(s + 3) = s^3 + s^2 + 2s + 24. \quad (5.9)$$

The polynomial satisfies all the necessary conditions since all the coefficients exist and are positive. Therefore, utilizing the Routh-Hurwitz array, we have

s^3	1	2
s^2	1	24
s^1	-22	0
s^0	24	0

Since two changes in sign appear in the first column, we find that two roots of $q(s)$ lie in the right-hand plane, and our prior knowledge is confirmed.

CASE 2. *Zeros in the first column while some other elements of the row containing a zero in the first column are nonzero.*

If only one element in the array is zero, it may be replaced with a small positive number ϵ which is allowed to approach zero after completing the array. For example, consider the following characteristic equation:

$$q(s) = s^5 + 2s^4 + 2s^3 + 4s^2 + 11s + 10. \quad (5.10)$$

The Routh-Hurwitz array is then

s^5	1	2	11
s^4	2	4	10
s^3	ϵ	6	0
s^2	c_1	10	0
s^1	d_1	0	0
s^0	10	0	0

where

$$c_1 = \frac{4\epsilon - 12}{\epsilon} = \frac{-12}{\epsilon} \quad \text{and} \quad d_1 = \frac{6c_1 - 10\epsilon}{c_1} \to 6.$$

There are two sign changes due to the large negative number in the first column, $c_1 = -12/\epsilon$. Therefore the system is unstable and two roots lie in the right-half of the plane.

Example 5.3. As a final example of the type of Case 2, consider the characteristic equation

$$q(s) = s^4 + s^3 + s^2 + s + K, \quad (5.11)$$

where it is desired to determine the gain K which results in borderline stability. The Routh-Hurwitz array is then

$$
\begin{array}{c|ccc}
s^4 & 1 & 1 & K \\
s^3 & 1 & 1 & 0 \\
s^2 & \epsilon & K & 0 \\
s^1 & c_1 & 0 & 0 \\
s^0 & K & 0 & 0
\end{array}
$$

where

$$
c_1 = \frac{\epsilon - K}{\epsilon} \rightarrow \frac{-K}{\epsilon}.
$$

Therefore, for any value of K greater than zero, the system is unstable. Also, since the last term in the first column is equal to K, a negative value of K will result in an unstable system. Therefore, the system is unstable for all values of gain K.

CASE 3. *Zeros in the first column, and the other elements of the row containing the zero are also zero.*

Case 3 occurs when all the elements in one row are zero or when the row consists of a single element which is zero. This condition occurs when the polynomial contains singularities that are symmetrically located about the origin of the s-plane. Therefore, Case 3 occurs when factors such as $(s + \sigma)(s - \sigma)$, or $(s + j\omega)(s - j\omega)$, occur. This problem is circumvented by utilizing the *auxiliary equation* which immediately precedes the zero entry in the Routh array. The order of the auxiliary equation is always even and indicates the number of symmetrical root pairs.

In order to illustrate this approach, let us consider a third-order system with a characteristic equation:

$$
q(s) = s^3 + 2s^2 + 4s + K, \tag{5.12}
$$

where K is an adjustable loop gain. The Routh array is then

$$
\begin{array}{c|cc}
s^3 & 1 & 4 \\
s^2 & 2 & K \\
s^1 & \dfrac{8 - K}{2} & 0 \\
s^0 & K & 0
\end{array}
$$

Therefore, for a stable system, we require that

$$
0 \leqslant K \leqslant 8.
$$

When $K = 8$, we have two roots on the $j\omega$-axis and a borderline stability case. Note that we obtain a row of zeros (Case 3) when $K = 8$. The auxiliary equations, $U(s)$, is the equation of the row preceding the row of zeros. The equation of the row preceding the row of zeros is, in this case, obtained from the s^2-row. We recall that this row contains the coefficients of the even powers of s and therefore in this case, we have

$$U(s) = 2s^2 + Ks^0 = 2s^2 + 8 = 2(s^2 + 4) = 2(s + j2)(s - j2). \qquad (5.13)$$

In order to show that the auxiliary equation, $U(s)$ is indeed a factor of the characteristic equation, we divide $q(s)$ by $U(s)$ to obtain

$$
\begin{array}{r}
\frac{1}{2}s \;\; + 1 \\
2s^2 + 8 \overline{)\; s^3 + 2s^2 + 4s + 8} \\
\underline{s^3 \qquad\quad + 4s} \\
2s^2 \qquad\;\; + 8 \\
\underline{2s^2 \qquad\;\; + 8} \\
0
\end{array}
$$

Therefore, when $K = 8$, the factors of the characteristic equation are

$$q(s) = (s + 2)(s + j2)(s - j2). \qquad (5.14)$$

Example 5.4. As an example, consider the polynomial

$$q(s) = s^5 + s^4 + 4s^3 + 24s^2 + 3s + 63. \qquad (5.15)$$

The Routh-Hurwitz array is

s^5	1	4	3
s^4	1	24	63
s^3	-20	-60	0
s^2	21	63	0
s^1	0	0	0

Therefore, the auxiliary equation is

$$U(s) = 21s^2 + 63 = 21(s^2 + 3) = 21(s + j\sqrt{3})(s - j\sqrt{3}), \qquad (5.16)$$

which indicates that two roots are on the imaginary axis. In order to examine the remaining roots, we divide by the auxiliary equation to obtain

$$\frac{q(s)}{s^2 + 3} = s^3 + s^2 + s + 21.$$

Establishing a Routh-Hurwitz array for this equation, we have

s^3	1	1
s^2	1	21
s^1	-20	0
s^0	21	0

The two changes in sign in the first column indicate the presence of two roots in the right-hand plane, and the system is unstable. The roots in the right-hand plane are $s = +1 \pm j\sqrt{6}$.

Example 5.5. Large disk storage devices are used with today's computers. The data head is moved to different positions on the spinning disk and rapid, accurate response is required [6]. A block diagram of a disk storage data head positioning system is shown in Fig. 5.4. It is desired to determine the range of K and a for which the system is stable. The characteristic equation is

$$1 + G(s) = 1 + \frac{K(s + a)}{s(s + 1)(s + 2)(s + 3)} = 0.$$

Therefore $q(s) = s^4 + 6s^3 + 11s^2 + (K + 6)s + Ka = 0$.
Establishing the Routh array we have

$$
\begin{array}{c|ccc}
s^4 & 1 & 11 & Ka \\
s^3 & 6 & (K + 6) & \\
s^2 & b_3 & Ka & \\
s^1 & c_3 & & \\
s^0 & Ka & &
\end{array}
$$

where $b_3 = \dfrac{60 - K}{6}$ and $c_3 = \dfrac{b_3(K + 6) - 6Ka}{b_3}$.

The coefficient c_3 sets the acceptable range of K and a, while b_3 requires that K be less than 60. Setting $c_3 = 0$ we obtain

$$(K - 60)(K + 6) + 36Ka = 0.$$

The required relationship between K and a is then

$$a \leqslant \frac{(60 - K)(K + 6)}{36K} \qquad \text{when } a \text{ is positive.}$$

Therefore, if $K = 40$, we require $a \leqslant 0.639$.

5.3 THE RELATIVE STABILITY OF FEEDBACK CONTROL SYSTEMS

The verification of stability by the Routh-Hurwitz criterion provides only a partial answer to the question of stability. The Routh-Hurwitz criterion ascertains the absolute stability of a system by determining if any of the roots of the characteristic equation lie in the right-half of the s-plane. However, if the system satisfies the Routh-Hurwitz criterion and is absolutely stable, it is desirable to determine the *relative stability;* that is, it is necessary to investigate the relative damping of each root of the characteristic equation. The relative stability of a system may be defined as the property which is measured by the relative settling times of each root or pair of roots. Therefore, relative stability is represented by the real part of each root.

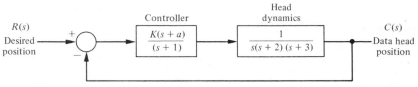

Figure 5.4

Thus, root r_2 is relatively more stable than the roots r_1, $\hat{r}_1$ as shown in Fig. 5.5. The relative stability of a system can also be defined in terms of the relative damping coefficients ζ of each complex root pair and therefore in terms of the speed of response and overshoot instead of settling time.

Hence the investigation of the relative stability of each root is clearly necessary since, as we found in Chapter 4, the location of the closed-loop poles in the s-plane determines the performance of the system. Thus it is imperative that we reexamine the characteristic equation $q(s)$ and consider several methods for the determination of relative stability.

Since the relative stability of a system is dictated by the location of the roots of the characteristic equation, a first approach using an s-plane formulation is to extend the Routh-Hurwitz criterion to ascertain relative stability. This can be simply accomplished by utilizing a change of variable which shifts the s-plane axis in order to utilize the Routh-Hurwitz criterion. Examining Fig. 5.5, we notice that a shift of the vertical axis in the s-plane to $-\sigma_1$ will result in the roots r_1, $\hat{r}_1$ appearing on the shifted axis. The correct magnitude to shift the vertical axis must be obtained on a trial-and-error basis. Then, without solving the fifth-order polynomial $q(s)$, one may determine the real part of the dominant roots r_1, $\hat{r}_1$.

Example 5.6.   Consider the simple third-order characteristic equation

$$q(s) = (s + 1 + j1)(s + 1 - j1)(s + 2) = s^3 + 4s^2 + 6s + 4. \qquad (5.17)$$

Initially, one might shift the axis other than one unit and obtain a Routh-Hurwitz array without a zero occurring in the first column. However, upon setting the shifted variable s_n equal to $s + 1$, we obtain

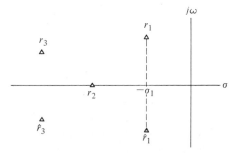

Fig. 5.5.   Root locations in the s-plane.

$$(s_n - 1)^3 + 4(s_n - 1)^2 + 6(s_n - 1) + 4 = s_n^3 + s_n^2 + s_n + 1. \qquad (5.18)$$

Then the Routh array is established as

$$
\begin{array}{c|cc}
s_n^3 & 1 & 1 \\
s_n^2 & 1 & 1 \\
s_n^1 & 0 & 0 \\
s_n^0 & 1 & 0
\end{array}
$$

Clearly, there are roots on the shifted imaginary axis and the roots can be obtained from the auxiliary equation, which is

$$
\begin{aligned}
U(s_n) &= s_n^2 + 1 = (s_n + j)(s_n - j) \\
&= (s + 1 + j)(s + 1 - j).
\end{aligned}
\qquad (5.19)
$$

The shifting of the s-plane axis in order to ascertain the relative stability of a system is a very useful approach, particularly for higher-order systems with several pairs of closed-loop complex conjugate roots.

5.4 THE DETERMINATION OF ROOT LOCATIONS IN THE s-PLANE

The relative stability of a feedback control system is directly related to the location of the closed-loop roots of the characteristic equation in the s-plane. Therefore, it is often necessary and easiest to simply determine the values of the roots of the characteristic equation. This approach has become particularly attractive today due to the availability of digital computer programs for determining the roots of polynomials. However, this approach may even be the most logical when using manual calculations if the order of the system is relatively low. For a fifth-order system or lower, it is usually simpler to utilize manual calculation methods.

The determination of the roots of a polynomial can be obtained by utilizing *synthetic division* which is based on the remainder theorem; that is, upon dividing the polynomial by a factor, the remainder is zero when the factor is a root of the polynomial. Synthetic division is commonly used to carry out the division process. The relations for the roots of a polynomial as obtained in Eq. (5.6) are utilized to aid in the choice of a first estimate of a root.

Example 5.7. Let us determine the roots of the polynomial

$$q(s) = s^3 + 4s^2 + 6s + 4. \qquad (5.20)$$

Establishing a table of synthetic division, we have

$$
\begin{array}{rrrr|l}
1 & 4 & 6 & 4 & \underline{-1} = \text{trial root} \\
  & -1 & -3 & -3 & \\
\hline
1 & 3 & 3 \mid & 1 = \text{remainder}
\end{array}
$$

for a trial root of $s = -1$. In this table, we multiply by the trial root and successively add in each column. With a remainder of one, we might try $s = -2$ which results in the form

$$
\begin{array}{rrrr|r}
1 & 4 & 6 & 4 & \underline{-2} \\
& -2 & -4 & -4 & \\
\hline
1 & 2 & 2 & | & 0
\end{array}
$$

Since the remainder is zero, one root is equal to -2 and the remaining roots may be obtained from the remaining polynomial $(s^2 + 2s + 2)$ by using the quadratic root formula.

The search for a root of the polynomial can be aided considerably by utilizing the rate of change of the polynomial at the estimated root in order to obtain a new estimate. The Newton-Raphson method is a rapid method utilizing synthetic division to obtain the value of

$$
\frac{dq(s)}{ds}\bigg|_{s=s_1},
$$

where s_1 is a first estimate of the root. The Newton-Raphson method is an iteration approach utilized in many digital computer root-solving programs. A new estimate s_{n+1} of the root is based on the last estimate as [12, 13]

$$
s_{n+1} = s_n - \frac{q(s_n)}{q'(s_n)}, \tag{5.21}
$$

where

$$
q'(s_n) = \frac{dq(s)}{ds}\bigg|_{s=s_n}.
$$

The synthetic division process may be utilized to obtain $q(s_n)$ and $q'(s_n)$. The synthetic division process for a trial root may be written for the polynomial

$$
q(s) = a_m s^m + a_{m-1} s^{m-1} + \cdots + a_1 s + a_0
$$

as

$$
\begin{array}{ccccc|c}
a_m & a_{m-1} & \cdots & a_1 & a_0 & \underline{s_n} \\
& b_m s_n & \cdots & & b_1 s_n & \\
\hline
b_m & b_{m-1} & \cdots & b_1 & b_0 &
\end{array}
$$

where $b_0 = q(s_n)$, the remainder of the division process. When s_n is a root of $q(s)$, the remainder is equal to zero and the remaining polynomial

$$
b_m s^{m-1} + b_{m-1} s^{m-2} + \cdots + b_1
$$

may itself be factored. The derivative evaluated at the nth estimate of the root, $q'(s_n)$, may also be obtained by repeating the synthetic division process on the b_m, $b_{m-1}, \ldots, b_1$ coefficients. The value of $q'(s_n)$ is the remainder of this repeated synthetic division process. This process converges as the square of the absolute error. The Newton-Raphson method, using synthetic division, is readily illustrated as can be seen by repeating Example 5.7.

Example 5.8. For the polynomial $q(s) = s^3 + 4s^2 + 6s + 4$, we establish a table of synthetic division for a first estimate as follows:

$$
\begin{array}{cccc|l}
1 & 4 & 6 & 4 & \underline{-1} \\
  & -1 & -3 & -3 & \\
\hline
1 & 3 & 3 & 1 & = q(s_1) \\
  & -1 & -2 & & \\
\hline
1 & 2 & 1 & & = q'(s_1)
\end{array}
$$

The derivative of $q(s)$ evaluated at s_1 is determined by continuing the synthetic division as shown. Then the second estimate becomes

$$
s_2 = s_1 - \frac{q(s_1)}{q'(s_1)} = -1 - \left(\frac{1}{1}\right) = -2.
$$

As we found in Example 5.7, s_2 is, in fact, a root of the polynomial and results in a zero remainder.

Example 5.9. Let us consider the polynomial

$$
q(s) = s^3 + 3.5s^2 + 6.5s + 10. \tag{5.22}
$$

From Eq. (5.6), we note that the sum of all the roots is equal to -3.5 and that the product of all the roots is -10. Therefore, as a first estimate, we try $s_1 = -1$ and obtain the following table:

$$
\begin{array}{cccc|l}
1 & 3.5 & 6.5 & 10 & \underline{-1} \\
  & -1 & -2.5 & -4 & \\
\hline
1 & 2.5 & 4 & 6 & = q(s_1) \\
  & -1 & -1.5 & & \\
\hline
1 & 1.5 & 2.5 & & = q'(s_1)
\end{array}
$$

Therefore, a second estimate is

$$
s_2 = -1 - \left(\frac{6}{2.5}\right) = -3.40.
$$

Now let us use a second estimate which is convenient for these manual calculations. Therefore, on the basis of the calculation of $s_2 = -3.40$, we will choose a second estimate $s_2 = -3.00$. Establishing a table for $s_2 = -3.00$ and completing the synthetic division, we find that

$$
s_3 = -s_2 - \frac{q(s_2)}{q'(s_2)} = -3.00 - \frac{(-5)}{12.5} = -2.60.
$$

Finally, completing a table for $s_3 = -2.50$, we find that the remainder is zero and the polynomial factors are $q(s) = (s + 2.5)(s^2 + s + 4)$.

The availability of a digital computer or a programmable calculator enables one to readily determine the roots of a polynomial by using root determination programs, which usually use the Newton-Raphson algorithm, Eq. (5.21). This, as is the case for time-shared computers with remote consoles, is a particularly useful approach when a computer is readily available for immediate access. The availability of a console connected in a time-shared manner to a digital computer is particularly advantageous for a control engineer, since the ability to perform on-line calculations aids in the iterative analysis and design process.

5.5 SUMMARY

In this chapter we have considered the concept of the stability of a feedback control system. A definition of a stable system in terms of a bounded system response was outlined and related to the location of the poles of the system transfer function in the s-plane.

The Routh-Hurwitz stability criterion was introduced and several examples were considered. The relative stability of a feedback control system was also considered in terms of the location of the poles and zeros of the system transfer function in the s-plane. Finally, the determination of the roots of the characteristic equation was considered and the Newton-Raphson method was illustrated.

PROBLEMS

5.1. Utilizing the Routh-Hurwitz criterion, determine the stability of the following polynomials:

(a) $s^2 + 3s + 1$ (b) $s^3 + 4s^2 + 5s + 6$

(c) $s^3 + 3s^2 - 6s + 10$ (d) $s^4 + s^3 + 2s^2 + 6s + 8$

(e) $s^4 + s^3 + 2s^2 + 2s + K$ (f) $s^5 + s^4 + 2s^3 + s + 3$

(g) $s^5 + s^4 + 2s^3 + s^2 + s + K$

For all cases, determine the number of roots, if any, in the right-hand plane. Also, when it is adjustable, determine the range of K which results in a stable system.

5.2. An antenna control system was analyzed in Problem 3.5 and it was determined that in order to reduce the effect of wind disturbances, the gain of the magnetic amplifier k_a should be as large as possible. (a) Determine the limiting value of gain for maintaining a stable system. (b) It is desired to have a system settling time equal to two seconds. Using a shifted axis and the Routh-Hurwitz criterion, determine the value of gain which satisfies this requirement. Assume that the complex roots of the closed-loop system dominate the transient response. (Is this a valid approximation in this case?)

5.3. A simplified model of the human speech system was discussed in Problem 3.8. When the speech control system becomes unstable, stuttering occurs. Using the Routh-Hurwitz

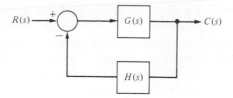

Figure P5.4

criterion and assuming $K_2 = 0.01$, $K_3 = 0.1$, and $K_4 = 0.1$, determine the value of gain K_5 when stuttering occurs. The characteristic equation is a fourth-order equation and is correctly obtained by evaluating all the factors of the denominator of $T(s) = C(s)/R(s)$.

5.4. A feedback control system is shown in Fig. P5.4. The process transfer function is

$$G(s) = \frac{K(s + 40)}{s(s + 10)},$$

and the feedback transfer function is $H(s) = 1/(s + 20)$. (a) Determine the limiting value of gain K for a stable system. (b) For the gain which results in borderline stability, determine the magnitude of the imaginary roots. (c) Reduce the gain to one-half the magnitude of the borderline value and determine the relative stability of the system (1) by shifting the axis and using the Routh-Hurwitz criterion, and (2) by determining the root locations. Show the roots are between -1 and -2.

5.5. Determine the relative stability of the systems with the following characteristic equations (a) by shifting the axis in the s-plane and using the Routh-Hurwitz criterion, and (b) by determining the location of the roots in the s-plane:

1. $s^3 + s^2 + 2s + 0.5 = 0$
2. $s^4 + 9s^3 + 30s^2 + 42s + 20 = 0$
3. $s^3 + 19s^2 + 110s + 200 = 0$

5.6. A unity-feedback control system is shown in Fig. P5.6. Determine the relative stability of the system with the following transfer functions by determining the location of the roots in the s-plane:

(a) $G(s) = \dfrac{65 + 33s}{s^2(s + 9)}$

(b) $G(s) = \dfrac{24}{s(s^3 + 10s^2 + 35s + 50)}$

(c) $G(s) = \dfrac{2(s + 4)(s + 8)}{s(s + 5)^2}$

5.7. The linear model of a phase detector (phase-lock loop) can be represented by Fig. P5.7. [7, 8] The phase-lock systems are designed to maintain zero difference in phase between the

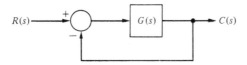

Figure P5.6

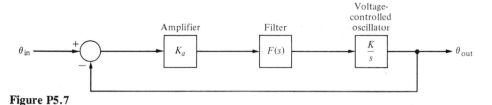

Figure P5.7

input carrier signal and a local voltage controlled oscillator. Phase-lock loops find application in color television, missile tracking, and space telemetry. The filter for a particular application is chosen as

$$F(s) = \frac{(s + 10)}{(s + 0.1)(s + 100)}.$$

It is desired to minimize the steady-state error of the system for a ramp change in the phase information signal. (a) Determine the limiting value of the gain $K_a K = K_v$ in order to maintain a stable system. (b) It is decided that a steady-state error equal to 1° is acceptable for a ramp signal of 100 rad/sec. For that value of gain K_v, determine the location of the roots of the system.

5.8. A very interesting and useful velocity control system has been designed for a wheelchair control system [14]. It is desirable to enable people paralyzed from the neck down to drive themselves about in motorized wheelchairs. A proposed system utilizing velocity sensors mounted in a headgear is shown in Fig. P5.8. The headgear sensor provides an output proportional to the magnitude of the head movement. There is a sensor mounted at 90° intervals so that forward, left, right, or reverse can be commanded. Typical values for the time constants are $\tau_1 = 0.5$ sec, $\tau_3 = 1$ sec and $\tau_4 = 1/3$ sec. (a) Determine the limiting gain $K = K_1 K_2 K_3 K_4$ for a stable system. (b) When the gain K is set equal to one-third of the limiting value, determine if the settling time of the system is less than 4 sec. (c) Determine the value of gain which results in a system with a settling time of 4 sec. Also, obtain the value of the roots of the characteristic equation when the settling time is equal to 4 sec.

5.9. The need for modern rapid transit systems which operate at high speeds results in exacting control system requirements. The Bay Area Rapid Transit District experienced difficulties with several designs near San Francisco, California. The rail transit system for the San Francisco area cost nearly two billion dollars and commenced in 1972. It is desired to operate the electronically controlled trains at 80 mph with headway between trains as little as 90 sec.

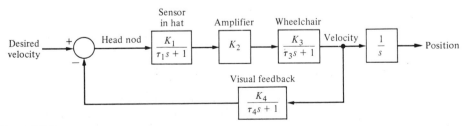

Figure P5.8

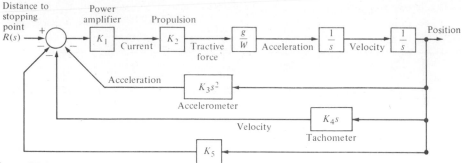

Figure P5.9

In addition, it is desirable that the trains be stopped within several inches of the selected station position in order to use doors that open in walls at the stations. In Fig. P5.9, the factor g/W represents the train, where g equals the acceleration of gravity and W equals the train weight. (a) Determine the steady-state error to (1) a velocity command $R(s) = A/s^2$ and (2) a final position command $R(s) = B/s$ when $K_5 = 1$. (b) Determine the limiting forward gain $K = K_1 K_2(g/W)$ for a stable system when K_3 and K_4 are less than one and $K_5 = 1$. (c) Determine suitable values for K_3 and K_4 so that the overshoot of the system is less than 0.1% and the settling time of the system is equal to 4 seconds.

5.10. Feedback is used by humans to control their posture and retain their balance [9, 10]. Humans use visual signals, otoliths, and signals from their semicircular canals for feedback. Study of these balance systems is important for medical, rehabilitative purposes and for the design of robots. The open-loop transfer function of a posture control system may be approximated by

$$GH(s) = \frac{K(s + 2)}{s(s + 5)(s^2 + 2s + 5)}.$$

(a) Determine the value of gain when a person's balance is lost and the system oscillates. (b) Calculate the roots when the system oscillates for K in part (a). Assume unity feedback.

5.11. A feedback control system has a characteristic equation:

$$s^3 + (4 + K)s^2 + 6s + 16 + 8K = 0.$$

The parameter K must be positive. What is the maximum value K can assume before the system becomes unstable? When K is equal to the maximum value, the system oscillates. Determine the frequency of oscillation.

5.12. A feedback control system has a characteristic equation:

$$s^6 + 2s^5 + 5s^4 + 8s^3 + 8s^2 + 8s + 4 = 0.$$

Determine if the system is stable and determine the values of the roots.

5.13. The stability of a motorcycle and rider is an important area for study since many motorcycle designs result in vehicles that are difficult to control [15]. The handling characteristics of a motorcycle must include a model of the rider as well as the vehicle. The dynamics of one motorcycle and an average rider can be represented by an open-loop transfer function

$$GH(s) = \frac{K(s^2 + 30s + 1125)}{s(s + 20)(s^2 + 10s + 125)(s^2 + 60s + 3400)}.$$

(a) An approximation, calculate the acceptable range of K for a stable system when the numerator polynomial (zeroes) and the denominator polynomial ($s^2 + 60s + 3400$) are neglected.

(b) Calculate the actual range of acceptable K accounting for all zeroes and poles.

REFERENCES

1. W. J. Cunningham, "The Concept of Stability," *American Scientist,* December 1963; pp. 425–436.

2. A. Hurwitz, "On the Conditions Under Which an Equation Has Only Roots With Negative Real Parts," *Mathematische Annalen,* **46,** 1895; pp. 273–284. Also in *Selected Papers on Mathematical Trends in Control Theory,* Dover, New York, 1964; pp. 70–82.

3. E. J. Routh, *Dynamics of a System of Rigid Bodies,* Macmillan, New York, 1892.

4. J. J. D'Azzo and C. H. Houpis, *Linear Control System Analysis and Design,* McGraw-Hill, New York, 1975, Ch. 6.

5. F. H. Raven, *Automatic Control Engineering,* 3rd ed., McGraw-Hill, New York, 1978.

6. H. P. Stickel, "New 50-Megabyte Disc Drive," *Hewlett-Packard Journal,* August 1977; pp. 2–16.

7. A. B. Przedpelski, "Optimize Phase Lock Loops to Meet Your Needs," *Electronic Design,* September 13, 1978; pp. 134–137.

8. W. C. Lindsey, *Phase-Locked Loops and Their Application,* IEEE Press, New York, 1978.

9. P. C. Camana, H. Hemami, and C. W. Stockwell, "Determination of Feedback for Human Positive Control Without Physical Intervention," *J. of Cybernetics,* **7,** 1977; pp. 199–225.

10. H. Hemami, "Reduced Order Models for Biped Locomotion," *IEEE Transactions on Systems, Man and Cybernetics,* April 1978; pp. 321–325.

11. F. B. Farquharson, "Aerodynamic Stability of Suspension Bridges, With Special Reference to the Tacoma Narrows Bridge," *Bulletin 116, Part I,* The Engineering Experiment Station, University of Washington, 1950.

12. R. C. Dorf, *Introduction to Computers and Computer Science,* 2nd ed., Boyd and Fraser Publishing Co., San Francisco, 1977.

13. G. E. Forsythe, M. A. Malcolm, and C. Moler, *Computer Methods for Mathematical Computations,* Prentice-Hall, Englewood Cliffs, N.J., 1977.

14. D. Selwyn, "Head-mounted Internal Servo Control for Handicapped," *Proceedings of*

the 6th IEEE Symposium on Human Factors in Electronics, May 1965, Boston, Mass.; pp. 1–8.

15. D. H. Weir and J. W. Zellner, "Lateral-Directional Motorcycle Dynamics and Rider Control," *Proceedings of the 1978 Congress, Society of Automotive Engineers,* 1978; pp. 7–23.

6 / The Root Locus Method

6.1 INTRODUCTION

The relative stability and the transient performance of a closed-loop control system are directly related to the location of the closed-loop roots of the characteristic equation in the s-plane. Also, it is frequently necessary to adjust one or more system parameters in order to obtain suitable root locations. Therefore, it is worthwhile to determine how the roots of the characteristic equation of a given system migrate about the s-plane as the parameters are varied; that is, it is useful to determine the locus of roots in the s-plane as a parameter is varied. The *root locus method* was introduced by Evans in 1948 and has been developed and utilized extensively in control engineering practice [1, 2, 3]. The root locus technique is a graphical method for drawing the locus of roots in the s-plane as a parameter is varied. In fact, the root locus method provides the engineer with a measure of the sensitivity of the roots of the system to a variation in the parameter being considered. The root locus technique may be used to great advantage in conjunction with the Routh-Hurwitz criterion and the Newton-Raphson method.

Since the root locus method provides graphical information, an approximate sketch can be used to obtain qualitative information concerning the stability and performance of the system. Furthermore, the locus of the roots of the characteristic equation of a multiloop system may be investigated as readily as for a single-loop. If the root locations are not satisfactory, the necessary parameter adjustments can often be readily ascertained from the root locus.

6.2 THE ROOT LOCUS CONCEPT

The dynamic performance of a closed-loop control system is described by the closed-loop transfer function

$$T(s) = \frac{C(s)}{R(s)} = \frac{p(s)}{q(s)}, \tag{6.1}$$

165

where $p(s)$ and $q(s)$ are polynomials in s. The roots of the characteristic equation $q(s)$ determine the modes of response of the system. For a closed-loop system, we found in Section 2.7 that by using Mason's signal-flow gain formula, we had

$$\Delta(s) = 1 - \sum_{n=1}^{N} L_n + \sum_{m,q}^{M,N} L_m L_q - \sum L_r L_s L_t + \ldots, \tag{6.2}$$

where L_q equals the value of the qth self-loop transmittance. Clearly, we have a characteristic equation which may be written as

$$q(s) = \Delta(s) = 1 + F(s). \tag{6.3}$$

In order to find the roots of the characteristic equation we set Eq. (6.3) equal to zero and obtain

$$1 + F(s) = 0. \tag{6.4}$$

Of course, Eq. (6.4) may be rewritten as

$$F(s) = -1, \tag{6.5}$$

and the roots of the characteristic equation must also satisfy this relation. In the case of the simple single-loop system, as shown in Fig. 6.1, we have the characteristic equation

$$1 + GH(s) = 0, \tag{6.6}$$

where $F(s) = G(s)H(s)$. The characteristic roots of the system must satisfy Eq. (6.5), where the roots lie in the s-plane. Since s is a complex variable, Eq. (6.5) may be rewritten in polar form as

$$|F(s)| \, \underline{/F(s)} = -1, \tag{6.7}$$

and therefore it is necessary that

$$|F(s)| = 1$$

and

$$\underline{/F(s)} = 180° \pm k360°, \tag{6.8}$$

where $k = 0, \pm 1, \pm 2, \pm 3, \ldots$ The graphical computation required for Eq. (6.8) is readily accomplished by using a protractor for estimating angles.

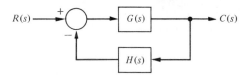

Fig. 6.1. Closed-loop control system.

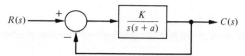

Fig. 6.2. Unity feedback control system.

    The simple second-order system considered in the previous chapters is shown in Fig. 6.2. The characteristic equation representing this system is

$$\Delta(s) = 1 + G(s) = 1 + \frac{K}{s(s+a)} = 0$$

or, alternatively,

$$q(s) = s^2 + as + K = s^2 + 2\zeta\omega_n s + \omega_n^2 = 0. \tag{6.9}$$

The locus of the roots as the gain K is varied is found by requiring that

$$|G(s)| = \left|\frac{K}{s(s+a)}\right| = 1 \tag{6.10}$$

and

$$\underline{/G(s)} = \pm 180°, \pm 540°, \ldots \tag{6.11}$$

The gain K may be varied from zero to an infinitely large positive value. For a second-order system, the roots are

$$s_1, s_2 = -\zeta\omega_n \pm \omega_n\sqrt{\zeta^2 - 1}, \tag{6.12}$$

and for $\zeta < 1$, we know that $\theta = \cos^{-1}\zeta$. Graphically, for two open-loop poles as shown in Fig. 6.3, the locus of roots is a vertical line for $\zeta \leq 1$ in order to satisfy the

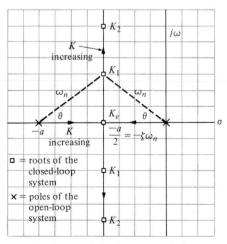

Fig. 6.3. Root locus for a second-order system.

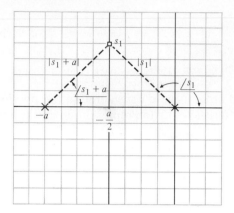

Fig. 6.4. Evaluation of the angle and gain at s_1.

angle requirement, Eq. (6.11). For example, as shown in Fig. 6.4, at a root s_1, the angles are

$$\left/ \frac{K}{s(s+a)} \right|_{s=s_1} = -\underline{/s_1} = \underline{/(s_1+a)}$$

$$= -[(180° - \theta) + \theta] = -180°. \tag{6.13}$$

This angle requirement is satisfied at any point on the vertical line which is a perpendicular bisector of the line 0 to a. Furthermore, the gain K at the particular point s_1 is found by using Eq. (6.9) as

$$\left| \frac{K}{s(s+a)} \right|_{s=s_1} = \frac{K}{|s_1|\,|s_1+a|} = 1 \tag{6.14}$$

and thus

$$K = |s_1|\,|s_1+a|, \tag{6.15}$$

where $|s_1|$ is the magnitude of the vector from the origin to s_1, and $|s_1 + a|$ is the magnitude of the vector from $-a$ to s_1.

In general, the function $F(s)$ may be written as

$$F(s) = \frac{K(s+z_1)(s+z_2)(s+z_3) \cdots (s+z_m)}{(s+p_1)(s+p_2)(s+p_3) \cdots (s+p_n)}. \tag{6.16}$$

Then the magnitude and angle requirement for the root locus are

$$|F(s)| = \frac{K|s+z_1|\,|s+z_2| \cdots}{|s+p_1|\,|s+p_2| \cdots} = 1 \tag{6.17}$$

and

$$\underline{/F(s)} = \underline{/s+z_1} + \underline{/s+z_2} + \cdots - (\underline{/s+p_1} + \underline{/s+p_2} + \cdots). \tag{6.18}$$

The magnitude requirement, Eq. (6.17), enables one to determine the value of K for a given root location s_1. A test point in the s-plane, s_1, is verified as a root location when Eq. (6.18) is satisfied. The angles are all measured in a counter-clockwise direction from a horizontal line.

In order to further illustrate the root locus procedure, let us reconsider the second-order system of Fig. 6.2. The effect of varying the parameter, a, can be effectively portrayed by rewriting the characteristic equation for the root locus form with a as the multiplying factor in the numerator. Then the characteristic equation is

$$1 + F(s) = 1 + \frac{K}{s(s + a)} = 0$$

or, alternatively,

$$s^2 + as + K = 0.$$

Dividing by the factor $(s^2 + K)$, we obtain

$$1 + \frac{as}{s^2 + K} = 0. \tag{6.19}$$

Then the magnitude criterion is satisfied when

$$\frac{a|s_1|}{|s_1^2 + K|} = 1 \tag{6.20}$$

at the root s_1. The angle criterion is

$$\underline{/s_1} - (\underline{/s_1 + j\sqrt{K}} + \underline{/s_1 - j\sqrt{K}}) = \pm 180°, \pm 540°, \ldots$$

In order to construct the root locus, we find the points in the s-plane that satisfy the angle criterion. The points in the s-plane that satisfy the angle criterion are located on a trial-and-error basis by searching in an orderly manner for a point with a total angle of $\pm 180°$, $\pm 540°$, or in general,

$$\frac{\pm(2q + 1)180°}{n_p - n_z}.$$

The algebraic sum of the angles from the poles and zeros is measured with a protractor, or, more readily, with a Spirule (a protractor with a hinged arm which facilitates the addition of the angles).* Using a protractor or a Spirule, we find the locus of roots as shown in Fig. 6.5. Specifically, at the root s_1, the magnitude of the parameter, a, is found from Eq. (6.20) as

$$a = \frac{|s_1 - j\sqrt{K}| \, |s_1 + j\sqrt{K}|}{|s_1|}. \tag{6.21}$$

---

* Available from the Spirule Company, 9728 El Venado Drive, Whittier, Calif.

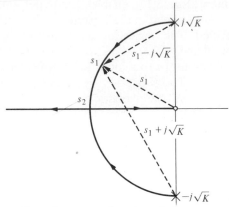

Fig. 6.5. Root locus as a function of the parameter a.

The roots of the system merge on the real axis at the point s_2 and provide a critically damped response to a step input. The parameter, a, has a magnitude at the critically damped roots $s_2 = \sigma_2$ equal to

$$a = \frac{|\sigma_2 - j\sqrt{K}|\,|\sigma_2 + j\sqrt{K}|}{\sigma_2}$$

$$= \frac{1}{\sigma_2}(\sigma_2^2 + K), \tag{6.22}$$

where σ_2 is evaluated from the s-plane vector lengths as $\sigma_2 = \sqrt{K}$. As a increases beyond the critical value, the roots are both real and distinct; one root is larger than σ_2 and one is smaller.

In general, an orderly process for locating the locus of roots as a parameter varies is desirable. In the following section, we shall develop such an orderly approach to obtaining a root locus diagram.

6.3 THE ROOT LOCUS PROCEDURE

The characteristic equation of a system provides a valuable insight concerning the response of the system when the roots of the equation are determined. In order to locate the roots of the characteristic equation in a graphical manner on the s-plane, we shall develop an orderly procedure which facilitates the rapid sketching of the locus. *First* we write the characteristic equation as

$$1 + F(s) = 0 \tag{6.23}$$

and rearrange the equation, if necessary, so that the parameter of interest, k, appears as the multiplying factor in the form

$$1 + kP(s) = 0. \tag{6.24}$$

Secondly, we factor $P(s)$, if necessary, and write the polynomial in the form of poles and zeros as follows:

$$1 + k \frac{\prod_{i=1}^{M} (s + z_i)}{\prod_{j=1}^{n} (s + p_j)} = 0. \tag{6.25}$$

Then we locate the poles and zeros on the s-plane with appropriate markings. Now, we are usually interested in determining the locus of roots as k varies as

$$0 \leq k \leq \infty.$$

Rewriting Eq. (6.24), we have

$$\prod_{j=1}^{n} (s + p_j) + k \prod_{i=1}^{M} (s + z_i) = 0. \tag{6.26}$$

Therefore, when $k = 0$, the roots of the characteristic equation are simply the poles of $P(s)$. Furthermore, when k approaches infinity, the roots of the characteristic equation are simply the zeros of $P(s)$. Therefore, we note that *the locus of the roots of the characteristic equation $1 + kP(s) = 0$ begins at the poles of $P(s)$ and ends at the zeros of $P(s)$ as k increases from 0 to infinity.* For most functions, $P(s)$, which we will encounter, several of the zeros of $P(s)$ lie at infinity in the s-plane.

The root locus on the real axis always lies in a section of the real axis to the left of an odd number of poles and zeros. This fact is clearly ascertained by examining the angle criterion of Eq. (6.18). These useful steps in plotting a root locus will be illustrated by a suitable example.

Example 6.1. A single-loop feedback control system possesses the following characteristic equation:

$$1 + GH(s) = 1 + \frac{K(1/2s + 1)}{s(1/4s + 1)} = 0. \tag{6.27}$$

First, the transfer function $GH(s)$ is rewritten in terms of poles and zeros as follows:

$$1 + \frac{2K(s + 2)}{s(s + 4)} = 0, \tag{6.28}$$

and the multiplicative gain parameter is $k = 2K$. In order to determine the locus of roots for the gain $0 \leq K \leq \infty$, we locate the poles and zeros on the real axis as shown in Fig. 6.6(a). Clearly, the angle criterion is satisfied on the real axis between the points 0 and -2, since the angle from pole p_1 is 180° and the angle from the zero and pole p_2 is zero degrees. Since the locus begins at the poles and ends at the zeros, the locus of roots appears as shown in Fig. 6.6(b), where the direction of the locus as K is increasing ($K \uparrow$) is shown by an arrow. We note that since the system

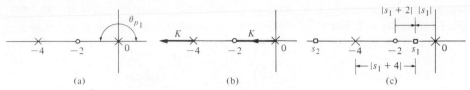

Fig. 6.6. Root locus of a second-order system with a zero.

has two real poles and one real zero, the second locus segment ends at a zero at negative infinity. In order to evaluate the gain K at a specific root location on the locus, we utilize the magnitude criterion, Eq. (6.17). For example, the gain K at the root $s = s_1 = -1$ is found from (6.17) as

$$\frac{(2K)|s_1 + 2|}{|s_1| \, |s_1 + 4|} = 1$$

or

$$K = \frac{|-1| \, |-1 + 4|}{2|-1 + 2|} = \frac{3}{2}. \tag{6.29}$$

This magnitude can also be evaluated graphically as is shown in Fig. 6.6(c). Finally, for the gain of $K = \frac{3}{2}$, one other root exists, located on the locus to the left of the pole at -4. The location of the second root is found graphically to be located at $s = -6$ as shown in Fig. 6.6(c).

Since the loci begin at the poles and end at the zeros, the *number of separate loci* is equal to the number of poles if the number of poles is greater than the number of zeros. In the unusual case when the number of zeros is greater than the number of poles, the number of separate loci would be the number of zeros. Therefore, as we found in Fig. 6.6, the number of separate loci is equal to two since there are two poles and one zero.

The root loci must be symmetrical with respect to the horizontal real axis since the complex roots must appear as pairs of complex conjugate roots.

When the number of finite zeros of $P(s)$, n_z, is less than the number of poles, n_p, by the number $N = n_p - n_z$, then N sections of loci must end at zeros at infinity. These sections of loci proceed to the zeros at infinity along *asymptotes* as k approaches infinity. These linear *asymptotes are centered* at a point on the real axis given by

$$\sigma_A = \frac{\sum \text{poles of } P(s) - \sum \text{zeros of } P(s)}{n_p - n_z} = \frac{\sum_{j=1}^{n} p_j - \sum_{i=1}^{M} z_i}{n_p - n_z}. \tag{6.30}$$

The angle of the asymptotes with respect to the real axis is

$$\phi_A = \frac{(2q + 1)}{n_p - n_z} 180°, \quad q = 0, 1, 2, \ldots, (n_p - n_z - 1), \tag{6.31}$$

where q is an integer index [3]. The usefulness of this rule is obvious for sketching the approximate form of a root locus. Equation (6.31) can be readily derived by considering a point on a root locus segment at a remote distance from the finite poles and zeros in the s-plane. The net phase angle at this remote point is 180°, since it is a point on a root locus segment. Since the finite poles and zeros of $P(s)$ are a great distance from the remote point, the angle from each pole and zero, ϕ, is essentially equal and therefore the net angle is simply $\phi(n_p - n_z)$, where n_p and n_z are the number of finite poles and zeros, respectively. Thus, we have

$$\phi(n_p - n_z) = 180°,$$

or, alternatively,

$$\phi = \frac{180°}{n_p - n_z}.$$

Accounting for all possible root locus segments at remote locations in the s-plane, we obtain Eq. (6.31).

The center of the linear asymptotes, often called the asymptote centroid, is determined by considering the characteristic equation $1 + GH(s) = 0$ (Eq. 6.25). For large values of s, only the higher-order terms need be considered so that the characteristic equation reduces to

$$1 + \frac{ks^M}{s^n} = 0.$$

However, this relation, which is an approximation, indicates that the centroid of $(n - M)$ asymptotes is at the origin, $s = 0$. A better approximation is obtained if we consider a characteristic equation of the form

$$1 + \frac{k}{(s - \sigma_A)^{n-M}} = 0$$

with a centroid at σ_A.

The centroid is determined by considering the first two terms of Eq. (6.25), which may be found from the relation

$$1 + \frac{k\prod_{i=1}^{M}(s + z_i)}{\prod_{j=1}^{n}(s + p_j)} = 1 + k\frac{(s^M + b_{M-1}s^{M-1} + \cdots + b_0)}{(s^n + a_{n-1}s^{n-1} + \cdots + a_0)}.$$

From Chapter 5, especially Eq. (5.5), we note that

$$b_{M-1} = \sum_{i=1}^{M} z_i \quad \text{and} \quad a_{n-1} = \sum_{j=1}^{n} p_j.$$

Considering only the first two terms of this expansion we have

$$1 + \frac{k}{s^{n-M} + (a_{n-1} - b_{M-1})s^{n-M-1}} = 0.$$

The first two terms of

$$1 + \frac{k}{(s - \sigma_A)^{n-M}} = 0$$

are

$$1 + \frac{k}{s^{n-M} + (n - M)\sigma_A s^{n-M-1}} = 0.$$

Equating the term for s^{n-M-1} we obtain

$$(a_{n-1} - b_{M-1}) = (n - M)\sigma_A,$$

which is equivalent to Eq. (6.30).

For example, reexamine the system shown in Fig. 6.2 and discussed in Section 6.2. The characteristic equation is written as

$$1 + \frac{K}{s(s + a)} = 0.$$

Since $n_p - n_z = 2$, we expect two loci to end at zeros at infinity. The asymptotes of the loci are located at a center

$$\sigma_A = \frac{-a}{2},$$

and at angles of

$$\phi_A = 90°, \quad q = 0,$$

and

$$\phi_A = 270°, \quad q = 1.$$

Therefore the root locus is readily sketched and the locus as shown in Fig. 6.3 is obtained. An example will further illustrate the process of utilizing the asymptotes.

Example 6.2.   A feedback control system has a characteristic equation as follows:

$$1 + F(s) = 1 + \frac{K(s + 1)}{s(s + 2)(s + 4)^2}. \tag{6.32}$$

We wish to sketch the root locus in order to determine the effect of the gain K. The poles and zeros are located in the s-plane as shown in Fig. 6.7(a). The root loci on the real axis must be located to the left of an odd number of poles and zeros and are

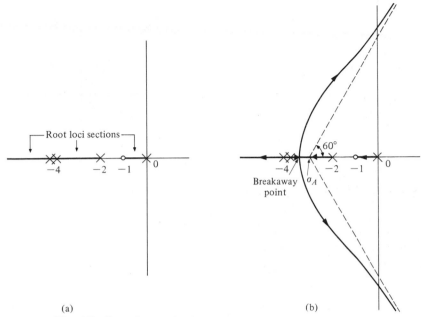

(a)   (b)

Fig. 6.7. Root locus of a fourth-order system with a zero.

therefore located as shown in Fig. 6.7(a) as heavy lines. The intersection of the asymptotes is

$$\sigma_A = \frac{(-2) + 2(-4) - (-1)}{4 - 1} = \frac{-9}{3} = -3. \tag{6.33}$$

The angles of the asymptotes are

$$\phi_A = +60°, \quad q = 0,$$
$$\phi_A = 180°, \quad q = 1,$$
$$\phi_A = 300°, \quad q = 2,$$

where there are three asymptotes since $n_p - n_z = 3$. Also, we note that the root loci must begin at the poles, and therefore two loci must leave the double pole at $s = -4$. Then, with the asymptotes drawn in Fig. 6.7(b), we may sketch the form of the root locus as shown in Fig. 6.7(b). The actual shape of the locus in the area near σ_A would be graphically evaluated, if necessary. The actual point at which *the root locus crosses the imaginary axis* is readily evaluated by utilizing the *Routh-Hurwitz criterion*.

The root locus in the previous example left the real axis at a *breakaway point*. The locus breakaway from the real axis occurs where the net change in angle caused by a small vertical displacement is zero. The locus leaves the real axis where there are a multiplicity of roots, typically two roots. The breakaway point for a simple

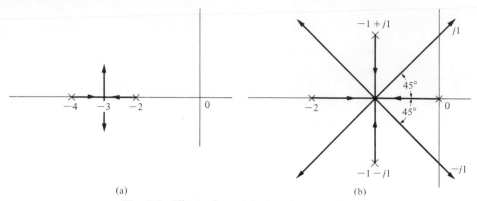

(a)   (b)

Fig. 6.8. Illustration of the breakaway point.

second-order system is shown in Fig. 6.8(a) and, for a special case of a fourth-order system, in Fig. 6.8(b). In general, due to the phase criterion, *the tangents to the loci at the breakaway point are equally spaced over 360 degrees.* Therefore, in Fig. 6.8(a), we find that the two loci at the breakaway point are spaced 180° apart, while in Fig. 6.8(b), the four loci are spaced 90° apart.

The breakaway point on the real axis can be evaluated graphically or analytically. The most straightforward method of evaluating the breakaway point involves the rearranging of the characteristic equation in order to isolate the multiplying factor K. Then the characteristic equation is written as

$$p(s) = K. \tag{6.34}$$

For example, consider a unity feedback closed-loop system with an open-loop transfer function

$$G(s) = \frac{K}{(s + 2)(s + 4)},$$

which has a characteristic equation as follows:

$$1 + G(s) = 1 + \frac{K}{(s + 2)(s + 4)} = 0. \tag{6.35}$$

Alternatively, the equation may be written as

$$K = p(s) = -(s + 2)(s + 4). \tag{6.36}$$

The root loci for this system is shown in Fig. 6.8(a). We expect the breakaway point to be near $s = \sigma = -3$ and plot $p(s)|_{s=\sigma}$ near that point as shown in Fig. 6.9. In this case, $p(s)$ equals zero at the poles $s = -2$ and $s = -4$. The plot of $p(s)$ versus σ is symmetrical and the maximum point occurs at $s = \sigma = -3$, the breakaway point. Analytically, the very same result may be obtained by determining the maximum

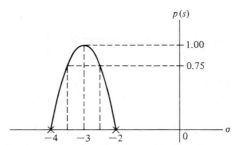

Fig. 6.9. A graphical evaluation of the breakaway point.

of $K = p(s)$. In order to find the maximum analytically, we differentiate, set the differentiated polynomial equal to zero and determine the roots of the polynomial. Therefore one may evaluate

$$\frac{dK}{ds} = \frac{dp(s)}{ds} = 0 \tag{6.37}$$

in order to find the breakaway point. Clearly, Eq. (6.37) is an analytical expression of the graphical procedure outlined in Fig. 6.9, and will result in an equation of only one order less than the total number of poles and zeros ($n_p + n_z - 1$). In almost all cases, we will prefer to use the graphical method of locating the breakaway point when it is necessary to do so.

The proof of Eq. (6.37) is obtained from a consideration of the characteristic equation

$$1 + F(s) = 1 + \frac{KY(s)}{X(s)} = 0,$$

which may be written as

$$1 + F(s) = X(s) + KY(s) = 0. \tag{6.38}$$

For a small increment in K, we have

$$X(s) + (K + \Delta K)Y(s) = 1 + \frac{\Delta K Y(s)}{X(s) + KY(s)} = 0. \tag{6.39}$$

Since the denominator is the original characteristic equation, at a breakaway point a multiplicity of roots exists and

$$\frac{Y(s)}{X(s) + KY(s)} = \frac{C_i}{(s - s_i)^n} = \frac{C_i}{(\Delta s_i)^n}. \tag{6.40}$$

Then we may write Eq. (6.39) as

$$1 + \frac{\Delta K C_i}{(\Delta s_i)^n} = 0 \tag{6.41}$$

or, alternatively,

$$\frac{\Delta K}{\Delta s} = \frac{-(\Delta s)^{n-1}}{C_i}. \tag{6.42}$$

Therefore, as we let Δs approach zero, we obtain

$$\frac{dK}{ds} = 0 \tag{6.43}$$

at the breakaway points.

Now, reconsidering the specific case where

$$G(s) = \frac{K}{(s + 2)(s + 4)},$$

we obtain $p(s)$ as

$$K = p(s) = -(s + 2)(s + 4) = -(s^2 + 6s + 8). \tag{6.44}$$

Then, differentiating, we have

$$\frac{dK}{ds} = -(2s + 6) = 0 \tag{6.45}$$

or the breakaway point occurs at $s = -3$. A more complicated example will illustrate the approach and exemplify the advantage of the graphical technique.

Example 6.3. A feedback control system is shown in Fig. 6.10. The characteristic equation is

$$1 + G(s)H(s) = 1 + \frac{K(s + 1)}{s(s + 2)(s + 3)} = 0. \tag{6.46}$$

Since the number of poles, n_p, minus the number of zeros, n_z, is equal to two, we have two asymptotes at $\pm 90°$ with a center at $\sigma_A = -2$. The asymptotes and the sections of loci on the real axis is shown in Fig. 6.11(a). A breakaway point occurs between $s = -2$ and $s = -3$. In order to evaluate the breakaway point, we rewrite the characteristic equation so that K is separated as follows:

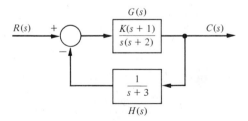

Fig. 6.10. Closed-loop system.

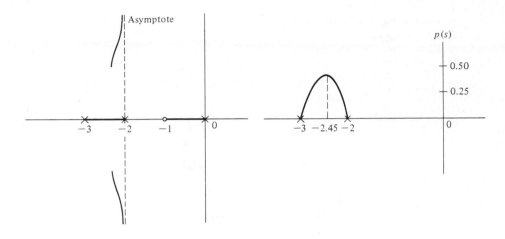

(a)   (b)

Fig. 6.11. Evaluation of the asymptotes and breakaway point.

$$s(s + 2)(s + 3) + K(s + 1) = 0$$

or

$$p(s) = \frac{-s(s + 2)(s + 3)}{(s + 1)} = K. \tag{6.47}$$

Then, evaluating $p(s)$ at various values of s between $s = -2$ and $s = -3$, we obtain the results of Table 6.1 as shown in Fig. 6.11(b). Alternatively, we differentiate Eq. (6.47) and set equal to zero to obtain

$$\frac{d}{ds}\left(\frac{-s(s + 2)(s + 3)}{(s + 1)}\right) = \frac{(s^3 + 5s^2 + 6s) - (s + 1)(3s^2 + 10s + 6)}{(s + 1)^2} = 0$$

$$= 2s^3 + 8s^2 + 10s + 6 = 0. \tag{6.48}$$

Now, in order to locate the maximum of $p(s)$, we locate the roots of Eq. (6.48) by synthetic division or by the Newton-Raphson method to obtain $s = -2.45$. It is evident from this one example that the evaluation of $p(s)$ near the expected breakaway point will result in the simplest method of evaluating the breakaway point. As the order of the characteristic equation increases, the usefulness of the graphical (or tabular) evaluation of the breakaway point will increase in contrast to the analytical approach.

The *angle of departure of the locus from a pole* and the *angle of arrival at the*

Table 6.1

$p(s)$	0	+0.412	+0.420	+0.417	+0.390	0
s	−2.00	−2.40	−2.45	−2.50	−2.60	−3.0

locus at a zero can be determined from the phase angle criterion. The angle of locus departure from a pole is the difference between the net angle due to all other poles and zeros and the criterion angle of $\pm 180° (2q + 1)$, and similarly for the locus angle of arrival at a zero. The angle of departure (or arrival) is particularly of interest for complex poles (and zeros) since the information is helpful in completing the root locus. For example, consider the third-order open-loop transfer function

$$F(s) = G(s)H(s) = \frac{K}{(s + p_3)(s^2 + 2\zeta\omega_n s + \omega_n^2)}. \tag{6.49}$$

The pole locations and the vector angles at one complex pole p_1 are shown in Fig. 6.12(a). The angles at a test point s_1, an infinitesimal distance from p_1, must meet the angle criterion. Therefore, since $\theta_2 = 90°$, we have

$$\theta_1 + \theta_2 + \theta_3 = \theta_1 + 90° + \theta_3 = +180°,$$

or the angle of departure at pole p_1 is

$$\theta_1 = 90° - \theta_3$$

as shown in Fig. 6.12(b). The departure at pole p_2 is the negative of that at p_1 since p_1 and p_2 are complex conjugates. Another example of a departure angle is shown in Fig. 6.13. In this case, the departure angle is found from

$$\theta_2 - (\theta_1 + \theta_3 + 90°) = 180°.$$

Since $(\theta_2 - \theta_3) = \gamma$, we find that the departure angle is $\theta_1 = 90° + \gamma$.

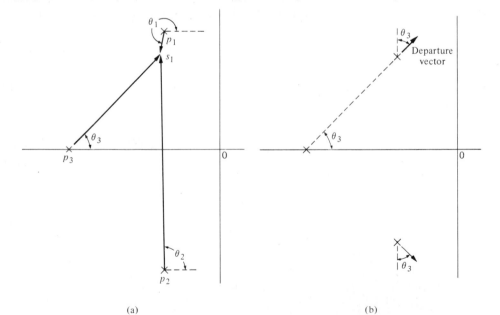

(a)   (b)

Fig. 6.12. Illustration of the angle of departure.

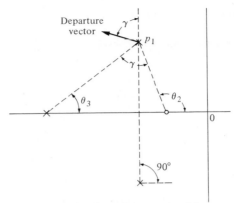

Fig. 6.13. Evaluation of the angle of departure.

It is worthwhile at this point to summarize the steps utilized in the root locus method and then illustrate their use in a complete example. The steps utilized in evaluating the locus of roots of a characteristic equation are as follows:

1. Write the characteristic equation in pole-zero form so that the parameter of interest k appears as $1 + kF(s) = 0$.
2. Locate the open-loop poles and zeros of $F(s)$ in the s-plane.
3. Locate the segments of the real axis which are root loci.
4. Determine the number of separate loci.
5. Locate the angles of the asymptotes and the center of the asymptotes.
6. Determine the breakaway point on the real axis (if any).
7. By utilizing the Routh-Hurwitz criterion, determine the point at which the locus crosses the imaginary axis (if it does so).
8. Estimate the angle of locus departure from complex poles and the angle of locus arrival at complex zeros.

Example 6.4.

1. We desire to plot the root locus for the characteristic equation of a system when

$$1 + \frac{K}{s(s + 4)(s + 4 + j4)(s + 4 - j4)} = 0 \qquad (6.50)$$

as K varies from zero to infinity.

2. The poles are located on the s-plane as shown in Fig. 6.14(a).
3. A segment of the root locus exists on the real axis between $s = 0$ and $s = -4$.
4. Since the number of poles n_p is equal to four, we have 4 separate loci.
5. The angles of the asymptotes are

$$\phi_A = \frac{(2q + 1)}{4} 180°, \quad q = 0, 1, 2, 3,$$

$$\phi_A = +45°, \ 135°, \ 225°, \ 315°$$

The center of the asymptotes is

$$\sigma_A = \frac{-4 - 4 - 4}{4} = -3.$$

Then the asymptotes are drawn as shown in Fig. 6.14(a).

6. The breakaway point is estimated by evaluating

$$K = p(s) = -s(s + 4)(s + 4 + j4)(s + 4 - j4)$$

between $s = -4$ and $s = 0$. We expect the breakaway point to lie between $s = -3$ and $s = -1$ and, therefore, we search for a maximum value of $p(s)$ in that region. The resulting values of $p(s)$ for several values of s are given in Table 6.2. The maximum of $p(s)$ is found to lie at approximately $s = -1.5$ as indicated in the table. A more accurate estimate of the breakaway point is

Table 6.2

$p(s)$	0	51	68.5	80	85	75	0
s		−4.0	−3.0	−2.5	−2.0	−1.5	−1.0	0

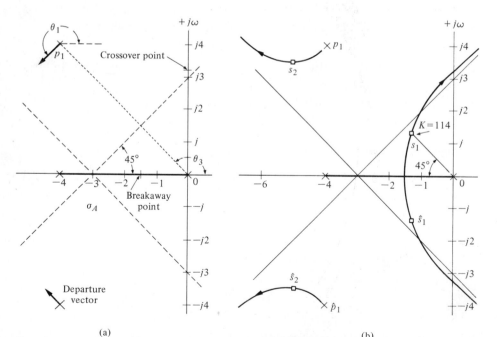

(a)   (b)

Fig. 6.14. The root locus for Example 6.4.

normally not necessary or worthwhile. The breakaway point is then indicated on Fig. 6.14(a).

7. The characteristic equation is rewritten as

$$s(s + 4)(s^2 + 8s + 32) + K = s^4 + 12s^3 + 64s^2 + 128s + K = 0. \quad (6.51)$$

Therefore, the Routh-Hurwitz array is

$$
\begin{array}{c|ccc}
s^4 & 1 & 64 & K \\
s^3 & 12 & 128 & \\
s^2 & b_1 & K & \\
s & c_1 & & \\
s^0 & K & &
\end{array}
$$

where

$$b_1 = \frac{12(64) - 128}{12} = 53.33 \quad \text{and} \quad c_1 = \frac{53.33(128) - 12K}{53.33}.$$

Hence the limiting value of gain for stability is $K = 570$ and the roots of the auxiliary equation are

$$53.33s^2 + 570 = 53.33(s^2 + 10.6) = 53.33(s + j3.25)(s - j3.25). \quad (6.52)$$

The points where the locus crosses the imaginary axis is shown in Fig. 6–14(a).

8. The angle of departure at the complex pole p_1 can be estimated by utilizing the angle criterion as follows:

$$\theta_1 + 90° + 90° + \theta_3 = 180°,$$

where θ_3 is the angle subtended by the vector from pole p_3. The angles from the pole at $s = -4$ and $s = -4 - j4$ are each equal to 90°. Since $\theta_3 = 135°$, we find that

$$\theta_1 = -135° = +225°$$

as shown in Fig. 6.14(a).

Utilizing all the information obtained from the 8 steps of the root locus method, the complete root locus is plotted by using a protractor or Spirule to locate points that satisfy the angle criterion. The root locus for this system is shown in Fig. 6.14(b). When complex roots near the origin have a damping ratio of $\zeta = 0.707$, the gain K can be determined graphically as shown in Fig. 6.14(b). The vector lengths to the root location s_1 from the open-loop poles are evaluated and result in a gain at s_1 of

$$
\begin{aligned}
K &= |s_1| |s_1 + 4| |s_1 - p_1| |s_1 - \hat{p}_1| \\
&= (1.9)(3.0)(3.8)(6.0) = 130.
\end{aligned} \quad (6.53)
$$

The remaining pair of complex roots occurs at s_2 and $\hat{s}_2$ when $K = 180$. The effect of the complex roots at s_2 and $\hat{s}_2$ on the transient response will be negligible compared to the roots s_1 and $\hat{s}_1$. This fact can be ascertained by considering the damping of the response due to each pair of roots. The damping due to s_1 and $\hat{s}_1$ is

$$e^{-\zeta_1 \omega n_1 t} = e^{-\sigma_1 t},$$

and the damping factor due to s_2 and $\hat{s}_2$ is

$$e^{-\zeta_2 \omega n_2 t} = e^{-\sigma_2 t},$$

where σ_2 is approximately five times as large as σ_1. Therefore, the transient response term due to s_2 will decay much more rapidly than the transient response term due to s_1. Thus the response to a unit step input may be written as

$$c(t) = 1 + c_1 e^{-\sigma_1 t} \sin (\omega_1 t + \theta_1) + c_2 e^{-\sigma_2 t} \sin (\omega_2 t + \theta_2)$$
$$\cong 1 + c_1 e^{-\sigma_1 t} \sin (\omega_1 t + \theta_1). \tag{6.54}$$

The complex conjugate roots near the origin of the s-plane relative to the other roots of the closed-loop system are labeled the *dominant roots* of the system since they represent or dominate the transient response. The relative dominance of the roots is determined by the ratio of the real parts of the complex roots and will result in reasonable dominance for ratios exceeding five.

Of course, the dominance of the second term of Eq. (6.54) also depends upon the relative magnitudes of the coefficients c_1 and c_2. These coefficients, which are the residues evaluated at the complex roots, in turn depend upon the location of the zeros in the s-plane. Therefore, the concept of dominant roots is useful for estimating the response of a system but must be used with caution and with a comprehension of the underlying assumptions.

A computer program written in Fortran IV for the calculation of the root locus is available [4]. This program, called RTLOC, obtains the roots of the characteristic equation and plots them as a function of K. Also, college computer centers usually have available computer programs for the calculation of the roots of an equation for a series of values of a parameter such as K.

6.4 AN EXAMPLE OF A CONTROL SYSTEM ANALYSIS AND DESIGN UTILIZING THE ROOT LOCUS METHOD

The analysis and design of a control system can be accomplished by utilizing the Laplace transform, a signal-flow diagram, the s-plane, and the root locus method. It will be worthwhile at this point in the development to examine a control system and select suitable parameter values based on the root locus method.

An automatic self-balancing scale in which the weighing operation is controlled by the physical balance function through an electrical feedback loop is shown in Fig. 6.15 [5]. The balance is shown in the equilibrium condition, and x is the travel of the counterweight W_c from an unloaded equilibrium condition. The weight to be measured, W, is applied 5 cm from the pivot and the length of the beam to the viscous damper, l_i, is 20 cm. It is desired to accomplish the following items:

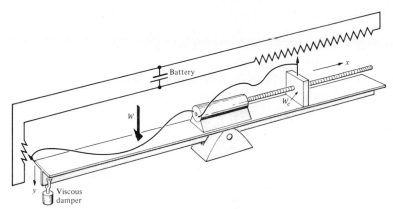

Fig. 6.15. An automatic self-balancing scale. (From J. H. Goldberg, *Automatic Controls,* Allyn and Bacon, Boston, 1964, with permission.)

1. Select the parameters and the specifications of the feedback system.
2. Obtain a model and signal-flow diagram representing the system.
3. Select the gain K based on a root locus diagram.
4. Determine the dominant mode of response.

The inertia of the beam will be chosen to be equal to 0.05 kg-m². We must select a battery voltage that is large enough to provide a reasonable position sensor gain, so let us choose E_{bb} = 24 volts. We will utilize a lead screw of 20 turns/cm and a potentiometer for x equal to 6 cm in length. Accurate balances are required, and therefore an input potentiometer for y will be chosen to be 0.5 cm in length. A reasonable viscous damper will be chosen with a damping constant $f = 10\sqrt{3}$ kg/ m/sec. Finally, a counterweight W_c is chosen so that the expected range of weights W can be balanced. Therefore, in summary, the *parameters* of the system are selected as listed in Table 6.3.

Specifications. A rapid and accurate response resulting in a small steady-state weight measurement error is desired. Therefore we will require the system be at least a type one system so that a zero measurement error is obtained. An under-damped response to a step change in the measured weight, W, is satisfactory, and

Table 6.3

W_c = 2 kg	Lead screw gain $K_s = \dfrac{1}{4000\pi}$ m/rad
I = 0.05 kg-m²	
l_w = 5 cm	Input potentiometer gain K_i = 4800 v/m
l_i = 20 cm	
$f = 10\sqrt{3}$ kg/m/sec	Feedback potentiometer gain K_f = 400 v/m

therefore a dominant response with $\zeta = 0.5$ will be specified. The settling time of the balance following the introduction of a weight to be measured should be less than two sec in order to provide a rapid weight-measuring device. The specifications are summarized in Table 6.4.

The derivation of a model of the electromechanical system may be accomplished by obtaining the equations of motion of the balance. For small deviations from balance, the deviation angle θ is

$$\theta \simeq \frac{y}{l_i}. \tag{6.55}$$

The motion of the beam about the pivot is represented by the torque equation:

$$I \frac{d^2\theta}{dt^2} = \Sigma \text{ torques}.$$

Therefore, in terms of the deviation angle, the motion is represented by

$$I \frac{d^2\theta}{dt^2} = l_w W - x W_c - l_i^2 f \frac{d\theta}{dt}. \tag{6.56}$$

The input voltage to the motor is

$$v_m(t) = K_i y - K_f x. \tag{6.57}$$

The transfer function of the motor is

$$\frac{\theta_m(s)}{V_m(s)} = \frac{K_m}{s(\tau s + 1)}, \tag{6.58}$$

where τ will be considered to be negligible with respect to the time constants of the overall system, and θ_m is the output shaft rotation. A signal-flow graph representing Eqs. (6.56) through (6.58) is shown in Fig. 6.16. Examining the forward path from W to $X(s)$ we find that the system is a type one system due to the integration preceding $Y(s)$. Therefore, the steady-state error of the system is zero.

The closed-loop transfer function of the system is obtained by utilizing Mason's flow-graph formula and is found to be

$$\frac{X(s)}{W(s)} = \frac{(l_w l_i K_i K_m K_s / I s^3)}{1 + (l_i^2 f / I s) + (K_m K_s K_f / s) + (l_i K_i K_m K_s W_c / I s^3) + (l_i^2 f K_m K_s K_f / I s^2)}, \tag{6.59}$$

where the numerator is the path factor from W to X, the second term in the denom-

Table 6.4 Specifications

Steady-state error	$K_p = \infty$
Underdamped response	$\zeta = 0.5$
Settling time	Less than 2 sec

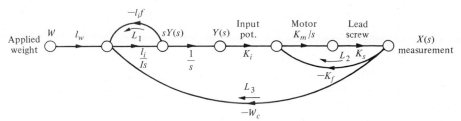

Fig.6.16. Signal flow graph model of the automatic self-balancing scale.

inator is the loop L_1, the third term is the loop factor L_2, the fourth term is the loop L_3, and the fifth term is the two nontouching loops $L_1 L_2$. Therefore, the closed-loop transfer function is

$$\frac{X(s)}{W(s)} = \frac{l_w l_i K_i K_m K_s}{s(Is + l_i^2 f)(s + K_m K_s K_f) + W_c K_m K_s K_i l_i}. \tag{6.60}$$

The steady-state gain of the system is then

$$\lim_{t \to \infty} \left(\frac{x(t)}{|W|}\right) = \lim_{s \to 0} s \left(\frac{X(s)}{W(s)}\right) = \frac{l_w}{W_c} = 2.5 \text{ cm/kg} \tag{6.61}$$

when $W(s) = |W|/s$. In order to obtain the root locus as a function of the motor constant K_m, we substitute the selected parameters into the characteristic equation which is the denominator of Eq. (6.60). Therefore, we obtain the following characteristic equation:

$$s(s + 8\sqrt{3}) \left(s + \frac{K_m}{10\pi}\right) + \frac{96 K_m}{10\pi} = 0. \tag{6.62}$$

Rewriting the characteristic equation in root locus form, we first isolate K_m as follows:

$$s^2(s + 8\sqrt{3}) + s(s + 8\sqrt{3}) \frac{K_m}{10\pi} + \frac{96 K_m}{10\pi} = 0. \tag{6.63}$$

Then, rewriting Eq. (6.63) in root locus form, we have

$$1 + kP(s) = 1 + \frac{(K_m/10\pi)[s(s + 8\sqrt{3}) + 96]}{s^2(s + 8\sqrt{3})} = 0$$

$$= 1 + \frac{(K_m/10\pi)(s + 6.93 + j6.93)(s + 6.93 - j6.93)}{s^2(s + 8\sqrt{3})}. \tag{6.64}$$

The root locus as K_m varies is shown in Fig. 6.17. The dominant roots can be placed at $\zeta = 0.5$ when $K = 25.3 = K_m/10\pi$. In order to achieve this gain,

$$K_m = 795 \frac{\text{rad/sec}}{\text{volt}} = 7600 \frac{\text{rpm}}{\text{volt}}, \tag{6.65}$$

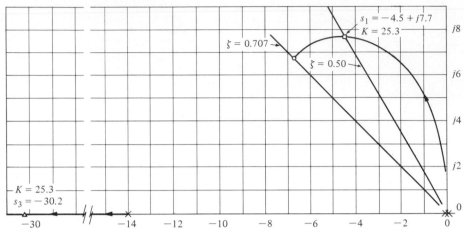

Fig. 6.17. Root locus as K_m varies.

an amplifier would be required to provide a portion of the required gain. The real part of the dominant roots is greater than four and therefore the settling time, $4/\sigma$, is less than 1 sec, and the settling time requirement is satisfied. The third root of the characteristic equation is a real root at $s = -30.2$ and the underdamped roots clearly dominate the response. Therefore the system has been analyzed by the root locus method and a suitable design for the parameter K_m has been achieved. The efficiency of the s-plane and root locus methods is clearly demonstrated by this example.

6.5 PARAMETER DESIGN BY THE ROOT LOCUS METHOD

The original development of the root locus method was concerned with the determination of the locus of roots of the characteristic equation as the system gain, K, is varied from zero to infinity. However, as we have seen, the effect of other system parameters may be readily investigated by using the root locus method. Fundamentally, the root locus method is concerned with a characteristic equation (Eq. 6.23), which may be written as

$$1 + F(s) = 0. \tag{6.66}$$

Then the standard root locus method we have studied may be applied. The question arises: How do we investigate the effect of two parameters, α and β? It appears that the root locus method is a single parameter method; however, fortunately it can be readily extended to the investigation of two or more parameters.

The characteristic equation of a dynamic system may be written as

$$a_n s^n + a_{n-1} s^{n-1} + \cdots + a_1 s + a_0 = 0. \tag{6.67}$$

Clearly, the effect of the coefficient a_1 may be ascertained from the root locus equation

$$1 + \frac{a_1 s}{a_n s^n + a_{n-1} s^{n-1} + \cdots + a_2 s^2 + a_0} = 0. \qquad (6.68)$$

If the parameter of interest, α, does not appear solely as a coefficient, the parameter is isolated as

$$a_n s^n + a_{n-1} s^{n-1} + \cdots + (a_{n-q} - \alpha) s^{n-q} + \alpha s^{n-q} + \cdots + a_1 s + a_0 = 0. \qquad (6.69)$$

Then, for example, a third-order equation of interest might be

$$s^3 + (3 + \alpha) s^2 + 3s + 6 = 0. \qquad (6.70)$$

In order to ascertain the effect of the parameter α, we isolate the parameter and rewrite the equation in root locus form as shown in the following steps:

$$s^3 + 3s^2 + \alpha s^2 + 3s + 6 = 0, \qquad (6.71)$$

$$1 + \frac{\alpha s^2}{s^3 + 3s^2 + 3s + 6} = 0. \qquad (6.72)$$

Then, in order to determine the effect of two parameters, we must repeat the root locus approach twice. Thus, for a characteristic equation with two variable parameters, α and β, we have

$$a_n s^n + a_{n-1} s^{n-1} + \cdots + (a_{n-q} - \alpha) s^{n-q} + \alpha s^{n-q} + \cdots$$
$$+ (a_{n-r} - \beta) s^{n-r} + \beta s^{n-r} + \cdots + a_1 s + a_0 = 0. \qquad (6.73)$$

The two variable parameters have been isolated and the effect of α will be determined, followed by the determination of the effect of β. For example, for a certain third-order characteristic equation with α and β as parameters, we obtain

$$s^3 + s^2 + \beta s + \alpha = 0. \qquad (6.74)$$

In this particular case, the parameters appear as the coefficients of the characteristic equation. The effect of varying β from zero to infinity is determined from the root locus equation

$$1 + \frac{\beta s}{s^3 + s^2 + \alpha} = 0. \qquad (6.75)$$

One notes that the denominator of Eq. (6.75) is the characteristic equation of the system with $\beta = 0$. Therefore, one first evaluates the effect of varying α from zero to infinity by utilizing the equation

$$s^3 + s^2 + \alpha = 0,$$

rewritten as

$$1 + \frac{\alpha}{s^2(s + 1)} = 0, \qquad (6.76)$$

where β has been set equal to zero in Eq. (6.74). Then, upon evaluating the effect of α, a value of α is selected and used with Eq. (6.75) to evaluate the effect of β.

This two-step method of evaluating the effect of α and then β may be carried out as a two-root locus procedure. First, we obtain a locus of roots as α varies, and we select a suitable value of α; the results are satisfactory root locations. Then we obtain the root locus for β by noting that the poles of Eq. (6.75) are the roots evaluated by the root locus of Eq. (6.76). A limitation of this approach is that one will not always be able to obtain a characteristic equation which is linear in the parameter under consideration, for example, α.

In order to effectively illustrate this approach, let us obtain the root locus for α and then β for Eq. (6.74). A sketch of the root locus as α varies for Eq. (6.76) is shown in Fig. 6.18(a), where the roots for two values of gain α are shown. If the gain α is selected as α_1, then the resultant roots of Eq. (6.76) become the poles of Eq. (6.75). The root locus of Eq. (6.75) as β varies is shown in Fig. 6.18(b), and a suitable β can be selected on the basis of the desired root locations.

Using the root locus method, we will further illustrate this parameter design approach by a specific design example.

Example 6.5. A feedback control system is to be designed to satisfy the following specifications:

(a) steady-state error for a ramp input $\leqslant 10\%$ of input magnitude,

(b) damping ratio of dominant roots $\geqslant 0.707$,

(c) settling time of the system $\leqslant 3$ sec.

The structure of the feedback control system is shown in Fig. 6.19, where the amplifier gain K_1 and the derivative feedback gain K_2 are to be selected. The steady-state error specification can be written as follows:

$$e_{ss} = \lim_{t \to \infty} e(t) = \lim_{s \to 0} sE(s) = \lim_{s \to 0} \frac{s(|R|/s^2)}{1 + GH(s)}. \qquad (6.77)$$

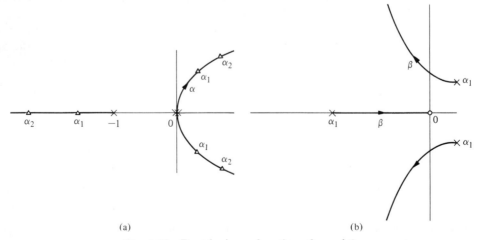

(a)   (b)

Fig. 6.18. Root loci as a function of α and β.

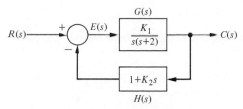

Fig. 6.19. Closed-loop system.

Therefore, the steady-state error requirement is

$$\frac{e_{ss}}{|R|} = \lim_{s \to 0} \frac{1}{sGH(s)} \leq 0.10. \tag{6.78}$$

Since

$$GH(s) = \frac{K_1(1 + K_2 s)}{s(s + 2)},$$

it is required that $K_1 \geq 20$. Of course, in this case,

$$K_v = \frac{|R|}{e_{ss}} = \frac{K_1}{2}.$$

The damping ratio specification requires that the roots of the closed-loop system be below the line at 45° in the left-hand s-plane. The settling time specification can be rewritten in terms of the real part of the dominant roots as

$$T_s = \frac{4}{\sigma} \leq 3 \text{ sec.} \tag{6.79}$$

Therefore it is necessary that $\sigma \geq \frac{4}{3}$ and this area in the left-hand s-plane is indicated along with the ζ-requirement in Fig. 6.20. In order to satisfy the specifications, all the roots must lie within the shaded area of the left-hand plane.

The parameters to be selected are $\alpha = K_1$ and $\beta = K_2 K_1$. The characteristic equation is

$$1 + GH(s) = s^2 + 2s + \beta s + \alpha = 0. \tag{6.80}$$

The locus of roots as $\alpha = K_1$ varies is determined from the following equation:

$$1 + \frac{\alpha}{s(s + 2)} = 0. \tag{6.81}$$

(See Fig. 6.21a.) For a gain of $K_1 = \alpha = 20$, the roots are indicated on the locus. Then the effect of varying $\beta = 20K_2$ is determined from the locus equation

$$1 + \frac{\beta s}{s^2 + 2s + \alpha} = 0, \tag{6.82}$$

where the poles of this root locus are the roots of the locus of Fig. 6.21(a). The root

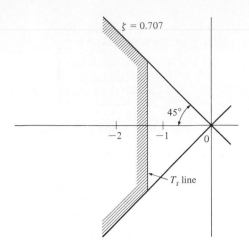

Fig. 6.20. A region in the s-plane for desired root location.

locus for Eq. (6.82) is shown in Fig. 6.21(b) and roots with $\zeta = 0.707$ are obtained when $\beta = 4.3 = 20K_2$ or when $K_2 = 0.215$. The real part of these roots is $\sigma = 3.15$, and therefore the settling time is equal to 1.27 sec, which is considerably less than the specification of 3 sec.

The root locus method may be extended to more than two parameters by extending the number of steps in the method outlined in this section. Furthermore, a family of root loci can be generated for two parameters in order to determine the

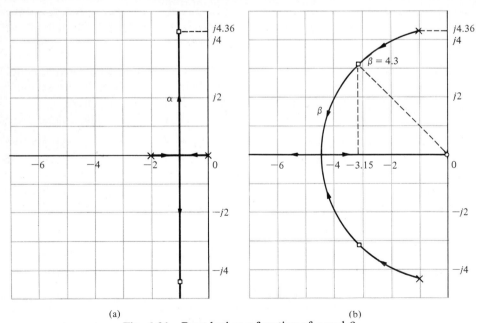

(a)   (b)

Fig. 6.21. Root loci as a function of α and β.

total effect of varying two parameters. For example, let us determine the effect of varying α and β of the following characteristic equation:

$$s^3 + 3s^2 + 2s + \beta s + \alpha = 0. \tag{6.83}$$

The root locus equation as a function of α is

$$1 + \frac{\alpha}{s(s + 1)(s + 2)} = 0. \tag{6.84}$$

The root locus as a function of β is

$$1 + \frac{\beta s}{s^3 + 3s^2 + 2s + \alpha} = 0. \tag{6.85}$$

The root locus for Eq. (6.84) as a function of α is shown in Fig. 6.22 with unbroken lines. The roots of this locus, indicated by a cross, become the poles for the locus of Eq. (6.85). Then the locus of Eq. (6.85) is continued on Fig. 6.22 with dotted lines where the locus for β is shown for several selected values of α. This family of loci, often called root contours, illustrates the effect of α and β on the roots of the characteristic equation of a system [3].

6.6 SENSITIVITY AND THE ROOT LOCUS

One of the prime reasons for the utilization of negative feedback in control systems is to reduce the effect of parameter variations. The effect of parameter variations, as we found in Section 3.2, can be described by a measure of the *sensitivity* of the system performance to specific parameter changes. In Section 3.2, we defined the *logarithmic sensitivity* originally suggested by Bode as

$$S_k^T = \frac{d \ln T}{d \ln k} = \frac{\partial T/T}{\partial k/k}, \tag{6.86}$$

where the system transfer function is $T(s)$ and the parameter of interest is k.

In recent years, with the increased utilization of the pole-zero (s-plane) approach, it has become useful to define a measure of sensitivity in terms of the positions of the roots of the characteristic equation [6,7]. Since the roots of the characteristic equation represent the dominant modes of transient response, the effect of parameter variations on the position of the roots is an important and useful measure of the sensitivity. The *root sensitivity* of a system $T(s)$ can be defined as

$$S_k^{r_i} = \frac{\partial r_i}{\partial \ln k} = \frac{\partial r_i}{\partial k/k}, \tag{6.87}$$

where r_i equals the ith root of the system so that

$$T(s) = \frac{K_1 \prod_{j=1}^{m} (s + Z_j)}{\prod_{i=1}^{n} (s + r_i)}, \tag{6.88}$$

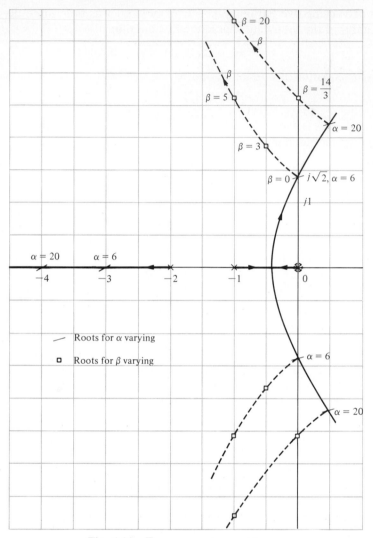

Fig. 6.22. Two-parameter root locus.

and k is the parameter. The root sensitivity relates the change in the location of the root in the s-plane to the change in the parameter. The root sensitivity is related to the Bode logarithmic sensitivity by the relation

$$S_k^T = \frac{\partial \ln K_1}{\partial \ln k} - \sum_{i=1}^{n} \frac{\partial r_i}{\partial \ln k} \cdot \frac{1}{(s + r_i)} \qquad (6.89)$$

when the zeros of $T(s)$ are independent of the parameter k so that

$$\frac{\partial Z_j}{\partial \ln k} = 0.$$

This logarithmic sensitivity can be readily obtained by determining the derivative of $T(s)$, Eq. (6.88), with respect to k. For the particular case when the gain of the system is independent of the parameter k, we have

$$S_k^T = - \sum_{i=1}^{n} S_k^{r_i} \cdot \frac{1}{(s + r_i)},$$
(6.90)

and the two sensitivity measures are directly related.

The evaluation of the root sensitivity for a control system can be readily accomplished by utilizing the root locus methods of the preceding section. The root sensitivity $S_k^{r_i}$ may be evaluated at root r_i by examining the root contours for the parameter k. An example will illustrate the process of evaluating the root sensitivity.

Example 6.6.   The characteristic equation of the feedback control system shown in Fig. 6.23 is

$$1 + \frac{K}{s(s + \beta)} = 0$$

or, alternatively,

$$s^2 + \beta s + K = 0.$$
(6.91)

The gain K will be considered to be the parameter α. Then the effect of a change in each parameter can be determined by utilizing the relations

$$\alpha = \alpha_0 \pm \Delta\alpha$$

and

$$\beta = \beta_0 \pm \Delta\beta,$$

where α_0 and β_0 are the nominal or desired values for the parameters α and β, respectively. We shall consider the case when the nominal pole value is $\beta_0 = 1$ and the desired gain is $\alpha_0 = K = 0.5$. Then the root locus as a function of $\alpha = K$ can be obtained by utilizing the root locus equation

$$1 + \frac{K}{s(s + \beta_0)} = 1 + \frac{K}{s(s + 1)} = 0$$
(6.92)

as shown in Fig. 6.24. The nominal value of gain $K = \alpha_0 = 0.5$ results in two complex roots, $r_1 = -0.5 + j0.5$ and $r_2 = \hat{r}_1$, as shown in Fig. 6.24. In order to

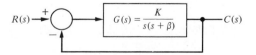

$R(s) \xrightarrow{+} \bigcirc \longrightarrow \boxed{G(s) = \dfrac{K}{s(s + \beta)}} \longrightarrow C(s)$

Fig. 6.23.   A feedback control system.

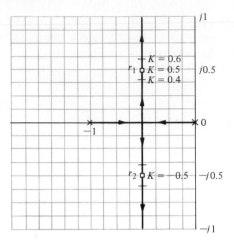

Fig. 6.24. The root locus for K.

evaluate the effect of unavoidable changes in the gain, the characteristic equation
with $\alpha = \alpha_0 \pm \Delta\alpha$ becomes

$$s^2 + s + \alpha_0 \pm \Delta\alpha = s^2 + s + 0.5 \pm \Delta\alpha$$

or

$$1 + \frac{\pm\Delta\alpha}{s^2 + s + 0.5} = 1 + \frac{\pm\Delta\alpha}{(s + r_1)(s + \hat{r}_1)} = 0. \tag{6.93}$$

Therefore, the effect of changes in the gain can be evaluated from the root locus of
Fig. 6.24. For a 20% change in α, we have $\pm\Delta\alpha = \pm 0.1$. The root locations for a
gain $\alpha = 0.4$ and $\alpha = 0.6$ are readily determined by root locus methods, and the
root locations for $\pm\Delta\alpha = \pm 0.1$ are shown on Fig. 6.24. When $\alpha = K = 0.6$, the root
in the second quadrant of the s-plane is

$$r_1 + \Delta r_1 = -0.5 + j0.59,$$

and the change in the root is $\Delta r_1 = +j0.09$. When $\alpha = K = 0.4$, the root in the
second quadrant is

$$r_1 + \Delta r_1 = -0.5 + j0.387,$$

and the change in the root is $\Delta r = -j0.11$. Thus the root sensitivity for r_1 is

$$s^{r_1}_{+\Delta K} = S^{r_1}_{K+} = \frac{\Delta r_1}{\Delta K/K} = \frac{+j0.09}{+0.2} = j0.45 = 0.45\underline{/+90°} \tag{6.94}$$

for positive changes of gain. For negative increments of gain, the sensitivity is

$$S^{r_1}_{-\Delta K} = S^{r_1}_{K-} = \frac{\Delta r_1}{\Delta K/K} = \frac{-j0.11}{+.0.2} = -j0.55 = 0.55\underline{/-90°}. \tag{6.95}$$

Of course, for infinitesimally small changes in the parameter ∂K, the sensitivity will be equal for negative or positive increments in K. The angle of the root sensitivity indicates the direction the root would move as the parameter varies. The angle of movement for $+\Delta\alpha$ is always $180°$ minus the angle of movement for $-\Delta\alpha$ at the point $\alpha = \alpha_0$.

The pole β varies due to environmental changes, and it may be represented by $\beta = \beta_0 + \Delta\beta$, where $\beta_0 = 1$. Then the effect of variation of the poles is represented by the characteristic equation

$$s^2 + s + \Delta\beta s + K = 0,$$

or, in root locus form, we have

$$1 + \frac{\Delta\beta s}{s^2 + s + K} = 0. \tag{6.96}$$

Again, the denominator of the second term is the unchanged characteristic equation when $\Delta\beta = 0$. The root locus for the unchanged system ($\Delta\beta = 0$) is shown in Fig. 6.24 as a function of K. For a design specification requiring $\zeta = 0.707$, the complex roots lie at

$$r_1 = -0.5 + j0.5 \qquad \text{and} \qquad r_2 = \hat{r}_1 = -0.5 - j0.5.$$

Then, since the roots are complex conjugates, the root sensitivity for r_1 is the conjugate of the root sensitivity for $\hat{r}_1 = r_2$. Using the parameter root locus techniques discussed in the preceding section, we obtain the root locus for $\Delta\beta$ as shown in Fig. 6.25. We are normally interested in the effect of a variation of the parameter so that $\beta = \beta_0 \pm \Delta\beta$, for which the locus as $\Delta\beta$ decreases is obtained from the root locus equation

$$1 + \frac{-(\Delta\beta)s}{s^2 + s + K} = 0. \tag{6.97}$$

Examining Eq. (6.97), we note that the equation is of the form

$$1 - kP(s) = 0.$$

Comparing this equation with Eq. (6.24) (Section 6.3), we find that the sign preceding the gain k is negative in this case. In a manner similar to the development of the root locus method in Section 6.3, we require that the root locus satisfy the equations

$$|kP(s)| = 1, \quad \underline{/P(s)} = 0° \pm q360°, \tag{6.98}$$

where $q = 0, 1, 2, \ldots$. The locus of roots follows a zero-degree locus (Eq. 6.98) in contrast with the $180°$ locus considered previously. However, the root locus rules of Section 6.3 may be altered to account for the zero-degree phase angle requirement, and then the root locus may be obtained as in the preceding sections. Therefore, in order to obtain the effect of reducing β, one determines the zero-degree

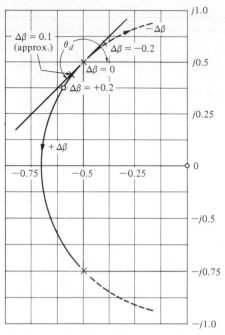

Fig. 6.25. The root locus for the parameter β.

locus in contrast to the $180°$ locus as shown by a dotted locus in Fig. 6.25. Therefore, to find the effect of a 20% change of the parameter β, we evaluate the new roots for $\pm\Delta\beta = \pm 0.20$ as shown in Fig. 6.25. The root sensitivity is readily evaluated graphically and, for a positive change in β, is

$$S_{\beta+}^{r_1} = \frac{\Delta r_1}{(\Delta\beta/\beta)} = \frac{0.16\underline{/-131°}}{0.20} = 0.80\underline{/-131°}. \tag{6.99}$$

The root sensitivity for a negative change in β is

$$S_{\beta-}^{r_1} = \frac{\Delta r_1}{(\Delta\beta/\beta)} = \frac{0.125\underline{/38°}}{0.20} = 0.625\underline{/+38°}. \tag{6.100}$$

As the percentage change $(\Delta\beta/\beta)$ decreases, the sensitivity measures, $S_{\beta+}^{r_1}$ and $S_{\beta-}^{r_1}$, will approach equality in magnitude and a difference in angle of $180°$. Thus, for small changes when $\Delta\beta/\beta \leq 0.10$, the sensitivity measures are related as

$$|S_{\beta+}^{r_1}| = |S_{\beta-}^{r_1}| \tag{6.101}$$

and

$$\underline{/S_{\beta+}^{r}} = 180° + \underline{/S_{\beta-}^{r}}. \tag{6.102}$$

Often the desired root sensitivity measure is for small changes in the parameter. When the relative change in the parameter is such that $\Delta\beta/\beta = 0.10$, a root locus approximation is satisfactory. The root locus for Eq. (6.98) when $\Delta\beta$ is varying leaves the pole at $\Delta\beta = 0$ at an angle of departure θ_d. Since θ_d is readily evaluated, one can estimate the increment in the root change by approximating the root locus with the line at θ_d. This approximation is shown in Fig. 6.25 and is accurate for only relatively small changes in $\Delta\beta$. However, the use of this approximation allows the analyst to avoid drawing the complete root locus diagram. Therefore, for Fig. 6.25, the root sensitivity may be evaluated for $\Delta\beta/\beta = 0.10$ along the departure line, and one obtains

$$S_{\beta+}^{r_1} = \frac{0.074 / -135°}{0.10} = 0.74 / -135°. \tag{6.103}$$

The root sensitivity measure for a parameter variation is useful for comparing the sensitivity for various design parameters and at different root locations. Comparing Eq. (6.103) for β with Eq. (6.94) for α, we find that the sensitivity for β is greater in magnitude by approximately 50% and the angle for $S_{\beta}^{r_1}$ indicates that the approach of the root toward the $j\omega$-axis is more sensitive for changes in β. Therefore, the tolerance requirements for β would be more stringent than for α. This information provides the designer with a comparative measure of the required tolerances for each parameter.

Example 6.7. A unity feedback control system has a forward transfer function

$$G(s) = \frac{20.7(s + 3)}{s(s + 2)(s + \beta)}, \tag{6.104}$$

where $\beta = \beta_0 + \Delta\beta$ and $\beta_0 = 8$. The characteristic equation as a function of $\Delta\beta$ is

$$s(s + 2)(s + 8 + \Delta\beta) + 20.7(s + 3) = 0$$

or

$$s(s + 2)(s + 8) + \Delta\beta s(s + 2) + 20.7(s + 3) = 0. \tag{6.105}$$

When $\Delta\beta = 0$, the roots may be determined by the root locus method or the Newton-Raphson method, and thus we evaluate the roots as

$$r_1 = -2.36 + j2.48, \qquad r_2 = \hat{r}_1, \qquad r_3 = -5.27.$$

The root locus for $\Delta\beta$ is determined by using the root locus equation

$$1 + \frac{\Delta\beta s(s + 2)}{(s + r_1)(s + \hat{r}_1)(s + r_3)} = 0. \tag{6.106}$$

The poles and zeros of Eq. (6.106) are shown in Fig. 6.26. The angle of departure at r_1 is evaluated from the angles as follows:

$$180° = -(\theta_d + 90° + \theta_{p_3}) + (\theta_{z_1} + \theta_{z_2})$$
$$= -(\theta_d + 90° + 40°) + (133° + 98°). \tag{6.107}$$

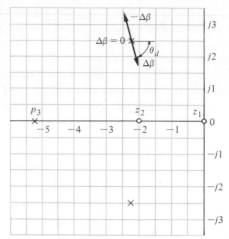

Fig. 6.26. Pole and zero diagram for the parameter β.

Therefore, $\theta_d = -80°$ and the locus is approximated near r_1 by the line at an angle of θ_d. For a change of $\Delta r_1 = 0.2/-80°$ along the departure line, the $+\Delta\beta$ is evaluated by determining the vector lengths from the poles and zeros. Then we have

$$+\Delta\beta = \frac{4.8(3.75)(0.2)}{(3.25)(2.3)} = 0.48. \tag{6.108}$$

Therefore, the sensitivity at r_1 is

$$S_\beta^{r_1} = \frac{\Delta r_1}{\Delta\beta/\beta} = \frac{0.2/-80°}{0.48/8} = 3.34/-80°, \tag{6.109}$$

which indicates that the root is quite sensitive to this 6% change in the parameter β. For comparison, it is worthwhile to determine the sensitivity of the root, r_1, to a change in the zero, $s = -3$. Then the characteristic equation is

$$s(s + 2)(s + 8) + 20.7(s + 3 + \Delta\gamma) = 0$$

or

$$1 + \frac{20.7\,\Delta\gamma}{(s + r_1)(s + \hat{r}_1)(s + r_3)} = 0. \tag{6.110}$$

The pole-zero diagram for Eq. (6.110) is shown in Fig. 6.27. The angle of departure at root r_1 is $180° = -(\theta_d + 90° + 40°)$ or

$$\theta_d = +50°. \tag{6.111}$$

For a change of $r_1 = 0.2/+50°$, the $\Delta\gamma$ is positive, and obtaining the vector lengths, we find

$$|\Delta\gamma| = \frac{5.22(4.18)(.2)}{20.7} = 0.21. \tag{6.112}$$

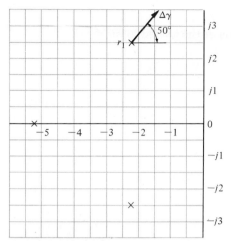

Fig. 6.27. Pole-zero diagram for the parameter γ.

Therefore, the sensitivity at r_1 for $+\Delta\gamma$ is

$$S_\gamma^{r_1} = \frac{\Delta r_1}{\Delta\gamma/\gamma} = \frac{0.2/\underline{+50°}}{0.21/3} = 2.84/\underline{+50°}. \qquad (6.113)$$

Thus we find that the magnitude of the root sensitivity for the pole β and the zero γ is approximately equal. However, the sensitivity of the system to the pole can be considered to be less than the sensitivity to the zero, since the angle of the sensitivity $S_\gamma^{r_1}$ is equal to $+50°$ and the direction of the root change is toward the $j\omega$-axis.

Evaluating the root sensitivity in the manner of the preceeding paragraphs, we find that for the pole $s = -\delta_0 = -2$ the sensitivity is

$$S_{\delta-}^{r_1} = 2.1/\underline{+27°}. \qquad (6.114)$$

Thus, for the parameter δ, the magnitude of the sensitivity is less than for the other parameters, but the direction of the change of the root is more important than for β and γ.

In order to utilize the root sensitivity measure for the analysis and design of control systems, a series of calculations must be performed for various selections of possible root configurations and the zeros and poles of the open-loop transfer function. Therefore, the use of the root sensitivity measure as a design technique is somewhat limited by the relatively large number of calculations required and by the lack of an obvious direction for adjusting the parameters in order to provide a minimized or reduced sensitivity. However, the root sensitivity measure can be utilized as an analysis measure which permits the designer to compare the sensitivity for several system designs based on a suitable method of design. The root sensitivity measure is a useful index of sensitivity of a system to parameter variations expressed in the s-plane. The weakness of the sensitivity measure is that it relies on the ability of the root locations to represent the performance of the system. As we have seen in the preceding chapters, the root locations represent the performance quite adequately for many systems, but due consideration must be given to the location of the zeros of the closed-loop transfer function and the dominancy of the pertinent roots. The root sensitivity measure is a very suitable measure of system performance sensitivity and can be used reliably for system analysis and design.

6.7 SUMMARY

The relative stability and the transient response performance of a closed-loop control system is directly related to the location of the closed-loop roots of the characteristic equation. Therefore we have investigated the movement of the characteristic roots on the s-plane as the system parameters are varied by utilizing the root locus method. The root locus method, a graphical technique, can be used to obtain an approximate sketch in order to analyze the initial design of a system and determine suitable alterations of the system structure and the parameter values. A summary of fifteen typical root locus diagrams is shown in Table 8.4.

Furthermore, we extended the root locus method for the design of several parameters for a closed-loop control system. Then the sensitivity of the characteristic roots was investigated for undesired parameter variations by defining a root sensitivity measure. It is clear that the root locus method is a powerful and useful approach for the analysis and design of modern control systems and will continue to be one of the most important procedures of control engineering.

PROBLEMS

6.1. Draw the root locus for the following open-loop transfer functions of the system shown in Fig. P6.1 when $0 < K < \infty$:

(a) $GH(s) = \dfrac{K}{s(s + 1)^2}$ (b) $GH(s) = \dfrac{K}{(s^2 + s + 1)(s + 1)}$

(c) $GH(s) = \dfrac{K(s + 1)}{s(s + 2)(s + 3)}$ (d) $GH(s) = \dfrac{K(s^2 + 4s + 8)}{s^2(s + 4)}$

6.2. The linear model of a phase detector was presented in Problem 5.7. Draw the root locus as a function of the gain $K_v = K_a K$. Determine the value of K_v attained if the complex roots have a damping ratio equal to 0.60.

6.3. The design of a ship stabilization system was presented in Problem 3.2. (a) Plot the root locus of the system for $K = K_a K_1$ when the fin actuator transfer function is

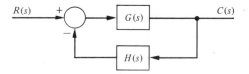

Figure P6.1

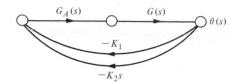

Figure P6.3

$$G_A(s) = \frac{K_a}{s + 2}.$$

(b) A roll rate sensor is added to the system as shown in Fig. P6.3. When $K_1 = 1$ and $K_2 = 2$, plot the root locus as a function of K_a. Determine the maximum damping ratio that can be attained and the gain K_a for this ζ_{max}.

6.4. The analysis of a large antenna was presented in Problem 3.5. Plot the root locus of the system as $0 < k_a < \infty$. Determine the maximum allowable gain of the amplifier for a stable system.

6.5. Automatic control of helicopters is necessary since, unlike fixed-wing aircraft which possess a fair degree of inherent stability, the helicopter is quite unstable. A helicopter control system which utilizes an automatic control loop plus a pilot stick control is shown in Fig. P6.5 [8]. When the pilot is not using the control stick, the switch may be considered to be open. The dynamics of the helicopter are represented by the transfer function

$$G_2(s) = \frac{25(s + 0.03)}{(s + 0.4)(s^2 - 0.36s + 0.16)}.$$

(a) With the pilot control loop open (hands-off control), plot the root locus for the automatic stabilization loop. Determine the gain K_2 which results in a damping for the complex roots equal to $\zeta = 0.707$. (b) For the gain K_2 obtained in part (a), determine the steady-state error due to a wind gust $T_d(s) = 1/s$. (c) With the pilot loop added, draw the root locus as K_1 varies from zero to ∞ when K_2 is set at the value calculated in (a). (d) Recalculate the steady-state error of part (b) when K_1 is equal to a suitable value based on the root locus.

6.6. An attitude control system for a satellite vehicle within the earth's atmosphere is shown in Fig. P6.6. The transfer functions of the system are

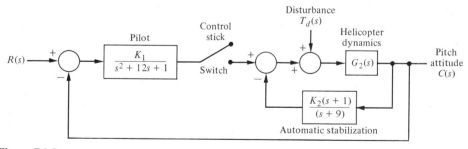

Figure P6.5

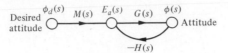

Figure P6.6

$$G(s) = \frac{K(s + 0.20)}{(s + 0.70)(s - 0.60)(s - 0.10)},$$

$$H(s) = \frac{(s + 1.20 + j1.4)(s + 1.20 - j1.4)}{(s + 2.5)}.$$

(a) Draw the root locus of the system as K varies from 0 to ∞. (b) Determine the gain K which results in a system with a 2.5% settling time less than 12 sec, and a damping ratio for the complex roots greater than 0.40.

6.7. The speed control system for an isolated power system is shown in Fig. P6.7 [9, 10]. The valve controls the steam flow input to the turbine in order to account for load changes, $\Delta L(s)$, within the power distribution network. The equilibrium speed desired results in a generator frequency equal to 60 cps. The effective rotary inertia, J, is equal to 4000 and the friction constant, f, is equal to 0.75. The steady-state speed regulation factor, R, is represented by the equation $R \cong (\omega_0 - \omega_r)/\Delta L$, where ω_r equals the speed at rated load and ω_0 equals the speed at no load. Clearly, it is desired to obtain a very small R, usually less than 0.10. (a) Using root locus techniques, determine the regulation, R, attainable when the damping ratio of the roots of the system must be greater than 0.60. (b) Verify that the steady-state speed deviation for a load torque change, $\Delta L(s) = \Delta L/s$, is, in fact, approximately equal to $R\Delta L$ when $R \leqslant 0.1$.

6.8. Reconsider the power control system of the preceding problem when the steam turbine is replaced by a hydroturbine. For hydroturbines, the large inertia of the water used as a source of energy causes a considerably larger time constant. The transfer function of a hydroturbine may be approximated by

$$G_t(s) = \frac{-\tau s + 1}{(\tau/2)s + 1},$$

where $\tau = 1$ sec. With the rest of the system remaining as given in Problem 6.7, repeat parts (a) and (b) of Problem 6.7.

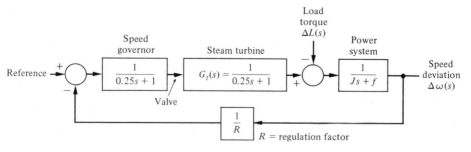

Figure P6.7

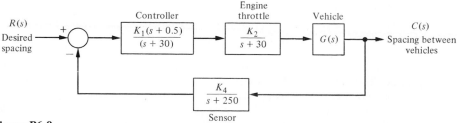

Figure P6.9

6.9. The achievement of safe, efficient control of the spacing of vehicles is an important part of future automatic ground transport [11, 12]. It is important that the system eliminate the effects of disturbances such as wind gusts as well as maintain accurate spacing between vehicles on a guideway. The system can be represented by the block diagram of Fig. P6.9. The vehicle dynamics can be represented by

$$G(s) = \frac{(s + 0.1)(s^2 + 2s + 289)}{s(s - 0.4)(s + 0.8)(s^2 + 1.45s + 361)}.$$

(a) Neglect the pole of the feedback sensor and draw the root locus of the system. (b) Determine all the roots when the loop gain $K = K_1 K_2 K_4 / 250$ is equal to 4000.

6.10. The supersonic transport airplane (SST) called the Concorde was put into service throughout the world in 1976. The Concorde, shown in Fig. P6.10(a), flies at 1400 miles per

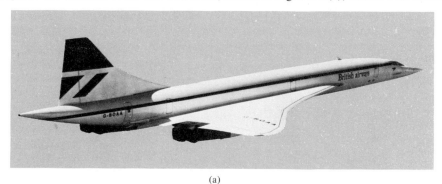

(a)

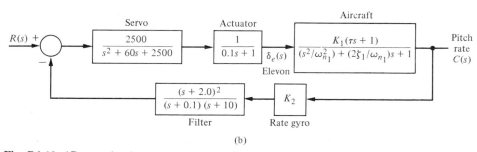

(b)

Fig. P6.10 (Concorde photo courtesy of British Airways)

hour, and has a trip time of three hours from London to New York. The flight control system requires good quality handling and comfortable flying conditions. An automatic flight control system can be designed for SST vehicles. The desired characteristics of the dominant roots of the control system shown in Fig. P6.10(b) are $\omega_n = 2.5$ rad/sec and $\zeta = 0.707$. The characteristics of the aircraft are $\omega_n = 2.5$, $\zeta_1 = 0.30$ and $\tau = 10$. Assume that the servo dynamics can be neglected. The gain factor K_1, however, will vary over the range 0.02 at medium-weight-cruise conditions to 0.20 at light-weight-descent conditions. (a) Draw the root locus as a function of the loop gain $K_1 K_2$. (b) Determine the necessary rate gyro gain K_2 in order to yield roots with $\zeta = 0.707$ when the aircraft is in the medium-cruise condition. (c) With the gain K_2 as found in (b), determine the ζ of the roots when the gain K_1 results from the condition of light-descent.

6.11. A computer system for a military application required a high performance magnetic tape transport system. However, the environmental conditions imposed on a military system result in a severe test of control engineering design. A direct-drive dc motor system for the magnetic tape reel system is shown in Fig. P6.11, where r equals the reel radius and J equals the reel and rotor inertia [13]. A complete reversal of the tape reel direction is required in 6 msec, and the tape reel must follow a step command in 3 msec or less. The tape is normally operating at a speed of 100 in/sec. The motor and components selected for this system possess the following characteristics:

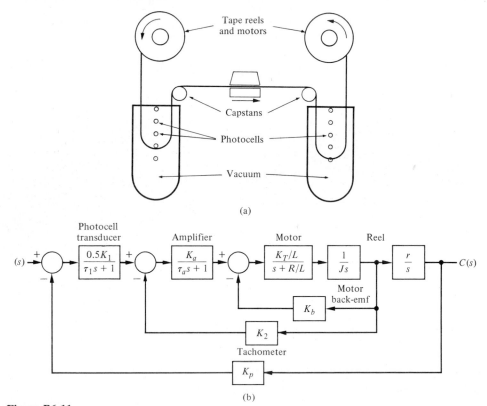

(a)

(b)

Figure P6.11

$$K_b = 0.40, \qquad\qquad K_T/LJ = 2.0,$$
$$K_p = 1, \qquad\qquad\qquad r = 0.2,$$
$$\tau_1 = \tau_a = 1 \text{ msec}, \qquad\quad K_1 = 2.0,$$
$$K_2 \text{ is adjustable}$$

The inertia of the reel and motor rotor is 2.5×10^{-3} when the reel is empty, and 5.0×10^{-3} when the reel is full. A series of photocells is used for an error sensing device. The time constant of the motor is $L/R = 0.5$ msec. (a) Draw the root locus for the system when $K_2 = 10$ and $J = 5.0 \times 10^{-3}$, and $0 < K_a < \infty$. (b) Determine the gain K_a which results in a well-damped system so that the zeta of all the roots is greater than or equal to 0.60. (c) With the K_a determined from part (b), draw a root locus for $0 < K_2 < \infty$.

6.12. A precision speed control system (Fig. P6.12) is required for a platform used in gyroscope and inertial system testing where a variety of closely controlled speeds is necessary. A direct-drive dc torque motor system was utilized in order to provide (1) a speed range of 0.01°/sec to 600°/sec, and (2) 0.1% steady-state error maximum [14]. The direct-drive dc torque motor avoids the use of a gear train with its attendant backlash and friction. Also the direct drive motor has a high torque capability, high efficiency, and low motor time constants. The motor gain constant is nominally $K_m = 0.10$, but is subject to variations up to 50%. The amplifier gain K_a is normally greater than 1,000 and subject to a variation of 25%. (a) Determine the minimum loop gain necessary to satisfy the steady-state error requirement. (b) Determine the limiting value of gain for stability. (c) With $K_m = 0.10$, draw the root locus as K_a varies from 0 to ∞. (d) With $K_a = 2000$, draw the root locus as K_m varies from 0 to ∞.

6.13. The control of large chemical processes is an important part of the field of control engineering. The signal-flow graph of the temperature control loop for a Xylene chemical process is shown in Fig. P6.13 [15]. The transfer functions for the components are

$$\text{Process:} \quad G_p(s) = \frac{0.03}{s^2 + 0.25s + 0.01}, \qquad \text{Sensor:} \quad G_s(s) = \frac{2}{3s + 1},$$
$$\text{Controller:} \quad G_c(s) = \frac{K(s^2 + s + 0.1)}{s}, \qquad \text{Valve:} \quad G_v(s) = \frac{2}{2s + 1}.$$

(a) Draw the root locus for the system for $0 < K < \infty$. (b) Determine the gain K when the dominant complex roots have a damping ratio equal to 0.707. (c) Determine and discuss the relative dominance of the complex roots close to the origin.

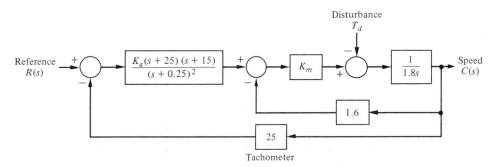

Figure P6.12

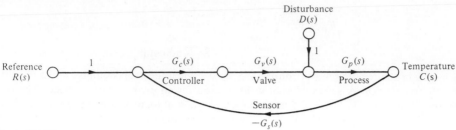

Figure P6.13

6.14. The open-loop transfer function of a single-loop negative feedback system is

$$GH(s) = \frac{K(s + 2)(s + 3)}{s^2(s + 1)(s + 24)(s + 30)}.$$

This system is called a *conditionally stable* system since the system is stable for only a range of the gain K as follows: $k_1 < K < k_2$. Using the Routh-Hurwitz criteria and the root locus method, determine the range of the gain for which the system is stable. Sketch the root locus for $0 < K < \infty$.

6.15. Let us again consider the stability and ride of a rider and high performance motorcycle as outlined in Problem 5.13 [16]. The dynamics of the motorcycle and rider can be represented by the open-loop transfer function

$$GH(s) = \frac{K(s^2 + 30s + 1125)}{s(s + 20)(s^2 + 10s + 125)(s^2 + 60s + 3400)}.$$

Draw the root locus for the system; determine the ζ of the dominant roots when $K = 3 \times 10^4$.

6.16. Control systems for maintaining constant tension on strip steel in a hot strip finishing mill are called "loopers." A typical system is shown in Fig. P6.16. The looper is an arm 2–3

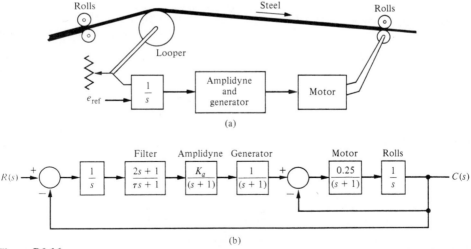

Figure P6.16

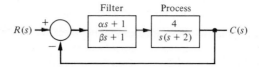

Figure P6.18

ft long with a roller on the end and is raised and pressed against the strip by a motor. The typical speed of the strip passing the looper is 2000 ft/min. A voltage proportional to the looper position is compared with a reference voltage and integrated where it is assumed that a change in looper position is proportional to a change in the steel strip tension. The time constant of the filter, τ, is negligible relative to the other time constants in the system. (a) Draw the root locus of the control system for $0 < K_a < \infty$. (b) Determine the gain K_a which results in a system whose roots have a damping ratio of $\zeta = 0.707$ or greater. (c) Determine the effect of τ as τ increases from a negligible quantity.

6.17. Reconsider the vibration absorber discussed in Problems 2.2 and 2.10 as a design problem. Using the root locus method, determine the effect of the parameters M_2 and k_{12}. Determine the specific values of the parameters M_2 and k_{12} so that the mass M_1 does not vibrate when $F(t) = a \sin \omega_0 t$. Assume that $M_1 = 1$, $k_1 = 1$, and $f_1 = 1$. Also assume that $k_{12} < 1$ and the term k_{12}^2 may be neglected.

6.18. A feedback control system is shown in Fig. P6.18. The filter $G_c(s)$ is often called a compensator and the design problem is that of selecting the parameters α and β. Using the root locus method, determine the effect of varying the parameters. Select a suitable filter so that the settling time is less than 4 sec and the damping ratio of the dominant roots is greater than 0.60.

6.19. In recent years, many automatic control systems for automatic highways have been proposed. One such system uses a guidance cable imbedded in the roadway to guide the vehicle along the desired lane [17]. An error detector is composed of two coils mounted on the front of the automobile which senses a magnetic field produced by the current in the buried guidance cable. The error signal is amplified and is used to automatically actuate the steering mechanism. The closed-loop system is shown in Fig. P6.19. The transient response of a guideway vehicle was recently investigated with a guidance cable system [17]. The transfer function for the Volkswagen system was found to be

$$G(s) = \frac{K_a(s^2 + 3.6s + 81)}{s(s + 1)(s + 5)}$$

for a velocity of 90 Km/hr when K_a equals the amplifier gain. (a) Draw a root locus and determine a suitable gain K_a so that the damping ratio of the complex roots is 0.707. (b) Determine the root sensitivity of the system for the complex root r_1 as a function of (1) K_a, and (2) the pole of $G(s)$, $s = -1$.

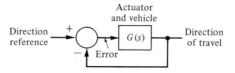

Figure P6.19

6.20. Determine the root sensitivity for the dominant roots of the design for Problem 6.18 for the gain $K = 4\alpha/\beta$ and the pole $s = -2$.

6.21. Determine the root sensitivity of the dominant roots of the power system of Problem 6.7. Evaluate the sensitivity for variations of (a) the poles at $s = -4$, and (b) the feedback gain, $1/R$.

6.22. Determine the root sensitivity of the dominant roots of Problem 6.1(a) when K is set so that the damping ratio of the unperturbed roots is 0.707. Evaluate and compare the sensitivity as a function of the poles and zeros of $GH(s)$.

6.23. Repeat Problem 6.22 for the open-loop transfer function $GH(s)$ of Problem 6.1(c).

6.24. For systems of relatively high degree, the form of the root locus can often assume an unexpected pattern. The root loci of four different feedback systems of third-order or higher are shown in Fig. P6.24 [15]. The open-loop poles and zeros of $KF(s)$ are shown and the form of the root loci as K varies from zero to infinity is presented. Verify, by constructing the root loci, the diagrams of Fig. P6.24.

6.25. Solid-state integrated electronic circuits are comprised of distributed R and C elements. Therefore, feedback electronic circuits in integrated circuit form must be investigated by obtaining the transfer function of the distributed RC networks. It has been shown that the slope of the attenuation curve of a distributed RC network is n 3db/octave, where n is the order of the RC filter [18]. This attenuation is in contrast with the normal n 6db/octave for the lumped parameter circuits. (The concept of the slope of an attenuation curve is considered in Chapter 7. If the reader is unfamiliar with this concept, this problem may be reexamined following the study of Chapter 7.) An interesting case arises when the distributed RC network occurs in a series to shunt feedback path of a transistor amplifier. Then the loop transfer function may be written as

$$GH(s) = \frac{K(s - 1)(s + 3)^{1/2}}{(s + 1)(s + 2)^{1/2}}.$$

(a) Using the root locus method, determine the locus of roots as K varies from zero to infinity. (b) Calculate the gain at borderline stability and the frequency of oscillation for this gain.

6.26. A single-loop negative feedback system has a loop transfer function

$$GH(s) = \frac{K(s + 1)^2}{s(s^2 + 1)(s + 4)}.$$

(a) Sketch the root locus for $0 \leqslant K \leqslant \infty$ to indicate the significant features of the locus. (b) Determine the range of the gain K for which the system is stable. (c) For what values of K in the range $K \geqslant 0$ do purely imaginary roots exist? What are the values of these roots? (d) Would the use of the dominant roots approximation for an estimate of settling time be justified in this case for a large magnitude of gain ($K > 10$)?

6.27. A unity negative feedback system has a transfer function

$$G(s) = \frac{K(s^2 + 0.105625)}{s(s^2 + 1)} = \frac{K(s + j0.325)(s - j0.325)}{s(s^2 + 1)}.$$

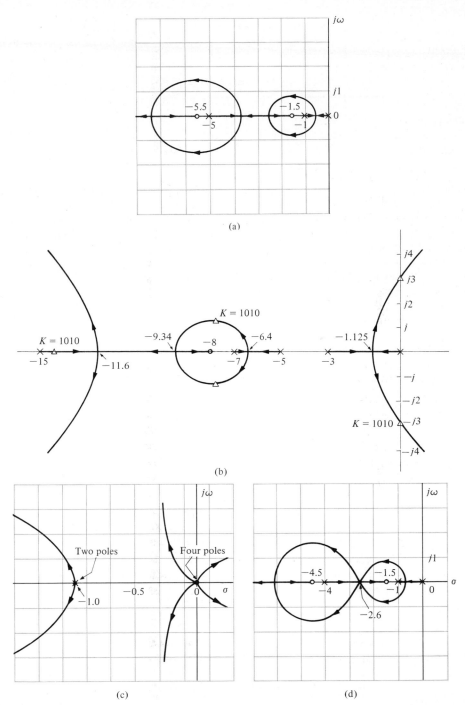

(a)

(b)

(c)

(d)

Figure P6.24

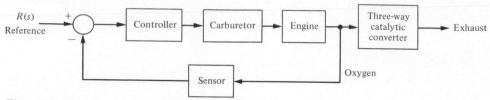

Figure P6.28

Calculate the root locus as a function of K. Carefully calculate where the segments of the locus enter and leave the real axis.

6.28. To meet current U.S. emissions standards for automobiles, hydrocarbon (HC) and carbon monoxide (CO) emissions are usually controlled by a catalytic converter in the automobile exhaust. Federal standards for nitrogen oxides (NO_x) emissions are met mainly by exhaust-gas recirculation (EGR) techniques. However, in 1981, when NO_x emissions standards are tightened from the current limit of 2.0 grams per mile to 1.0 grams per mile, these techniques alone may no longer be sufficient.

Although many schemes are under investigation for meeting the 1981 emissions standards for all three emissions, one of the most promising employs a three-way catalyst—for HC, CO, and NO_x emissions—in conjunction with a closed-loop engine-control system. The approach is to use a closed-loop engine control as shown in Fig. P6.28 [19]. The exhaust gas sensor gives an indication of a rich or lean exhaust and this is compared to a reference. The difference signal is processed by the controller and the output of the controller modulates the vacuum level in the carburetor to achieve the best air-fuel ratio for proper operation of the catalytic converter. The open-loop transfer function is represented by

$$GH(s) = \frac{K(s + 1)(s + 6)}{s(s + 2)(s + 3)}.$$

Calculate the root locus as a function of K. Carefully calculate where the segments of the locus enter and leave the real axis.

6.29. A unity feedback control system has a transfer function

$$G(s) = \frac{K(s^2 + 4s + 8)}{s^2(s + 4)}.$$

It is desired that the dominant roots have a damping ratio equal to 0.5. Find the gain K when this condition is satisfied. Show that at this gain the imaginary roots are $s = -1.3 \pm j2.2$.

6.30. An RLC network is shown in Fig. P6.30. The nominal values (normalized) of the net-

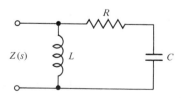

Figure P6.30

work elements are $L = C = 1$ and $R = 2.5$. Show that the root sensitivity of the two roots of the input impedance $Z(s)$ to a change in R is different by a factor of 4.

REFERENCES

1. W. R. Evans, "Graphical Analysis of Control Systems," *Transactions of the AIEE,* **67,** 1948; pp. 547–551. Also in *Automatic Control,* G. J. Thaler, ed., Dowden, Hutchinson, and Ross, Inc., Stroudsburg, Pa., 1974; pp. 417–421.

2. W. R. Evans, "Control System Synthesis by Root Locus Method," *Transactions of the AIEE,* **69,** 1950; pp. 1–4. Also in *Automatic Control,* G. J. Thaler, ed., Dowden, Hutchinson, and Ross, Inc., Stroudsburg, Pa., 1974; pp. 423–425.

3. W. R. Evans, *Control System Dynamics,* McGraw-Hill, New York, 1954.

4. J. L. Melsa and S. K. Jones, *Computer Programs for Computational Assistance in the Study of Linear Control Theory,* 2nd ed., McGraw-Hill, New York, 1973.

5. J. H. Goldberg, *Automatic Controls,* Allyn and Bacon, Boston, 1964.

6. P. M. Frank, *Introduction to System Sensitivity Theory,* Academic Press, New York, 1978.

7. H. Ur, "Root Locus Properties and Sensitivity Relations in Control Systems," *I.R.E. Transactions on Automatic Control,* January 1960; pp. 57–65.

8. A. R. S. Bramwell, *Helicopter Dynamics,* Halsted Press, New York, 1976.

9. P. J. Burrows and A. R. Daniels, "Digital Excitation Control of a.c. Turbogenerators Using a Dedicated Microprocessor," *Proceedings of the I.E.E.,* **125,** 3, March 1978; pp. 237–240.

10. D. V. Richardson, *Rotating Electric Machinery and Transformer Technology,* Reston Publishing Co., Reston, Va., 1978.

11. R. E. Fenton and P. M. Chu, "On Vehicle Automatic Longitudinal Control," *Transportation Science,* **11,** 1, February 1977; pp. 73–91.

12. G. M. Takasaki and R. E. Fenton, "On the Identification of Vehicle Longitudinal Dynamics," *IEEE Transactions on Automatic Control,* August 1977; pp. 610–615.

13. R. E. Matick, *Computer Storage Systems and Technology,* Wiley, New York, 1977.

14. K. Magnus, *Gyrodynamics,* Springer-Verlag, New York, 1974.

15. C. D. Johnson, *Process Control Instrumentation Technology,* Wiley, New York, 1977.

16. D. H. Weir and J. W. Zellner, "Lateral-Directional Motorcycle Dynamics and Rider Control," *Society of Automotive Engineers, 1978 Congress;* pp. 1–23.

17. S. E. Shladover, "Steering Controller Design for Automated Guideway Transit Vehicles," *J. of Dynamic Systems, Measurement, and Control,* March 1978; pp. 1–7.

18. F. K. Manasse, *Semiconductor Electronics Design,* Prentice-Hall, Englewood Cliffs, N.J., 1977.

19. G. W. Niepoth and S. P. Stonestreet, "Closed-Loop Engine Control," *IEEE Spectrum,* November 1977; pp. 53–55.

20. F. H. Raven, *Automatic Control Engineering,* 3rd ed., McGraw-Hill, New York, 1978; Ch. 7.

7 / Frequency Response Methods

7.1 INTRODUCTION

In the preceding chapters the response and performance of a system has been described in terms of the complex frequency variable s, and the location of the poles and zeros on the s-plane. A very practical and important alternative approach to the analysis and design of a system is the frequency response method. *The frequency response of a system is defined as the steady-state response of the system to a sinusoidal input signal.* The sinusoid is a unique input signal, and the resulting output signal for a linear system, as well as signals throughout the system, is sinusoidal in the steady-state; it differs from the input waveform only in amplitude and phase angle.

One advantage of the frequency response method is the ready availability of sinusoid test signals for various ranges of frequencies and amplitudes. Thus the experimental determination of the frequency response of a system is easily accomplished and is the most reliable and uncomplicated method for the experimental analysis of a system. Often, as we shall find in Section 7.4, the unknown transfer function of a system can be deduced from the experimentally determined frequency response of a system [1, 2]. Furthermore, the design of a system in the frequency domain provides the designer with control of the bandwidth of a system and some measure of the response of the system to undesired noise and disturbances.

A second advantage of the frequency response method is that the transfer function describing the sinusoidal steady-state behavior of a system can be obtained by replacing s with $j\omega$ in the system transfer function $T(s)$. The transfer function representing the sinusoidal steady-state behavior of a system is then a function of the complex variable $j\omega$ and is itself a complex function $T(j\omega)$ which possesses a magnitude and phase angle. The magnitude and phase angle of $T(j\omega)$ are readily represented by graphical plots which provide a significant insight for the analysis and design of control systems.

The basic disadvantage of the frequency response method for analysis and design is the indirect link between the frequency and time domain. Direct correlations between the frequency response and the corresponding transient response characteristics are somewhat tenuous, and in practice the frequency response characteristic is adjusted by using various design criteria which will normally result in a satisfactory transient response.

The Laplace transform pair was given in Section 2.4 and is written as [3]:

$$F(s) = \mathcal{L}\{f(t)\} = \int_0^\infty f(t)e^{-st}\, dt \tag{7.1}$$

and

$$f(t) = \mathcal{L}^{-1}\{F(s)\} = \frac{1}{2\pi j}\int_{\sigma-j\infty}^{\sigma+j\infty} F(s)e^{st}\, ds, \tag{7.2}$$

where the complex variable $s = \sigma + j\omega$. Similarly, the *Fourier transform* pair is written as

$$F(j\omega) = \mathcal{F}\{f(t)\} = \int_{-\infty}^\infty f(t)e^{-j\omega t}\, dt \tag{7.3}$$

and

$$f(t) = \mathcal{F}^{-1}\{\mathcal{F}(j\omega)\} = \frac{1}{2\pi}\int_{-\infty}^\infty F(j\omega)e^{j\omega t}\, d\omega. \tag{7.4}$$

The Fourier transform exists for $f(t)$ when

$$\int_{-\infty}^\infty |f(t)|\, dt < \infty.$$

The Fourier and Laplace transforms are closely related as we may see by examining Eqs. (7.1) and (7.3). When the function $f(t)$ is only defined for $t \geqslant 0$, as is often the case, the lower limits on the integrals are the same. Then, we note that the two equations differ only in the complex variable. Thus, if the Laplace transform of a function $f_1(t)$ is known to be $F_1(s)$, one can obtain the Fourier transform of this same time function $F_1(j\omega)$ by setting $s = j\omega$ in $F_1(s)$.

Again, one might ask, since the Fourier and Laplace transforms are so closely related, why not always use the Laplace transform? Why use the Fourier transform at all? The Laplace transform permits us to investigate the s-plane location of the poles and zeros of a transfer function $T(s)$ as in Chapter 6. However, the frequency response method allows us to consider the transfer function $T(j\omega)$ and concern ourselves with the amplitude and phase characteristics of the system. This ability to investigate and represent the character of a system by amplitude and phase equations and curves is an advantage for the analysis and design of control systems.

If we consider the frequency response of the closed-loop system, we might have an input $r(t)$ which has a Fourier transform, in the frequency domain, as follows:

$$R(j\omega) = \int_{-\infty}^{\infty} r(t)e^{-j\omega t} \, dt. \tag{7.5}$$

Then the output frequency response of a single-loop control system can be obtained by substituting $s = j\omega$ in the closed-loop system relationship, $C(s) = T(s)R(s)$, so that we have

$$C(j\omega) = T(j\omega)R(j\omega) = \frac{G(j\omega)}{1 + G(j\omega)H(j\omega)} R(j\omega). \tag{7.6}$$

Utilizing the inverse Fourier transform, the output transient response would be

$$c(t) = \mathcal{F}^{-1}\{C(j\omega)\} = \frac{1}{2\pi} \int_{-\infty}^{\infty} C(j\omega)e^{j\omega t} \, d\omega. \tag{7.7}$$

However, it is usually quite difficult to evaluate this inverse transform integral for any but the simplest systems, and a graphical integration may be used. Alternatively, as we will note in succeeding sections, several measures of the transient response can be related to the frequency characteristics and utilized for design purposes.

7.2 FREQUENCY RESPONSE PLOTS

The transfer function of a system $G(s)$ can be described in the frequency domain by the relation

$$G(j\omega) = G(s)|_{s=j\omega} = R(j\omega) + jX(j\omega), \tag{7.8}$$

where

$$R(j\omega) = \text{Re} \, (G(j\omega))$$

and

$$X(j\omega) = \text{Im} \, (G(j\omega)).$$

Alternatively, the transfer function can be represented by a magnitude $|G(j\omega)|$ and a phase $\phi(j\omega)$ as

$$G(j\omega) = |G(j\omega)|e^{j\phi(j\omega)} = |G(\omega)|\underline{/\phi(\omega)}, \tag{7.9}$$

where

$$\phi(\omega) = \tan^{-1} X(\omega)/R(\omega)$$

and

$$|G(\omega)|^2 = (R(\omega))^2 + (X(\omega))^2.$$

The graphical representation of the frequency response of the system $G(j\omega)$ can utilize either Eq. (7.8) or Eq. (7.9). The *polar plot* representation of the frequency

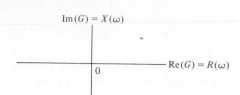

$\text{Im}(G) = X(\omega)$

0

$\text{Re}(G) = R(\omega)$

Fig. 7.1. The polar plane.

response is obtained by using Eq. (7.8). The coordinates of the polar plot are the real and imaginary parts of $G(j\omega)$ as shown in Fig. 7.1. An example of a polar plot will illustrate this approach.

Example 7.1. A simple RC filter is shown in Fig. 7.2. The transfer function of this filter is

$$G(s) = \frac{V_2(s)}{V_1(s)} = \frac{1}{RCs + 1}, \tag{7.10}$$

and the sinuoidal steady-state transfer function is

$$G(j\omega) = \frac{1}{j\omega(RC) + 1} = \frac{1}{j(\omega/\omega_1) + 1}, \tag{7.11}$$

where

$$\omega_1 = 1/RC.$$

Then the polar plot is obtained from the relation

$$G(j\omega) = R(\omega) + jX(\omega) = \frac{1 - j(\omega/\omega_1)}{(\omega/\omega_1)^2 + 1}$$

$$= \frac{1}{1 + (\omega/\omega_1)^2} - \frac{j(\omega/\omega_1)}{1 + (\omega/\omega_1)^2}. \tag{7.12}$$

The locus of the real and imaginary parts is shown in Fig. 7.3 and is easily shown to be a circle with the center at $(\tfrac{1}{2}, 0)$. When $\omega = \omega_1$, the real and imaginary parts are equal, and the angle $\phi(\omega) = -45°$. The polar plot can also be readily obtained from Eq. (7.9) as

$$G(j\omega) = |G(\omega)|/\underline{\phi(\omega)}, \tag{7.13}$$

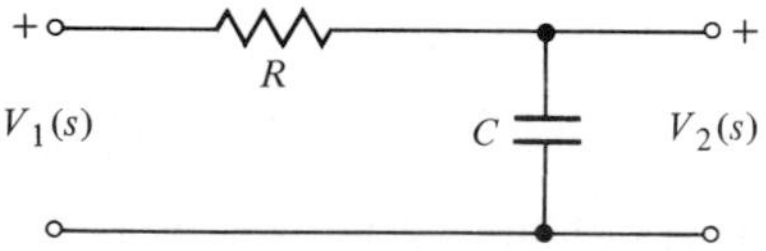

$+\!\circ$———$\bigvee\!\bigvee\!\bigvee$———$\circ\,+$
R

$V_1(s)$

C $V_2(s)$

Fig. 7.2. An RC filter.

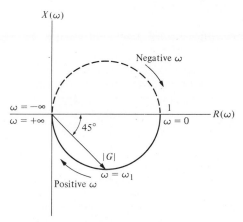

Fig. 7.3. Polar plot for RC filter.

where

$$|G(\omega)| = \frac{1}{(1 + (\omega/\omega_1)^2)^{1/2}} \quad \text{and} \quad \phi(\omega) = -\tan^{-1}(\omega/\omega_1).$$

Clearly, when $\omega = \omega_1$, magnitude is $|G(\omega_1)| = 1/\sqrt{2}$ and phase $\phi(\omega_1) = -45°$. Also, when ω approaches $+\infty$, we have $|G(\omega)| \to 0$ and $\phi(\omega) = -90°$. Similarly, when $\omega = 0$, we have $|G(\omega)| = 1$ and $\phi(\omega) = 0$.

Example 7.2. The polar plot of a transfer function will be useful for investigating system stability and will be utilized in Chapter 8. Therefore, it is worthwhile to complete another example at this point. Consider a transfer function

$$|G(s)|_{s=j\omega} = G(j\omega) = \frac{K}{j\omega(j\omega\tau + 1)} = \frac{K}{j\omega - \omega^2\tau}. \tag{7.14}$$

Then the magnitude and phase angle are written as

$$|G(\omega)| = \frac{K}{(\omega^2 + \omega^4\tau^2)^{1/2}} \tag{7.15}$$

and

$$\phi(\omega) = -\tan^{-1}\left(\frac{1}{-\omega\tau}\right).$$

The phase angle and magnitude are readily calculated at the frequencies $\omega = 0$, $\omega = 1/\tau$, and $\omega = +\infty$. The values of $|G(\omega)|$ and $\phi(\omega)$ are given in Table 7.1 and the polar plot of $G(j\omega)$ is shown in Fig. 7.4.

There are several possibilities for coordinates of a graph portraying the frequency response of a system. As we have seen, one may choose to utilize a polar

Table 7.1

ω	0	$1/2\tau$	$1/\tau$	∞
$|G(\omega)|$	∞	$4K\tau/\sqrt{5}$	$K\tau/\sqrt{2}$	0
$\phi(\omega)$	$-90°$	$-117°$	$-135°$	$-180°$

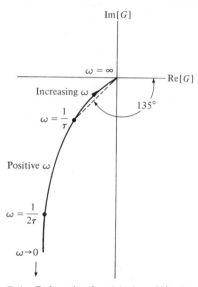

Fig. 7.4. Polar plot for $G(j\omega) = K/j\omega(j\omega\tau + 1)$.

plot to represent the frequency response (Eq. 7.8) of a system. However, the limitations of polar plots are readily apparent. The addition of poles or zeros to an existing system requires the recalculation of the frequency response as outlined in Examples 7.1 and 7.2. (See Table 7.1.) Furthermore, the calculation of the frequency response in this manner is tedious and does not indicate the effect of the individual poles or zeros.

Therefore, the introduction of *logarithmic plots*, often called *Bode plots*, simplifies the determination of the graphical portrayal of the frequency response. The logarithmic plots are called Bode plots in honor of H. W. Bode who used them extensively in his studies of feedback amplifiers [4, 5]. The transfer function in the frequency domain is

$$G(j\omega) = |G(\omega)|e^{j\phi(\omega)}. \tag{7.16}$$

The natural logarithm of Eq. (7.16) is

$$\ln G(j\omega) = \ln |G(\omega)| + j\phi(\omega), \tag{7.17}$$

where $\ln |G|$ is the magnitude in nepers. The logarithm of the magnitude is normally expressed in terms of the logarithm to the base 10 so that we use

$$\text{Logarithmic gain} = 20 \log_{10} |G(\omega)|,$$

where the units are decibels (db). A decibel conversion table is given in Appendix F. The logarithmic gain in db and the angle $\phi(\omega)$ can be plotted versus the frequency ω by utilizing several different arrangements. For a Bode diagram, the plot of logarithmic gain in db vs. ω is normally plotted on one set of axes and the phase $\phi(\omega)$ versus ω on another set of axes as shown on Fig. 7.5. For example, the Bode diagram of the transfer function of Example 7.1 can be readily obtained as we will find in the following example.

Example 7.3.   The transfer function of Example 7.1 is

$$G(j\omega) = \frac{1}{j\omega(RC) + 1} = \frac{1}{j\omega\tau + 1}, \tag{7.18}$$

where

$$\tau = RC,$$

the time constant of the network. The logarithmic gain is

$$20 \log |G| = 20 \log \left(\frac{1}{1 + (\omega\tau)^2}\right)^{1/2} = -10 \log (1 + (\omega\tau)^2). \tag{7.19}$$

For small frequencies, that is $\omega \ll 1/\tau$, the logarithmic gain is

$$20 \log |G| = -10 \log (1) = 0 \text{ db}, \quad \omega \ll 1/\tau. \tag{7.20}$$

For large frequencies, that is $\omega \gg 1/\tau$, the logarithmic gain is

$$20 \log |G| = -20 \log \omega\tau, \quad \omega \gg 1/\tau, \tag{7.21}$$

and at $\omega = 1/\tau$, we have

$$20 \log |G| = -10 \log 2 = -3.01 \text{ db}.$$

The magnitude plot for this network is shown in Fig. 7.5(a). The phase angle of this network is

$$\phi(j\omega) = -\tan^{-1} \omega\tau. \tag{7.22}$$

The phase plot is shown in Fig. 7.5(b). The frequency $\omega = 1/\tau$ is often called the *break frequency* or *corner frequency*.

Examining the Bode diagram of Fig. 7.5, we find that a linear scale of frequency is not the most convenient or judicious choice and we should consider the use of a logarithmic scale of frequency. The convenience of a logarithmic scale of frequency can be seen by considering Eq. (7.21) for large frequencies $\omega \gg 1/\tau$, as follows:

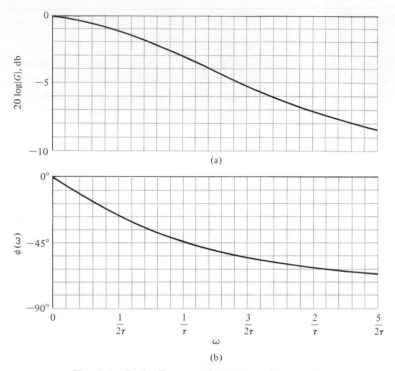

(a)

(b)

Fig. 7.5. Bode diagram of $G(j\omega) = 1/(j\omega\tau + 1)$.

$$20 \log |G| = -20 \log \omega\tau = -20 \log \tau - 20 \log \omega. \qquad (7.23)$$

Then, on a set of axes where the horizontal axis is $\log \omega$, the asymptotic curve for $\omega \gg 1/\tau$ is a straight line as shown in Fig. 7.6. The slope of the straight line can be ascertained from Eq. (7.21). An interval of two frequencies with a ratio equal to ten is called a decade so that the range of frequencies from ω_1 to ω_2, where $\omega_2 = 10\omega_1$,

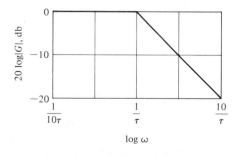

Fig. 7.6. Asymptotic curve for $(j\omega\tau + 1)^{-1}$.

is called a decade. Then, the difference between the logarithmic gains, for $\omega \gg$ $1/\tau$, over a decade of frequency is

$$20 \log |G(\omega_1)| - 20 \log |G(\omega_2)| = -20 \log \omega_1\tau - (-20 \log \omega_2\tau)$$

$$= -20 \log \frac{\omega_1\tau}{\omega_2\tau}$$

$$= -20 \log (\tfrac{1}{10}) = +20 \text{ db.} \qquad (7.24)$$

That is, the slope of the asymptotic line for this first-order transfer function is -20 db/decade, and the slope is shown for this transfer function in Fig. 7.6. Instead of using a horizontal axis of log ω and linear rectangular coordinates, it is simpler to use semilog paper with a linear rectangular coordinate for db and a logarithmic coordinate for ω. Alternatively, one could use a logarithmic coordinate for the magnitude as well as for frequency and avoid the necessity of calculating the logarithm of the magnitude.

The frequency interval $\omega_2 = 2\omega_1$ is often used and is called an octave of frequencies. The difference between the logarithmic gains for $\omega \gg 1/\tau$, for an octave, is

$$20 \log |G(\omega_1)| - 20 \log |G(\omega_2)| = -20 \log \frac{\omega_1\tau}{\omega_2\tau}$$

$$= -20 \log (\tfrac{1}{2}) = 6.02 \text{ db.} \qquad (7.25)$$

Therefore, the slope of the asymptotic line is -6 db/octave or -20 db/decade.

The primary advantage of the logarithmic plot is the conversion of multiplicative factors such as $(j\omega\tau + 1)$ into additive factors $20 \log (j\omega\tau + 1)$ by virtue of the logarithmic gain definition. This can be readily ascertained by considering a generalized transfer function as

$$G(j\omega) = \frac{K_b \prod_{i=1}^{Q} (1 + j\omega\tau_i)}{(j\omega)^N \prod_{m=1}^{M} (1 + j\omega\tau_m) \prod_{k=1}^{R} (1 + (2\zeta_k/\omega_{n_k})j\omega + (j\omega/\omega_{n_k})^2)}. \qquad (7.26)$$

This transfer function includes Q zeros, N poles at the origin, M poles on the real axis, and R pairs of complex conjugate poles. Clearly, obtaining the polar plot of such a function would be a formidable task indeed. However, the logarithmic magnitude of $G(j\omega)$ is

$$20 \log |G(\omega)| = 20 \log K_b + 20 \sum_{i=1}^{Q} \log |1 + j\omega\tau_i|$$

$$- 20 \log |(j\omega)^N| - 20 \sum_{m=1}^{M} \log |1 + j\omega\tau_m|$$

$$- 20 \sum_{k=1}^{R} \log \left| 1 + \left(\frac{2\zeta_k}{\omega_{n_k}}\right) j\omega + \left(\frac{j\omega}{\omega_{n_k}}\right)^2 \right|, \qquad (7.27)$$

and the Bode diagram can be obtained by adding the plot due to each individual factor. Furthermore, the separate phase angle plot is obtained as

$$\phi(\omega) = + \sum_{i=1}^{Q} \tan^{-1} \omega\tau_i - N(90°) - \sum_{m=1}^{M} \tan^{-1} \omega\tau_m$$
$$- \sum_{k=1}^{R} \tan^{-1} \left(\frac{2\zeta_k \omega_{n_k} \omega}{\omega_{n_k}^2 - \omega^2} \right), \tag{7.28}$$

which is simply the summation of the phase angles due to each individual factor of the transfer function.

Therefore, the four different kinds of factors which may occur in a transfer function are

1. constant gain K_b,
2. poles (or zeros) at the origin $(j\omega)$,
3. poles or zeros on the real axis $(j\omega\tau + 1)$,
4. complex conjugate poles (or zeros) $(1 + (2\zeta/\omega_n)j\omega + (j\omega/\omega_n)^2)$.

We can determine the logarithmic magnitude plot and phase angle for these four factors and then utilize them to obtain a Bode diagram for any general form of a transfer function. Typically, the curves for each factor are obtained and then added together graphically to obtain the curves for the complete transfer function. Furthermore, this procedure can be simplified by using the asymptotic approximations to these curves and obtaining the actual curves only at specific important frequencies.

Constant gain K_b. The logarithmic gain is

$$20 \log K_b = \text{constant in db},$$

and the phase angle is zero. The gain curve is simply a horizontal line on the Bode diagram.

Poles (or zeros) at the origin $(j\omega)$. A pole at the origin has a logarithmic magnitude

$$20 \log \left| \frac{1}{j\omega} \right| = -20 \log \omega \text{ db} \tag{7.29}$$

and a phase angle $\phi(\omega) = -90°$. The slope of the magnitude curve is -20 db/decade for a pole. Similarly for a multiple pole at the origin, we have

$$20 \log \left| \frac{1}{(j\omega)^N} \right| = -20 N \log \omega, \tag{7.30}$$

and the phase is $\phi(\omega) = -90°N$. In this case the slope due to the multiple pole is

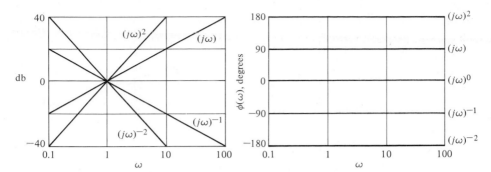

Fig. 7.7. Bode diagram for $(j\omega)^{\pm N}$.

$-20N$ db/decade. For a zero at the origin, we have a logarithmic magnitude

$$20 \log |j\omega| = +20 \log \omega, \tag{7.31}$$

where the slope is $+20$ db/decade and the phase angle is $+90°$. The Bode diagram of the magnitude and phase angle of $(j\omega)^{\pm N}$ is shown in Fig. 7.7 for $N = 1$ and $N = 2$.

Poles or zeros on the real axis. The pole factor $(1 + j\omega\tau)^{-1}$ has been considered previously and we found that

$$20 \log \left| \frac{1}{1 + j\omega\tau} \right| = -10 \log (1 + \omega^2\tau^2). \tag{7.32}$$

The asymptotic curve for $\omega \ll 1/\tau$ is $20 \log 1 = 0$ db and the asymptotic curve for $\omega \gg 1/\tau$ is $-20 \log \omega\tau$ which has a slope of -20 db/decade. The intersection of the two asymptotes occurs when

$$20 \log 1 = 0 \text{ db} = -20 \log \omega\tau$$

or when $\omega = 1/\tau$, the *break frequency*. The actual logarithmic gain when $\omega = 1/\tau$ is -3 db for this factor. The phase angle is $\phi(\omega) = -\tan^{-1} \omega\tau$ for the denominator factor. The Bode diagram of a pole factor $(1 + j\omega\tau)^{-1}$ is shown in Fig. 7.8.

The Bode diagram of a zero factor $(1 + j\omega\tau)$ is obtained in the same manner as the pole. However, the slope is positive at $+20$ db/decade, and the phase angle is $\phi(\omega) = +\tan^{-1} \omega\tau$.

A linear approximation to the phase angle curve can be obtained as shown in Fig. 7.8. This linear approximation, which passes through the correct phase at the break frequency, is within $6°$ of the actual phase curve for all frequencies. This approximation will provide a useful means for readily determining the form of the phase angle curves of a transfer function $G(s)$. However, often the accurate phase angle curves are required and the actual phase curve for the first-order factor must be drawn. Therefore, it is often worthwhile to prepare a cardboard (or plastic) template which may be utilized to repeatedly draw the phase curves for the individual

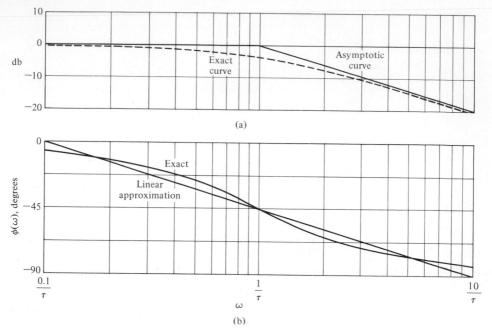

Fig. 7.8. Bode diagram of $(1 + j\omega\tau)^{-1}$.

factors. The exact values of the frequency response for the pole $(1 + j\omega\tau)^{-1}$ as well as the values obtained by using the approximation for comparison are given in Table 7.2.

Complex conjugate poles or zeros $[1 + (2\zeta/\omega_n)j\omega + (j\omega/\omega_n)^2]$. The quadratic factor for a pair of complex conjugate poles may be written in normalized form as

$$[1 + j2\zeta u - u^2]^{-1}, \tag{7.33}$$

where $u = \omega/\omega_n$. Then, the logarithmic magnitude is

$$20 \log |G(\omega)| = -10 \log ((1 - u^2)^2 + 4\zeta^2 u^2), \tag{7.34}$$

and the phase angle is

$$\phi(\omega) = -\tan^{-1}\left(\frac{2\zeta u}{1 - u^2}\right). \tag{7.35}$$

When $u \ll 1$, the magnitude is

$$db = -10 \log 1 = 0 \ db,$$

and the phase angle approaches $0°$. When $u \gg 1$, the logarithmic magnitude approaches

$$db = -10 \log u^4 = -40 \log u,$$

Table 7.2

$\omega\tau$	0.10	0.50	0.76	1	1.31	2	5	10
$20 \log |(1 + j\omega\tau)^{-1}|$, db	−0.04	−1.0	−2.0	−3.0	−4.3	−7.0	−14.2	−20.04
Asymptotic approximation, db	0	0	0	0	−2.3	−6.0	−14.0	−20.0
$\phi(\omega)$, degrees	−5.7	−26.6	−37.4	−45.0	−52.7	−63.4	−78.7	−84.3
Linear approximation, degrees	0	−31.5	−39.5	−45.0	−50.3	−58.5	−76.5	−90.0

which results in a curve with a slope of -40 db/decade. The phase angle, when $u \gg 1$, approaches $-180°$. The magnitude asymptotes meet at the 0-db line when $u = \omega/\omega_n = 1$. However, the difference between the actual magnitude curve and the asymptotic approximation is a function of the damping ratio and must be accounted for when $\zeta < 0.707$. The Bode diagram of a quadratic factor due to a pair of complex conjugate poles is shown in Fig. 7.9. The maximum value of the frequency response, M_{p_ω}, occurs at the *resonant frequency* ω_r. When the damping ratio approaches zero, then ω_r approaches ω_n, the natural frequency. The resonant frequency is determined by taking the derivative of the magnitude of Eq. (7.33) with respect to the normalized frequency, u, and setting it equal to zero. The resonant frequency is represented by the relation

$$\omega_r = \omega_n\sqrt{1 - 2\zeta^2}, \quad \zeta < 0.707, \tag{7.36}$$

and the maximum value of the magnitude $|G(\omega)|$ is

$$M_{p_\omega} = |G(\omega_r)| = (2\zeta\sqrt{1 - \zeta^2})^{-1}, \quad \zeta < 0.707, \tag{7.37}$$

for a pair of complex poles. The maximum value of the frequency response M_{p_ω}, and the resonant frequency ω_r are shown as a function of the damping ratio ζ for a pair of complex poles in Fig. 7.10. Assuming the dominance of a pair of complex conjugate closed-loop poles, we find that these curves are useful for estimating the damping ratio of a system from an experimentally determined frequency response.

The frequency response curves can be evaluated graphically on the s-plane by determining the vector lengths and angles at various frequencies ω along the ($s = +j\omega$)-axis. For example, considering the second-order factor with complex conjugate poles, we have

$$G(s) = \frac{1}{(s/\omega_n)^2 + 2\zeta s/\omega_n + 1} = \frac{\omega_n^2}{s^2 + 2\zeta\omega_n s + \omega_n^2}. \tag{7.38}$$

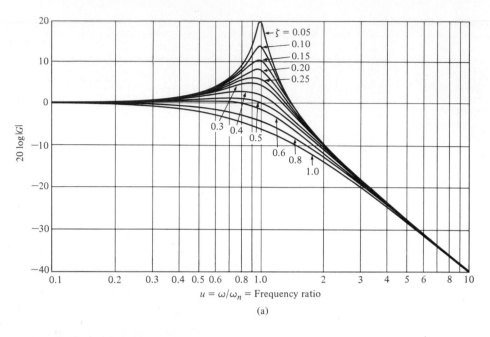

(a)

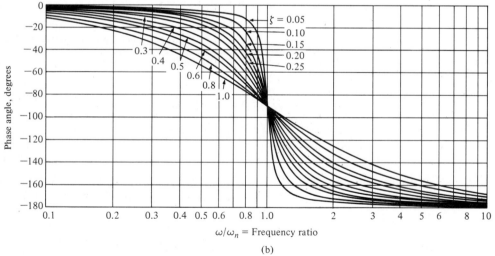

(b)

Fig. 7.9. Bode diagram for $G(j\omega) = [1 + (2\zeta/\omega_n)j\omega + (j\omega/\omega_n)^2]^{-1}$.

The poles lie on a circle of radius ω_n and are shown for a particular ζ in Fig. 7.11(a). The transfer function evaluated for real frequency $s = j\omega$ is written as

$$G(j\omega) = \left.\frac{\omega_n^2}{(s - s_1)(s - s_1^*)}\right|_{s=j\omega} = \frac{\omega_n^2}{(j\omega - s_1)(i\omega - s_1^*)}, \qquad (7.39)$$

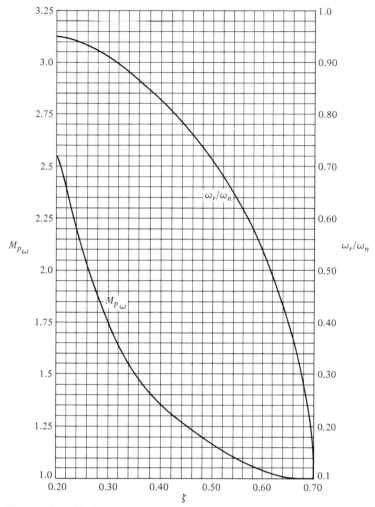

Fig. 7.10. The maximum of the frequency response, M_{p_ω}, and the resonant frequency, ω_r, versus ζ for a pair of complex conjugate poles.

where s_1 and s_1^* are the complex conjugate poles. The vectors $(j\omega - s_1)$ and $(j\omega - s_1^*)$ are the vectors from the poles to the frequency $j\omega$, as shown in Fig. 7.11(a). Then the magnitude and phase may be evaluated for various specific frequencies. The magnitude is

$$|G(\omega)| = \frac{\omega_n^2}{|j\omega - s_1| \, |j\omega - s_1^*|},$$ (7.40)

and the phase is

$$\phi(\omega) = -\underline{/(j\omega - s_1)} - \underline{/(j\omega - s_1^*)}.$$

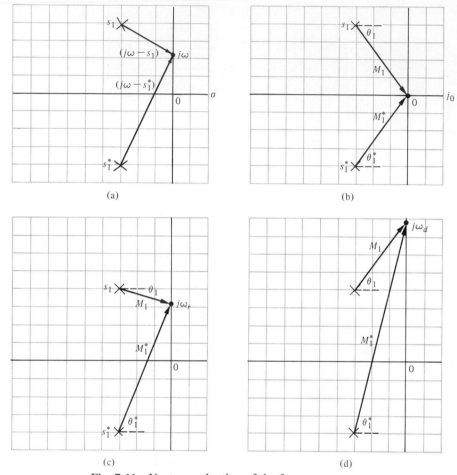

(a)

(b)

(c)

(d)

Fig. 7.11. Vector evaluation of the frequency response.

The magnitude and phase may be evaluated for three specific frequencies:

$$\omega = 0, \qquad \omega = \omega_r, \qquad \omega = \omega_d,$$

as shown in Figs. 7.11(b), 7.11(c), and 7.11(d), respectively. The magnitude and phase corresponding to these frequencies is shown in Fig. 7.12.

Example 7.4. As an example of the determination of the frequency response using the pole-zero diagram and the vectors to $j\omega$, consider the Twin-T network shown in Fig. 7.13 [6]. The transfer function of this network is

$$G(s) = \frac{E_{\text{out}}(s)}{E_{\text{in}}(s)} = \frac{(s\tau)^2 + 1}{(s\tau)^2 + 4s\tau + 1}, \tag{7.41}$$

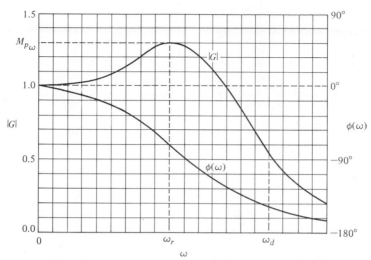

Fig. 7.12. Bode diagram for complex conjugate poles.

where $\tau = RC$. The zeros are at $\pm j1$ and the poles are at $-2 \pm \sqrt{3}$ in the $s\tau$-plane as is shown in Fig. 7.14(a). Clearly, at $\omega = 0$, we have $|G| = 1$ and $\phi(\omega) = 0°$. At $\omega = 1/\tau$, $|G| = 0$ and the phase angle of the vector from the zero at $s\tau = j1$ passes through a transition of 180°. When ω approaches ∞, $|G| = 1$ and $\phi(\omega) = 0$ again. Evaluating several intermediate frequencies, one can readily obtain the frequency response as shown in Fig. 7.14(b).

In the previous examples the poles and zeros of $G(s)$ have been restricted to the left-hand plane. However, a system may have zeros located in the right-hand s-plane and may still be stable. Transfer functions with zeros in the right-hand s-plane are classified as *nonminimum phase-shift* transfer functions. If the zeros of a transfer function are all reflected about the $j\omega$-axis, there is no change in the magnitude of the transfer function, and the only difference is in the phase-shift characteristics. If the phase characteristics of the two system functions are compared, it can be readily shown that the net phase shift over the frequency range from zero to infinity is less for the system with all its zeros in the left-hand s-plane. Thus, the transfer function $G_1(s)$, with all its zeros in the left-hand s-plane is called a *minimum phase* transfer function. The transfer function $G_2(s)$, with $|G_2(j\omega)| = |G_1(j\omega)|$ and all the

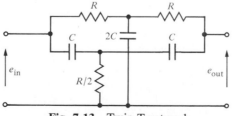

Fig. 7.13. Twin-T network.

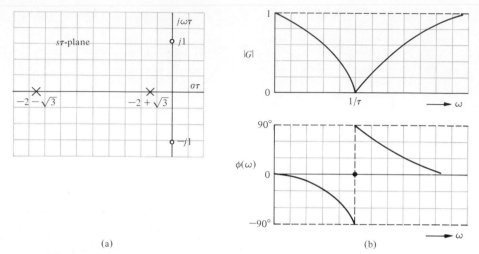

Fig. 7.14. Twin-T network. (a) Pole-zero pattern. (b) Frequency response.

zeros of $G_1(s)$ reflected about the $j\omega$-axis into the right-hand s-plane is called a nonminimum phase transfer function. Reflection of any zero or pair of zeros into the right-half plane results in a nonminimum phase transfer function.

The two pole-zero patterns shown in Fig. 7.15(a) and 7.15(b) have the same amplitude characteristics as can be deduced from the vector lengths. However, the phase characteristics are different for Fig. 7.15(a) and Fig. 7.15(b). The minimum phase characteristic of Fig. 7.15(a) and the nonminimum phase characteristic of Fig. 7.15(b) are shown in Fig. 7.16. Clearly, the phase shift of

$$G_1(s) = \frac{s + z}{s + p}$$

ranges over less than 80°, while the phase shift of

$$G_2(s) = \frac{s - z}{s + p}$$

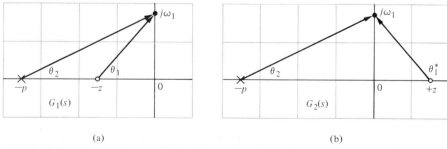

Fig. 7.15. Pole-zero patterns giving the same amplitude response and different phase characteristics.

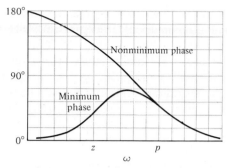

Fig. 7.16. The phase characteristics for the minimum phase and nonminimum phase transfer function.

ranges over 180°. The meaning of the term minimum phase is illustrated by Fig. 7.16. The range of phase shift of a minimum phase transfer function is the least possible or minimum corresponding to a given amplitude curve, while the range of the nonminimum phase curve is greater than the minimum possible for the given amplitude curve.

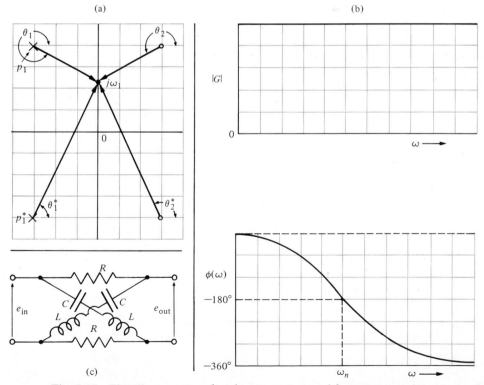

Fig. 7.17. The all-pass network pole-zero pattern and frequency response.

A particularly interesting nonminimum phase network is the *all-pass* network, which can be realized with a symmetrical lattice network [6]. A symmetrical pattern of poles and zeros is obtained as shown in Fig. 7.17(a). Again, the magnitude $|G|$ remains constant; in this case, it is equal to unity. However, the angle varies from $0°$ to $-360°$. Because $\theta_2 = 180° - \theta_1$ and $\theta_2^* = 180° - \theta_1^*$, the phase is given by $\phi(\omega) = -2(\theta_1 + \theta_1^*)$. The magnitude and phase characteristic of the all-pass network is shown in Fig. 7.17(b). A nonminimum phase lattice network is shown in Fig. 7.17(c).

7.3 EXAMPLE OF DRAWING THE BODE DIAGRAM

The Bode diagram of a transfer function $G(s)$, which contains several zeros and poles, is obtained by adding the plot due to each individual pole and zero. The simplicity of this method will be illustrated by considering a transfer function which possesses all the factors considered in the preceding section. The transfer function of interest is

$$G(j\omega) = \frac{5(1 + j0.1\omega)}{j\omega(1 + j0.5\omega)(1 + j0.6(\omega/50) + (j\omega/50)^2)}. \tag{7.42}$$

The factors, in order of their occurrence as frequency increases, are

1. a constant gain $K = 5$,
2. a pole at the origin,
3. a pole at $\omega = 2$,
4. a zero at $\omega = 10$,
5. a pair of complex poles at $\omega = \omega_n = 50$.

First, we plot the magnitude characteristic for each individual pole and zero factor and the constant gain.

1. The constant gain is $20 \log 5 = 14$ db as shown in Fig. 7.18.
2. The magnitude of the pole at the origin extends from zero frequency to infinite frequencies and has a slope of -20 db/decade intersecting the 0-db line at $\omega = 1$ as shown in Fig. 7.18.
3. The asymptotic approximation of the magnitude of the pole at $\omega = 2$ has a slope of -20 db/decade beyond the break frequency at $\omega = 2$. The asymptotic magnitude below the break frequency is 0 db as shown in Fig. 7.18.
4. The asymptotic magnitude for the zero at $\omega = +10$ has a slope of $+20$ db/decade beyond the break frequency at $\omega = 10$ as shown in Fig. 7.18.
5. The asymptotic approximation for the pair of complex poles at $\omega = \omega_n = 50$ has a slope of -40 db/decade due to the quadratic forms. The break frequency

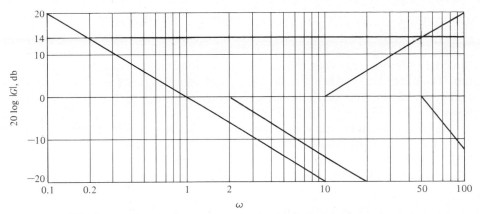

Fig. 7.18. Magnitude asymptotes of poles and zeros for example.

is $\omega = \omega_n = 50$ as shown in Fig. 7.18. This approximation must be corrected to the actual magnitude since the damping ratio is $\zeta = 0.3$ and the magnitude differs appreciably from the approximation, as shown in Fig. 7.19.

Therefore, the total asymptotic magnitude can be plotted by adding the asymptotes due to each factor, as shown by the solid line in Fig. 7.19. Examining the asymptotic curve of Fig. 7.19, one notes that the curve can be obtained directly by plotting each asymptote in order as frequency increases. Thus the slope is -20 db/decade due to $(j\omega)^{-1}$ intersecting 14 db at $\omega = 1$. Then at $\omega = 2$, the slope becomes -40 db/decade due to the pole at $\omega = 2$. The slope changes to -20 db/decade due

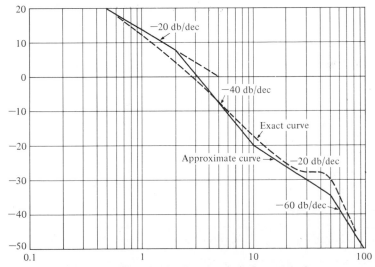

Fig. 7.19. Magnitude characteristic for example.

to the zero at $\omega = 10$. Finally, the slope becomes -60 db/decade at $\omega = 50$ due to the pair of complex poles at $\omega_n = 50$.

The exact magnitude curve is then obtained by utilizing Table 7.2, which provides the difference between the actual and asymptotic curves for a single pole or zero. The exact magnitude curve for the pair of complex poles is obtained by utilizing Fig. 7.9(a) for the quadratic factor. The exact magnitude curve for $G(j\omega)$ is shown by a dotted line in Fig. 7.19.

The phase characteristic may be obtained by adding the phase due to each individual factor. Usually, the linear approximation of the phase characteristic for a single pole or zero is suitable for the initial analysis or design attempt. Thus the individual phase characteristics for the poles and zeros are shown in Fig. 7.20.

1. The phase of the constant gain is, of course, zero degrees.

2. The phase of the pole at the origin is a constant -90 degrees.

3. The linear approximation of the phase characteristic for the pole at $\omega = 2$ is shown in Fig. 7.20, where the phase shift is $-45°$ at $\omega = 2$.

4. The linear approximation of the phase characteristic for the zero at $\omega = 10$ is also shown in Fig. 7.20, where the phase shift is $+45°$ at $\omega = 10$.

5. The actual phase characteristic for the pair of complex poles is obtained from Fig. 7.9 and is shown in Fig. 7.20.

Therefore, the total phase characteristic, $\phi(\omega)$, is obtained by adding the phase due to each factor as shown in Fig. 7.20. While this curve is an approximation, its

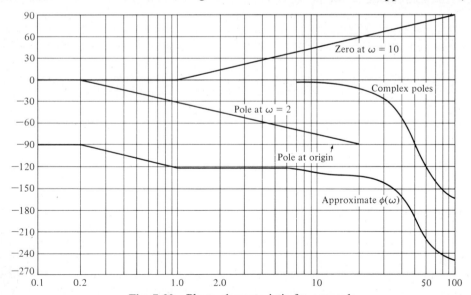

Fig. 7.20. Phase characteristic for example.

usefulness merits consideration as a first attempt to determine the phase characteristic. Thus, for example, a frequency of interest, as we shall note in the following
section, is that frequency for which $\phi(\omega) = -180°$. The approximate curve indicates
that a phase shift of $-180°$ occurs at $\omega = 46$. The actual phase shift at $\omega = 46$ can
be readily calculated as

$$\phi(\omega) = -90° - \tan^{-1} \omega\tau_1 + \tan^{-1} \omega\tau_2 - \tan^{-1} \frac{2\zeta u}{1 - u^2}, \qquad (7.43)$$

where

$$\tau_1 = 0.5, \qquad \tau_2 = 0.1, \qquad u = \omega/\omega_n = \omega/50.$$

Then, we find that

$$\phi(46) = -90° - \tan^{-1} 23 + \tan^{-1} 4.6 - \tan^{-1} 3.55$$
$$= -175°, \qquad (7.44)$$

and the approximate curve has an error of 5° at $\omega = 46$. However, once the approximate frequency of interest is ascertained from the approximate phase curve, the
accurate phase shift for the neighboring frequencies is readily determined by using
the exact phase shift relation (Eq. 7.43). This approach is usually preferable to the
calculation of the exact phase shift for all frequencies over several decades. In
summary, one may obtain approximate curves for the magnitude and phase shift of
a transfer function $G(j\omega)$ in order to determine the important frequency ranges.
Then, within the relatively small important frequency ranges, the exact magnitude
and phase shift can be readily evaluated by using the exact equations, such as Eq.
(7.43).

Computer programs written in Fortran IV for the calculation of the frequency
response of a transfer function are available [8, 15]. One program, called FRESP,
provides the values of the real and imaginary part of $G(j\omega)$ as well as the magnitude
and phase of $G(j\omega)$ for a selected frequency range. The program is also able to
provide a graphical output of the Bode diagram. College computer centers often
have this program or an equivalent program available to assist the student with the
calculation of the frequency response of a system.

7.4 FREQUENCY RESPONSE MEASUREMENTS

A sine wave may be used to measure the open-loop frequency response of a control
system. In practice a plot of amplitude versus frequency and phase versus frequency will be obtained [1, 2, 7]. From these two plots the open-loop transfer function $GH(j\omega)$ may be deduced. Similarly, the closed-loop frequency response of a
control system, $T(j\omega)$, may be obtained and the actual transfer function deduced.

A device called a wave analyzer can be used to measure the amplitude and
phase variations as the frequency of the input sine wave is altered. Also, a device

Fig. 7.21. (a) The Hewlett-Packard 5420A Digital Signal Analyzer serves as a transfer function analyzer. (Courtesy of Hewlett-Packard Corp.) (b—at right) The open-loop Bode plot for the speed control system of a tape drive for the HP5420A. (From T.L. Donahue and J.P. Oliverio, "Digital Signal Analyzer Applications," *Hewlett-Packard Journal*, October 1977; pp. 17–21.)

called a transfer funtion analyzer may be used to measure the open-loop and closed-loop transfer functions [9]. A transfer function analyzer is shown in Fig. 7.21(a) and an illustrative phase and magnitude plot provided by the analyzer is shown in Fig. 7.21(b).

As an example of determining the transfer function from the Bode plot, let us consider the plot shown in Fig. 7.22. The system is a stable circuit consisting of resistors and capacitors. Since the phase and magnitude decline as ω increases between 10 and 1000, and since the phase is $-45°$ and the gain -3 db at 370 rad/sec, we can then deduce that one factor is a pole near $\omega = 370$. Beyond 370 rad/sec

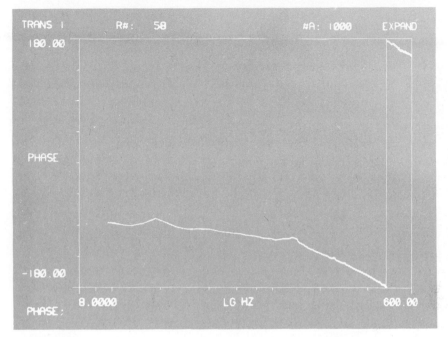

Fig. 7.21(b).

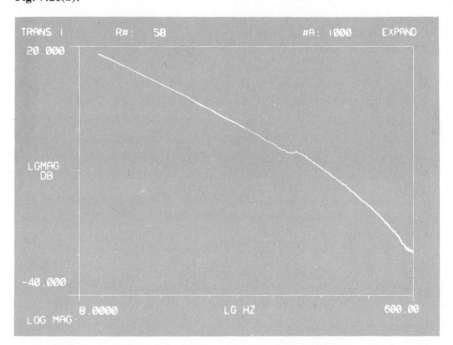

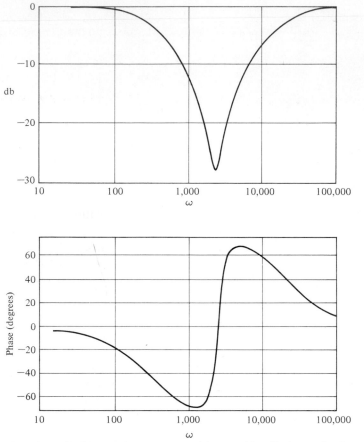

Fig. 7.22. A Bode diagram for a system with an unidentified transfer function.

the magnitude drops sharply at -40db/decade, indicating that another pole exists. However, the phase drops to $-55°$ at $\omega = 1100$ and then starts to rise again, eventually approaching zero degrees at large values of ω. Also, since the magnitude returns to 0db as ω exceeds 50,000, we determine that there are two zeros as well as two poles. We deduce the numerator is a quadratic factor with a small ζ yielding the sharp phase change. Therefore the transfer function is

$$T(s) = \frac{(s/\omega_n)^2 + (2\zeta/\omega_n)s + 1}{(\tau_1 s + 1)(\tau_2 s + 1)},$$

where we know $\tau_1 = 1/370$. Reviewing Fig. 7.9, we note that the phase passes through $+90°$ for a quadratic numerator at $\omega = \omega_n$. Since τ_1 yields $-90°$ by $\omega = 1000$, we deduce that $\omega_n = 2500$. Drawing the asymptotic curve for the pole $1/\tau_1$

and the numerator we estimate $\zeta = 0.15$. Finally the pole $p = 1/\tau_2$ yields a 45° phase shift from the asymptotic approximation so that $p = 20,000$. Therefore

$$T(s) = \frac{(s/2500)^2 + (0.3/2500)s + 1}{(s/370 + 1)(s/20,000 + 1)}.$$

This frequency response is actually obtained from a bridged-T network.

7.5 PERFORMANCE SPECIFICATIONS IN THE FREQUENCY DOMAIN

One must continually ask the question: how does the frequency response of a system relate to the expected transient response of the system? In other words, given a set of time-domain (transient performance) specifications, how does one specify the frequency response? For a simple second-order system we have already answered this question by considering the performance in terms of overshoot, settling time, and other performance criteria such as Integral Squared Error. For the second-order system which is shown in Fig. 7.23, the closed-loop transfer function is

$$T(s) = \frac{\omega_n^2}{s^2 + 2\zeta\omega_n s + \omega_n^2}. \tag{7.45}$$

The frequency response of this feedback system will appear as shown in Fig. 7.24. Since this is a second-order system, one relates the damping ratio of the system to the maximum magnitude M_{p_ω}. Furthermore, the resonant frequency ω_r and the -3-db bandwidth can be related to the speed of the transient response. Thus, as the bandwidth ω_B increases, the rise time of the step response of the system will decrease. Furthermore, the overshoot to a step input can be related to M_{p_ω} through the damping ratio ζ. The curves of Fig. 7.10 relate the resonance magnitude and frequency to the damping ratio of the second-order system. Then the step response overshoot may be estimated from Fig. 4.8 or may be calculated by utilizing Eq. (4.15). Thus, we find that as the resonant peak M_{p_ω} increases in magnitude, the overshoot to a step input increases. In general, the magnitude M_{p_ω} indicates the relative stability of a system.

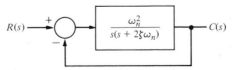

Fig. 7.23. A second-order closed-loop system.

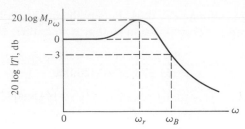

Fig. 7.24. Magnitude characteristic of the second-order system.

The bandwidth of a system, ω_B, as indicated on the frequency response can be approximately related to the natural frequency of the system. The response of the second-order system to a unit step input is

$$c(t) = 1 + e^{-\zeta\omega_n t} \cos(\omega_1 t + \theta). \tag{7.46}$$

The greater the magnitude of ω_n when ζ is constant, the more rapid the response approaches the desired steady-state value. Thus desirable frequency-domain specifications are

1. relatively small resonant magnitudes, $M_{p_\omega} < 1.5$, for example;

2. relatively large bandwidths so that the system time constant $\tau = 1/\zeta\omega_n$ is sufficiently small.

The usefulness of these frequency-response specifications and their relation to the actual transient performance depend upon the approximation of the system by a second-order pair of complex poles. This approximation was discussed in Section 6.3, and the second-order poles of $T(s)$ are called the *dominant roots*. Clearly, if the frequency response is dominated by a pair of complex poles the relationships between the frequency response and the time response discussed in this section will be valid. Fortunately, a large proportion of control systems satisfy this dominant second-order approximation in practice.

The steady-state error specification can also be related to the frequency response of a closed-loop system. As we found in Section 4.4, the steady-state error for a specific test input signal can be related to the gain and number of integrations (poles at the origin) of the open-loop transfer function. Therefore, for the system shown in Fig. 7.23, the steady-state error for a ramp input is specified in terms of K_v, the velocity constant. The steady-state error for the system is

$$\lim_{t\to\infty} e(t) = R/K_v, \tag{7.47}$$

where R = magnitude of the ramp input. The velocity constant for the closed-loop system of Fig. 7.23 is

$$K_v = \lim_{s\to 0} sG(s) = \lim_{s\to 0} s\left(\frac{\omega_n^2}{s(s + 2\zeta\omega_n)}\right) = \frac{\omega_n}{2\zeta}. \tag{7.48}$$

In Bode diagram form (in terms of time constants), the transfer function $G(s)$ is written as

$$G(s) = \frac{(\omega_n/2\zeta)}{s((1/2\zeta\omega_n)s + 1)} = \frac{K_v}{s(\tau s + 1)}, \tag{7.49}$$

and the gain constant is K_v for this type-one system. For example, reexamining the example of Section 7.3, we had a type-one system with an open-loop transfer function

$$G(j\omega) = \frac{5(1 + j\omega\tau_2)}{j\omega(1 + j\omega\tau_1)(1 + j0.6u - u^2)}, \tag{7.50}$$

where $u = \omega/\omega_n$. Therefore, in this case we have $K_v = 5$. In general, if the open-loop transfer function of a feedback system is written as

$$G(j\omega) = \frac{K\prod_{i=1}^{M} (1 + j\omega\tau_i)}{(j\omega)^N\prod_{k=1}^{Q} (1 + j\omega\tau_k)}, \tag{7.51}$$

then the system is type N and the gain K is the gain constant for the steady-state error. Thus, for a type-zero system which has an open-loop transfer function

$$G(j\omega) = \frac{K}{(1 + j\omega\tau_1)(1 + j\omega\tau_2)}, \tag{7.52}$$

$K = K_p$ (the position error constant) and appears as the low frequency gain on the Bode diagram.

Furthermore, the gain constant $K = K_v$ for the type-one system appears as the gain of the low frequency section of the magnitude characteristic. Considering only the pole and gain of the type-one sytem of Eq. (7.50), we have

$$G(j\omega) = \left(\frac{5}{j\omega}\right) = \left(\frac{K_v}{j\omega}\right), \quad \omega < 1/\tau_1, \tag{7.53}$$

and the K_v is equal to the frequency when this portion of the magnitude characteristic intersects the 0-db line. For example, the low frequency intersection of $(K_v/j\omega)$ in Fig. 7.19 is equal to $\omega = 5$ as we expect.

Therefore, the frequency response characteristics represent the performance of a system quite adequately, and with some experience they are quite useful for the analysis and design of feedback control systems.

7.6 LOG MAGNITUDE AND PHASE DIAGRAMS

There are several alternative methods of presenting the frequency response of a function $GH(j\omega)$. We have seen that suitable graphical presentations of the frequency response are (1) the polar plot and (2) the Bode diagram. An alternative approach to graphically portraying the frequency response is to plot the logarithmic magnitude in db versus the phase angle for a range of frequencies. Since this information is equivalent to that portrayed by the Bode diagram, it is normally easier to obtain the Bode diagram and transfer the information to the coordinates of the log magnitude versus phase diagram. Alternatively, one can construct templates for first and second-order factors and work directly on the log magnitude-phase diagram. The gain and phase of cascaded transfer functions may then be added vectorially directly on the diagram.

An illustration will best portray the use of the log-magnitude-phase diagram. The log-magnitude-phase diagram for a transfer function

$$GH_1(j\omega) = \frac{5}{j\omega(0.5j\omega + 1)((j\omega/6) + 1)} \tag{7.54}$$

is shown in Fig. 7.25. The numbers indicated along the curve are for values of frequency, ω.

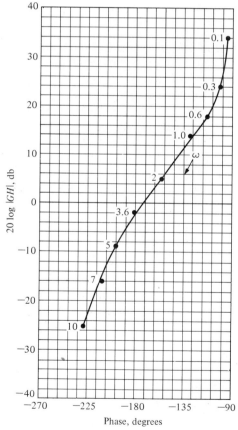

Fig. 7.25. Log magnitude-phase curve for $GH_1(j\omega)$.

The log-magnitude-phase curve for the transfer function

$$GH_2(j\omega) = \frac{5(0.1j\omega + 1)}{j\omega(0.5j\omega + 1)(1 + j0.6(\omega/50) + (j\omega/50)^2)} \tag{7.55}$$

considered in Section 7.3 is shown in Fig. 7.26. This curve is obtained most readily by utilizing the Bode diagrams of Figs. 7.19 and 7.20 to transfer the frequency response information to the log magnitude and phase coordinates. The shape of the locus of the frequency response on a log-magnitude-phase diagram is particularly important as the phase approaches 180° and the magnitude approaches 0 db. Clearly, the locus of Eq. (7.54) and Fig. 7.25 differs substantially from the locus of Eq. (7.55) and Fig. 7.26. Therefore, as the correlation between the shape of the locus and the transient response of a system is established, we will obtain another useful portrayal of the frequency response of a system. In the following chapter, we will establish a stability criterion in the frequency domain for which it will be useful

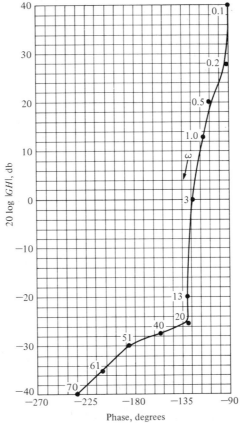

Fig. 7.26. Log magnitude-phase curve for $GH_2(j\omega)$.

to utilize the logarithmic magnitude-phase diagram to investigate the relative stability of closed-loop feedback control systems.

7.7 SUMMARY

In this chapter we have considered the representation of a feedback control system by its frequency response characteristics. The frequency response of a system was defined as the steady-state response of the system to a sinusoidal input signal. Several alternative forms of frequency response plots were considered. The polar plot of the frequency response of a system, $G(j\omega)$, was considered. Also, logarithmic plots, often called Bode plots, were considered and the value of the logarithmic measure was illustrated. The ease of obtaining a Bode plot for the various factors of $G(j\omega)$ was noted, and an example was considered in detail. The asymptotic approximation for drawing the Bode diagram simplifies the computation considerably. Several performance specifications in the frequency domain were discussed; among them were the maximum magnitude M_{p_ω} and the resonant frequency ω_r. The relationship between the Bode diagram plot and the system error constants (K_p and K_v) was noted. Finally, the log magnitude versus phase diagram was considered for graphically representing the frequency response of a system.

PROBLEMS

7.1. Sketch the polar plot of the frequency response for the following transfer functions:

(a) $GH(s) = \dfrac{1}{(1 + 0.5s)(1 + 2s)}$

(b) $GH(s) = \dfrac{(1 + 0.5s)}{s^2}$

(c) $GH(s) = \dfrac{(s + 3)}{(s^2 + 4s + 16)}$

(d) $GH(s) = \dfrac{30(s + 8)}{s(s + 2)(s + 4)}$

7.2. Draw the Bode diagram representation of the frequency response for the transfer functions given in Problem 7.1.

7.3. A rejection network which can be utilized instead of the Twin-T network of Example 7.4 is the Bridged-T network shown in Fig. P7.3. The transfer function of this network is

$$G(s) = \frac{s^2 + \omega_n^2}{s^2 + 2(\omega_n s/Q) + \omega_n^2}$$

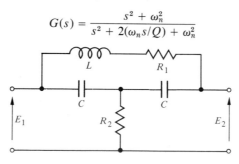

Figure P7.3

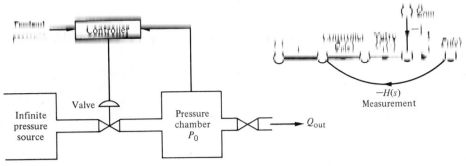

Figure P7.4

(Can you show this?), where $\omega_n^2 = 2/LC$ and $Q = \omega_n L/R_1$ and R_2 is adjusted so that $R_2 = (\omega_n L)^2/4R_1$ [6]. (a) Determine the pole-zero pattern and, utilizing the vector approach, evaluate the approximate frequency response. (b) Compare the frequency response of the Twin-T and Bridged-T networks when $Q = 10$.

7.4. A control system for controlling the pressure in a closed chamber is shown in Fig. P7.4. The transfer function for the measuring element is

$$H(s) = \frac{450}{s^2 + 90s + 900},$$

and the transfer function for the valve is

$$G_1(s) = \frac{1}{(0.1s + 1)(1/15s + 1)}.$$

The controller transfer function is

$$G_c(s) = (10 + 2s).$$

Obtain the frequency response characteristics for the loop transfer function

$$G_c(s)G_1(s)H(s) \cdot [1/s].$$

7.5. A simplified model of the human respiratory control system is shown in Fig. P7.5 [10]. This model provides an insight into the physiology of respiratory disorders. The respiratory control system controls the effective ventilation at the lungs, $V_a(s)$. The percentage of CO_2, P, in the circulated blood at the chemoreceptor (measuring element) is proportional to the

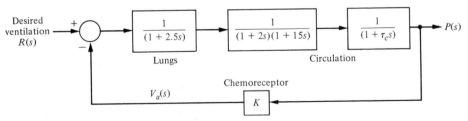

Figure P7.5

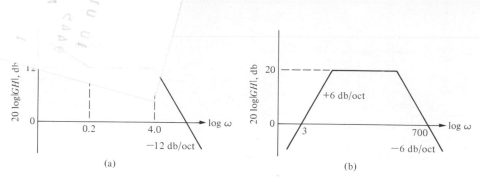

Figure P7.6

effective ventilation $V_a(s)$. The effective gain of the normal chemoreceptor will be assumed to be equal to 0.10. The delay time constant of the circulatory system τ_c is equal to 10 sec. Draw the approximate Bode diagram of the loop transfer function, $V_a(j\omega)/R(j\omega)$, for the model of the respiratory control system. Obtain an accurate phase and gain curve for the frequency range, where the phase is $-180°$.

7.6. The asymptotic logarithmic magnitude curves for two transfer functions are given in Fig. P7.6. Sketch the corresponding asymptotic phase shift curves for each system. Determine the transfer function for each system. Assume that the systems have minimum phase transfer functions.

7.7. A feedback control system is shown in Fig. P7.7. The specification for the closed-loop system requires that the overshoot to a step input is less than 30%. (a) Determine the corresponding specification in the frequency domain $M_{p\omega}$ for the closed-loop transfer function

$$\frac{C(j\omega)}{R(j\omega)} = T(j\omega).$$

(b) Determine the resonant frequency, ω_r. (c) Determine the bandwidth of the closed-loop system.

7.8. Driverless vehicles can be used in warehouses, airports, and many other applications. These vehicles follow a wire imbedded in the floor and adjust the steerable front wheels in order to maintain proper direction, as shown in Fig. P7.8(a) [14]. The sensing coils, mounted on the front wheel assembly, detect an error in the direction of travel and adjust the steering. The overall control system is shown in Fig. P7.8(b). The open-loop transfer function is

$$GH(s) = \frac{K}{s(s + \pi)^2} = \frac{K_v}{s(s/\pi + 1)^2}.$$

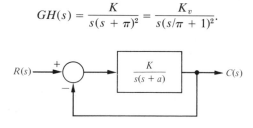

Figure P7.7

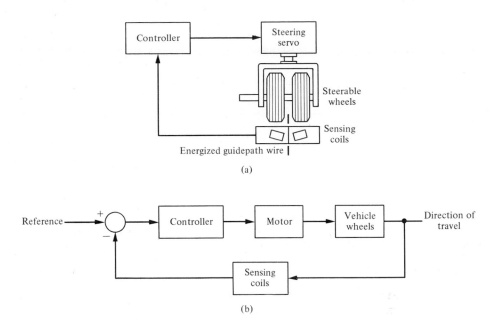

(a)

(b)

Figure P7.8

It is desired to have the bandwidth of the closed-loop system exceed 2π rad/sec. (a) Set $K_v = 2\pi$ and plot the Bode diagram. (b) Using the Bode diagram obtain the logarithmic magnitude versus phase angle curve.

7.9. Draw the logarithmic magnitude versus phase angle curves for the transfer functions a and b of Problem 7.1.

7.10. A linear actuator is utilized in the system shown in Fig. P7.10 to position a mass M. The actual position of the mass is measured by a slide wire resistor and thus $H(s) = 1.0$. The

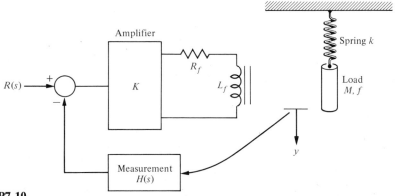

Figure P7.10

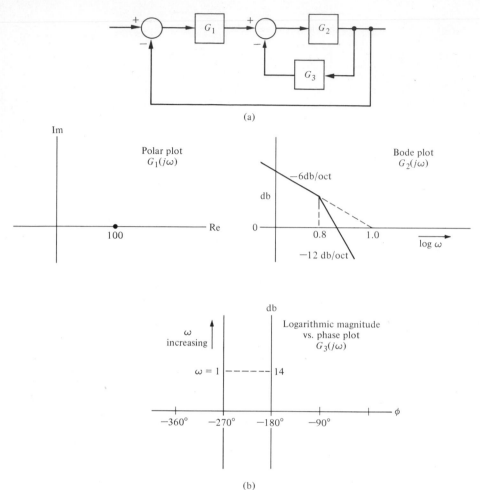

(a)

(b)

Figure P7.11

amplifier gain is to be selected so that the steady-state error of the system is less than 1% of the magnitude of the position reference $|R|$. The actuator has a field coil with a resistance $R_f = 0.1$ ohm and $L_f = 0.2$ henries. The mass of the load is 0.1 kg and the friction is 0.2 n-sec/m. The spring constant is equal to 0.4 n/m. (a) Determine the gain K necessary to maintain a steady-state error for a step input less than 1%. That is, K_p must be greater than 99. (b) Draw the Bode diagram of the loop transfer function $GH(s)$. (c) Draw the logarithmic magnitude versus phase angle curve for $GH(j\omega)$. (d) Draw the Bode diagram for the closed-loop transfer function $Y(j\omega)/R(j\omega)$. Determine $M_{p\omega}$, ω_r, and the bandwidth.

7.11. The block diagram of a feedback control system is shown in Fig. P7.11(a). The transfer functions of the blocks are represented by the frequency response curves shown in Fig. P7.11(b). (a) When G_3 is disconnected from the system, determine the damping ratio ζ of the

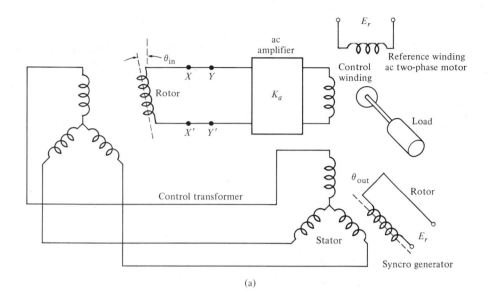

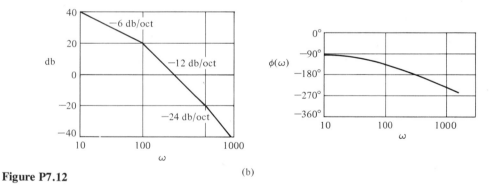

(b)

Figure P7.12

system. (b) Connect G_3 and determine the damping ratio ζ. Assume that the systems have minimum phase transfer functions.

7.12. A position control system may be constructed by using an ac motor and ac components as shown in Fig. P7.12. The syncro and control transformer may be considered to be a transformer with a rotating winding. The syncro position detector rotor turns with the load through an angle θ_0. The syncro motor is energized with an ac reference voltage, for example, 115 volts, 60 cps. The input signal or command is $R(s) = \theta_{in}(s)$ and is applied by turning the rotor of the control transformer. The ac two-phase motor operates as a result of the amplified error signal. The advantages of an ac control system are (1) freedom from dc drift effects and (2) the simplicity and accuracy of ac components. In order to measure the open-loop frequency response, one simply disconnects X from Y and X' from Y'. Then one applies a sinusoidal modulation signal generator to the $Y-Y'$ terminals and measures the response at $X-X'$. [The error $(\theta_0 - \theta_i)$ will be adjusted to zero before applying the ac generator.] The resulting frequency response of the loop, $GH(j\omega)$, is shown in Fig. P7.12(b). Determine the transfer function $GH(j\omega)$. Assume that the system has a minimum phase transfer function.

7.13. Automatic steering of a ship would be a particularly useful application of feedback control theory. In the case of heavily travelled seas, it is important to maintain the motion of the ship along an accurate track. An automatic system is able to maintain a much smaller error from the desired heading than is a helmsman who recorrects at infrequent intervals. A mathematical model of the steering system has been developed for a ship moving at a constant velocity and for small deviations from the desired track [11]. For a large tanker, the transfer function of the ship is

$$G(s) = \frac{E(s)}{\delta(s)} = \frac{0.164(s + 0.2)(-s + 0.32)}{s^2(s + 0.25(s - 0.009)}$$

where $E(s)$ is the Laplace transform of the deviation of the ship from the desired heading and $\delta(s)$ is the Laplace transform of the angle of deflection of the steering rudder.

Verify that the frequency response of the ship, $E(j\omega)/\delta(j\omega)$, is that shown in Fig. P7.13.

7.14. In order to determine the transfer function of a plant $G(s)$, the frequency response may be measured using a sinusoidal input. One system yields the following data [12]. Determine the transfer function $G(s)$.

ω, rad/sec	0.1	1	2	4	5	6.3	8	10	12.5	20	31
$|G(j\omega)|$	50	5.02	2.57	1.36	1.17	1.03	0.97	0.97	0.74	0.13	0.026
Phase, degrees	-90	-92.4	-96.2	-100	-104	-110	-120	-143	-169	-245	-258

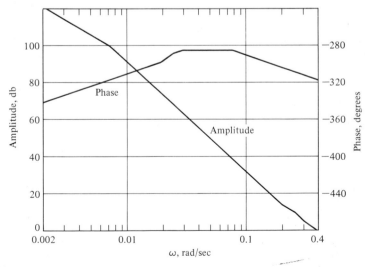

Figure P7.13

7.15. A system has a unity negative feedback structure with a plant transfer function

$$G(s) = \frac{2(s + 2)}{(s^2 - 1)}.$$

Construct the Bode diagram for this transfer function, $G(s)$.

REFERENCES

1. R. Keller, "Closed-loop Testing and Computer Analysis and Design of Control Systems," *Electronic Design,* November 22, 1978; pp. 132–138.

2. B. D. Pierce, "Measure Open-Loop Servo Response," *Electronic Design,* November 22, 1974; pp. 170–173.

3. T. N. Trick, *Introduction to Circuit Analysis*, Wiley, New York, 1977.

4. H. W. Bode, "Relations Between Attenuation and Phase in Feedback Amplifier Design," *Bell System Tech. J.,* July 1940; pp. 421–454; Also in *Automatic Control: Classical Linear Theory,* G. J. Thaler, ed., Dowden, Hutchinson and Ross, Inc., Stroudsburg, Pa., 1974; pp. 145–178.

5. M. D. Fagen, *A History of Engineering and Science in the Bell System,* Bell Telephone Laboratories, Inc., Murray Hill, N.J., 1978; Ch. 3.

6. A. Budak, *Circuit Theory Fundamentals and Applications,* Prentice-Hall, Englewood Cliffs, N.J., 1978.

7. R. R. Benedict, *Electronics for Scientists and Engineers,* 2nd ed., Prentice-Hall, Englewood Cliffs, N.J., 1976.

8. J. L. Melsa and S. K. Jones, *Computer Programs for Computational Assistance in the Study of Linear Control Theory,* McGraw-Hill, New York, 1973.

9. T. L. Donahue and J. P. Oliverio, "Digital Signal Analyzer Applications," *Hewlett-Packard Journal,* October 1977; pp. 17–21.

10. L. B. Korta, J. D. Horgan and R. L. Lange, "Stability Analysis of the Human Respiratory System," *Proceedings of the National Electronics Conference,* **21,** 1965; pp. 201–206.

11. F. A. Gross, "New Pathfinder Heavy Marine Radars," *Electronic Progress,* **20,** 2, 1978; pp. 2–9.

12. J. L. Melsa and D. G. Schultz, *Linear Control Systems,* McGraw-Hill, New York, 1969; Ch. 5.

13. J. Mazner, M. Nougaret, and C. Gaubert, "New Experiments for a Control System Laboratory," *IEEE Transactions on Education,* May 1976; pp. 59–62.

14. W. Hippler, "Programmed Driverless Vehicles Can Tie Your Operations Together," *Production Engineering,* December 1978; pp. 62–64.

15. O. Wing, *Circuit Theory With Computer Methods,* McGraw-Hill, New York, 1978; Chapter 8.

8 / Stability in the Frequency Domain

8.1 INTRODUCTION

For a control system, it is necessary to determine whether or not the system is stable. Furthermore, if the system is stable, it is often necessary to investigate the relative stability. In Chapter 5, we discussed the concept of stability and several methods of determining the absolute and relative stability of a system. The Routh-Hurwitz method discussed in Chapter 5 is a useful method for investigating the characteristic equation expressed in terms of the complex variable, $s = \sigma + j\omega$. Then in Chapter 6, we investigated the relative stability of a system utilizing the root-locus method, which is also in terms of the complex variable, s. In this chapter, we are concerned with investigating the stability of a system in the real frequency domain, that is, in terms of the frequency response discussed in Chapter 7.

The frequency response of a system represents the sinusoidal steady-state response of a system and provides sufficient information for the determination of the relative stability of the system. Since the frequency response of a system can readily be obtained experimentally by exciting the system with sinusoidal input signals, it can be utilized to investigate the relative stability of a system when the system parameter values have not been determined. Furthermore, a frequency-domain stability criterion would be useful for determining suitable approaches to altering a system in order to increase its relative stability.

A frequency domain stability criterion was developed by H. Nyquist in 1932 and remains a fundamental approach to the investigation of the stability of linear control systems [1, 2]. The *Nyquist stability criterion* is based upon a theorem in the theory of the function of a complex variable due to Cauchy. Cauchy's theorem is concerned with the *mapping* of *contours* in the complex s-plane, and fortunately the theorem can be understood without a formal proof, which uses complex variable theory.

In order to determine the relative stability of a closed-loop system, we must investigate the characteristic equation of the system:

$$F(s) = 1 + P(s) = 0. \tag{8.1}$$

For a multiloop system, we found in Section 2.7 that, in terms of signal-flow graphs, the characteristic equation is

$$F(s) = \Delta(s) = 1 - \Sigma L_n + \Sigma L_m L_q \cdots,$$

where $\Delta(s)$ is the graph determinant. Therefore we can represent the characteristic equation of single-loop or multiple-loop systems by Eq. (8.1), where $P(s)$ is a rational function of s. In order to assure stability, we must ascertain that all the zeros of $F(s)$ lie in the left-hand s-plane. Nyquist proposed, in order to investigate this, a mapping of the right-hand s-plane into the $F(s)$-plane. Therefore, to utilize and understand Nyquist's criterion, we shall first consider briefly the mapping of contours in the complex plane.

8.2 MAPPING OF CONTOURS IN THE s-PLANE

We are concerned with the mapping of contours in the s-plane by a function $F(s)$. Since s is a complex variable, $s = \sigma + j\omega$, the function $F(s)$ is itself complex and may be defined as $F(s) = u + jv$ and represented on a complex $F(s)$-plane with coordinates u and v. As an example, let us consider a function $F(s) = 2s + 1$ and a contour in the s-plane as shown in Fig. 8.1(a). The mapping of the s-plane unit square contour to the $F(s)$ plane is accomplished through the relation $F(s)$ so that

$$u + jv = F(s) = 2s + 1 = 2(\sigma + j\omega) + 1. \tag{8.2}$$

Therefore, in this case, we have

$$u = 2\sigma + 1 \tag{8.3}$$

and

$$v = 2\omega. \tag{8.4}$$

Thus the contour has been mapped by $F(s)$ into a contour of an identical form, a square, with the center shifted by one unit and the magnitude of a side multiplied by 2. This type of mapping, which retains the angles of the s-plane contour on the $F(s)$-plane, is called a *conformal mapping*. We also note that a closed contour in the s-plane results in a closed contour in the $F(s)$-plane.

The points A, B, C, and D, as shown in the s-plane contour, map into the points A, B, C, and D shown in the $F(s)$-plane. Furthermore, a direction of traversal of the s-plane contour may be indicated by the direction $ABCD$ and the arrows shown on the contour. Then a similar traversal occurs on the $F(s)$-plane contour as we pass $ABCD$ in order, as shown by the arrows. By convention, the area within a contour to the right of the traversal of the contour is considered to be the *area enclosed* by the contour. Therefore we will assume *clockwise traversal* of a contour to be positive and the area enclosed within the contour to be on the right. This

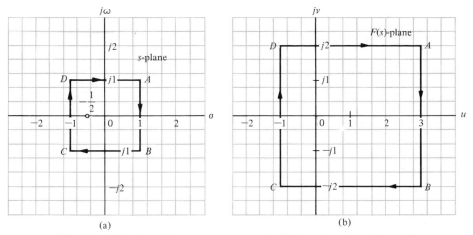

Fig. 8.1. Mapping of a square contour by $F(s) = 2s + 1 = 2(s + \frac{1}{2})$.

convention is opposite to that usually employed in complex variable theory but is equally applicable and is generally used in control system theory. The reader might consider the area on the right as he walks along the contour in a clockwise direction and call this rule "clockwise and eyes right."

Typically, we are concerned with an $F(s)$ which is a rational function of s. Therefore it will be worthwhile to consider another example of a mapping of a contour. Let us again consider the unit square contour for the function

$$F(s) = \frac{s}{s + 2}. \tag{8.5}$$

Several values of $F(s)$ as s traverses the square contour are given in Table 8.1, and the resulting contour in the $F(s)$-plane is shown in Fig. 8.2(b). The contour in the $F(s)$-plane encloses the origin of the $F(s)$-plane since the origin lies within the enclosed area of the contour in the $F(s)$-plane.

Cauchy's theorem is concerned with the mapping of a function $F(s)$, which has a finite number of poles and zeros within the contour so that we may express $F(s)$ as

$$F(s) = \frac{K\prod_{i=1}^{n} (s + s_i)}{\prod_{k=1}^{M} (s + s_k)}, \tag{8.6}$$

Table 8.1

$s = \sigma + j\omega$	Point A $1 + j1$	1	Point B $1 - j1$	$-j1$	Point C $-1 - j1$	-1	Point D $-1 + j1$	$j1$
$F(s) = u + jv$	$\dfrac{4 + 2j}{10}$	$\dfrac{1}{3}$	$\dfrac{4 - 2j}{10}$	$\dfrac{1 - 2j}{5}$	$-j$	-1	$+j$	$\dfrac{1 + 2j}{5}$

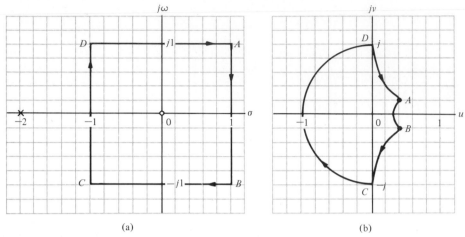

(a)   (b)

Fig. 8.2. Mapping for $F(s) = s/(s + 2)$.

where s_i are the zeros of the function $F(s)$ and s_k are the poles of $F(s)$. The function $F(s)$ is the characteristic equation so that

$$F(s) = 1 + P(s),\qquad(8.7)$$

where

$$P(s) = \frac{N(s)}{D(s)}.$$

Therefore we have

$$F(s) = 1 + \frac{N(s)}{D(s)} = \frac{D(s) + N(s)}{D(s)} = \frac{K \prod_{i=1}^{n}(s + s_i)}{\prod_{k=1}^{M}(s + s_k)},\qquad(8.8)$$

and the poles of $P(s)$ are the poles of $F(s)$. However, it is the zeros of $F(s)$ that are the characteristic roots of the system and indicate the response of the system. This is clear if we recall that the output of the system is

$$C(s) = T(s)R(s) = \frac{\Sigma P_k \Delta_k}{\Delta(s)} R(s) = \frac{\Sigma P_k \Delta_k}{F(s)} R(s),\qquad(8.9)$$

where P_k and Δ_k are the path factors and cofactors as defined in Section 2.7.

Now, reexamining the example when $F(s) = 2(s + \frac{1}{2})$, we have one zero of $F(s)$ at $s = -\frac{1}{2}$ as shown in Fig. 8.1. The contour which we chose (i.e., the unit square) enclosed and encircled once the zero within the area of the contour. Similarly, for the function $F(s) = s/(s + 2)$, the unit square encircled the zero at the origin but did not encircle the pole at $s = -2$. The encirclement of the poles and zeros of $F(s)$ can be related to the encirclement of the origin in the $F(s)$ plane by a theorem of Cauchy, commonly known as the *principle of the argument,* which states [3, 4]:

> If a contour Γ_s in the s-plane encircles Z zeros and P poles of $F(s)$ and does not pass through any poles or zeros of $F(s)$ as the traversal is in the clockwise direction along the contour, the corresponding contour Γ_F in the $F(s)$ plane encircles the origin of the $F(s)$ plane $N = Z - P$ times in the clockwise direction.

Thus for the examples shown in Figs. 8.1 and 8.2, the contour in the $F(s)$-plane encircles the origin once, since $N = Z - P = 1$, as we expect. As another example, consider the function $F(s) = s/(s + \frac{1}{2})$. For the unit square contour shown in Fig. 8.3(a), the resulting contour in the $F(s)$ plane is shown in Fig. 8.3(b). In this case, $N = Z - P = 0$ as is the case in Fig. 8.3(b), since the contour Γ_F does not encircle the origin.

Cauchy's theorem can be best comprehended by considering $F(s)$ in terms of the angle due to each pole and zero as the contour Γ_s is traversed in a clockwise direction. Thus let us consider the function

$$F(s) = \frac{(s + z_1)(s + z_2)}{(s + p_1)(s + p_2)}, \tag{8.10}$$

where z_i is a zero of $F(s)$ and p_k is a pole of $F(s)$. Equation (8.10) can be rewritten as

$$
\begin{aligned}
F(s) &= |F(s)| \underline{/F(s)} \\
&= \frac{|s + z_1| \, |s + z_2|}{|s + p_1| \, |s + p_2|} (\underline{/s + z_1} + \underline{/s + z_2} - \underline{/s + p_1} - \underline{/s + p_2}) \\
&= |F(s)| (\phi_{z_1} + \phi_{z_2} - \phi_{p_1} - \phi_{p_2}).
\end{aligned} \tag{8.11}
$$

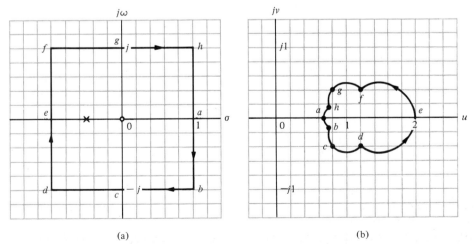

(a)   (b)

Fig. 8.3. Mapping for $F(s) = s/(s + \frac{1}{2})$.

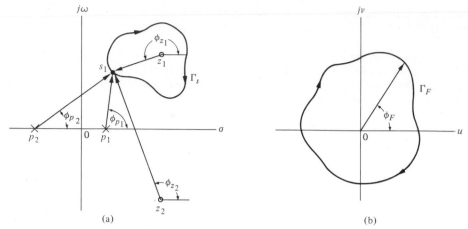

Fig. 8.4. Evaluation of the net angle of Γ_F.

Now, considering the vectors as shown for a specific contour Γ_s (Fig. 8.4a), one can determine the angles as s traverses the contour. Clearly, the net angle change, as s traverses along Γ_s a full rotation of 360° for ϕ_{p_1}, ϕ_{p_2} and ϕ_{z_2}, is zero degrees. However, for ϕ_{z_1} as s traverses 360° around Γ_s, the angle ϕ_{z_1} traverses a full 360° clockwise. Thus, as Γ_s is completely traversed, the net angle of $F(s)$ is equal to 360° since only one zero is enclosed. If Z zeros were enclosed within Γ_s, then the net angle would be equal to $\phi_Z = 2\pi(Z)$ rad. Following this reasoning, if Z zeros and P poles are encircled as Γ_s is traversed, then $2\pi(Z) - 2\pi(P)$ is the net resultant angle of $F(s)$. Thus the net angle of Γ_F of the contour in the $F(s)$-plane, ϕ_F, is simply

$$\phi_F = \phi_Z - \phi_P$$

or

$$2\pi N = 2\pi Z - 2\pi p, \qquad (8.12)$$

and the net number of encirclements of the origin of the $F(s)$-plane is $N = Z - P$. Thus, for the contour shown in Fig. 8.4(a) which encircles one zero, the contour Γ_F shown in Fig. 8.4(b) encircles the origin once in the clockwise direction.

As an example of the use of Cauchy's theorem, consider the pole-zero pattern shown in Fig. 8.5(a) with the contour Γ_s to be considered. The contour encloses and encircles three zeros and one pole. Therefore we obtain

$$N = 3 - 1 = +2,$$

and Γ_F completes two clockwise encirclements of the origin in the $F(s)$-plane as shown in Fig. 8.5(b).

For the pole and zero pattern shown and the contour Γ_s as shown in Fig. 8.6(a),

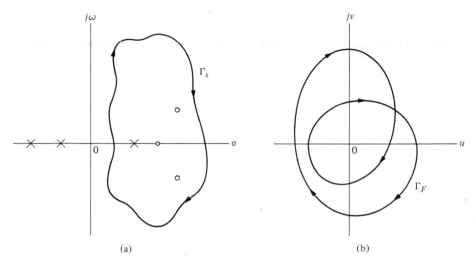

(a)   (b)

Fig. 8.5. Example of Cauchy's theorem.

one pole is encircled and no zeros are encircled. Therefore we have

$$N = Z - P = -1,$$

and we expect one encirclement of the origin by the contour Γ_F in the $F(s)$-plane. However, since the sign of N is negative, we find that the encirclement moves in the counterclockwise direction as shown in Fig. 8.6(b).

Now that we have developed and illustrated the concept of mapping of contours through a function $F(s)$, we are ready to consider the stability criterion proposed by Nyquist.

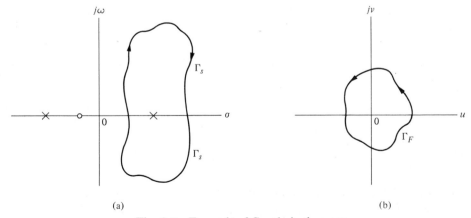

(a)   (b)

Fig. 8.6. Example of Cauchy's theorem.

8.3 THE NYQUIST CRITERION

In order to investigate the stability of a control system, we consider the characteristic equation, which is $F(s) = 0$, so that

$$F(s) = 1 + P(s) = \frac{K\prod_{i=1}^{n}(s + s_i)}{\prod_{k=1}^{M}(s + s_k)} = 0. \tag{8.13}$$

For a system to be stable, all the zeros of $F(s)$ must lie in the left-hand s-plane. Thus we find that the roots of a stable system [the zeros of $F(s)$] must lie to the left of the $j\omega$-axis in the s-plane. Therefore we chose a contour Γ_s in the s-plane which encloses the entire right-hand s-plane, and we determine whether any zeros of $F(s)$ lie within Γ_s by utilizing Cauchy's theorem. That is, we plot Γ_F in the $F(s)$-plane and determine the number of encirclements of the origin N. Then the number of zeros of $F(s)$ within the Γ_s contour [and therefore unstable zeros of $F(s)$] is

$$Z = N + P. \tag{8.14}$$

Thus if $P = 0$, as is usually the case, we find that the number of unstable roots of the system is equal to N, the number of encirclements of the origin of the $F(s)$ plane.

The Nyquist contour which encloses the entire right-hand s-plane is shown in Fig. 8.7. The contour Γ_s passes along the $j\omega$-axis from $-j\infty$ to $+j\infty$ and this part of the contour provides the familiar $F(j\omega)$. The contour is completed by a semicircular path of radius r where r approaches infinity.

Now, the Nyquist criterion is concerned with the mapping of the characteristic equation

$$F(s) = 1 + P(s) \tag{8.15}$$

and the number of encirclements of the origin of the $F(s)$-plane. Alternatively, we may define the function $F'(s)$ so that

$$F'(s) = F(s) - 1 = P(s). \tag{8.16}$$

The change of functions represented by Eq. (8.16) is very convenient since $P(s)$ is typically available in factored form, while $1 + P(s)$ is not. Then the mapping of Γ_s in the s-plane will be through the function $F'(s) = P(s)$ into the $P(s)$-plane. In this case the number of clockwise encirclements of the origin of the $F(s)$-plane becomes the number of clockwise encirclements of the -1 point in the $F'(s) = P(s)$ plane since $F'(s) = F(s) - 1$. Therefore the *Nyquist stability criterion* may be stated as follows:

A feedback system is stable if and only if the contour Γ_p in the $P(s)$-plane does not encircle the $(-1, 0)$ point when the number of poles of $P(s)$ in the right-hand s-plane is zero ($P = 0$).

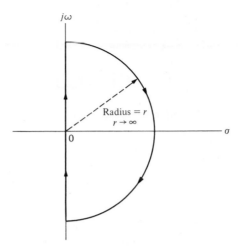

Fig. 8.7. The Nyquist contour.

When the number of poles of $P(s)$ in the right-hand s-plane is other than zero, the Nyquist criterion is:

> A feedback control system is stable if and only if, for the contour Γ_p, the number of counterclockwise encirclements of the $(-1, 0)$ point is equal to the number of poles of $P(s)$ with positive real parts.

The basis for the two statements is the fact that for the $F'(s) = P(s)$ mapping, the number of roots (or zeros) of $1 + P(s)$ in the right-hand s-plane is represented by the expression

$$Z = N + P.$$

Clearly, if the number of poles of $P(s)$ in the right-hand s-plane is zero ($P = 0$), we require for a stable system that $N = 0$ and the contour Γ_p must not encircle the -1 point. Also, if P is other than zero and we require for a stable system that $Z = 0$, then we must have $N = -P$, or P counterclockwise encirclements.

It is best to illustrate the use of the Nyquist criterion by completing several examples.

Example 8.1. A single-loop control system is shown in Fig. 8.8, where

$$GH(s) = \frac{K}{s(\tau s + 1)}. \tag{8.17}$$

In this single-loop case, $P(s) = GH(s)$ and we determine the contour $\Gamma_p = \Gamma_{GH}$ in the $GH(s)$-plane. The contour Γ_s in the s-plane is shown in Fig. 8.9(a), where an infinitesimal detour around the pole at the origin is effected by a small semicircle of

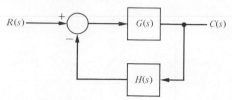

Fig. 8.8. Single-loop feedback control system.

radius ϵ, where $\epsilon \rightarrow 0$. This detour is a consequence of the condition of Cauchy's theorem which requires that the contour cannot pass through the pole at the origin. A sketch of the contour Γ_{GH} is shown in Fig. 8.9(b). Clearly, the portion of the contour Γ_{GH} from $\omega = 0^+$ to $\omega = +\infty$ is simply $GH(j\omega)$, the real frequency polar plot. Let us consider each portion of the Nyquist contour Γ_s in detail and determine the corresponding portions of the $GH(s)$-plane contour Γ_{GH}.

(a) *The origin of the s-plane.* The small semicircular detour around the pole at the origin can be represented by setting $s = \epsilon e^{j\phi}$ and allowing ϕ to vary from $-90°$ at $\omega = 0^-$ to $+90°$ at $\omega = 0^+$. Since ϵ approaches zero, the mapping $GH(s)$ is

$$\lim_{\epsilon \to 0} GH(s) = \lim_{\epsilon \to 0} \left(\frac{K}{\epsilon e^{j\phi}} \right) = \lim_{\epsilon \to 0} \left(\frac{K}{\epsilon} \right) e^{-j\phi}. \tag{8.18}$$

Therefore the angle of the contour in the $GH(s)$-plane changes from $90°$ at $\omega = 0^-$ to $-90°$ at $\omega = 0^+$, passing through $0°$ at $\omega = 0$. The radius of the contour in the $GH(s)$-plane for this portion of the contour is infinite, and this portion of the contour is shown in Fig. 8.9(b).

(b) *The portion from* $\omega = 0^+$ *to* $\omega = +\infty$. The portion of the contour Γ_s from $\omega = 0^+$ to $\omega = +\infty$ is mapped by the function $GH(s)$ as the real frequency polar plot

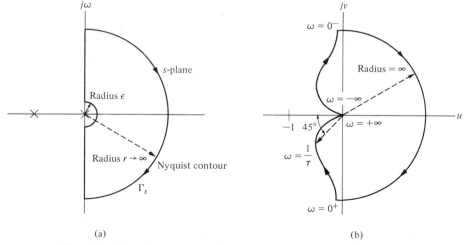

(a)   (b)

Fig. 8.9. Nyquist contour and mapping for $GH(s) = K/s(\tau s + 1)$.

since $s = j\omega$ and

$$GH(s)|_{s=j\omega} = GH(j\omega) \qquad (8.19)$$

for this part of the contour. This results in the real frequency polar plot shown in Fig. 8.9(b). When ω approaches $+\infty$, we have

$$\lim_{\omega \to +\infty} GH(j\omega) = \lim_{\omega \to +\infty} \frac{K}{+j\omega(j\omega\tau + 1)}$$

$$= \lim_{\omega \to \infty} \left| \frac{K}{\tau\omega^2} \right| \underline{/-(\pi/2) - \tan^{-1} \omega\tau}. \qquad (8.20)$$

Therefore the magnitude approaches zero at an angle of $-180°$.

(c) *The portion from* $\omega = +\infty$ *to* $\omega = -\infty$. The portion of Γ_s from $\omega = +\infty$ to $\omega = -\infty$ is mapped into the point zero at the origin of the $GH(s)$-plane by the function $GH(s)$. The mapping is represented by

$$\lim_{r \to \infty} GH(s)|_{s=re^{j\phi}} = \lim_{r \to \infty} \left| \frac{K}{r^2} \right| e^{-2j\phi} \qquad (8.21)$$

as ϕ changes from $\phi = +90°$ at $\omega = +\infty$ to $\phi = -90°$ at $\omega -\infty$. Thus the contour moves from an angle of $-180°$ at $\omega = +\infty$ to an angle of $+180°$ at $\omega = -\infty$. The magnitude of the $GH(s)$ contour when r is infinite is always zero or a constant.

(d) *The portion from* $\omega = -\infty$ *to* $\omega = 0^-$. The portion of the contour Γ_s from $\omega = -\infty$ to $\omega = 0^-$ is mapped by the function $GH(s)$ as

$$GH(s)|_{s=-j\omega} = GH(-j\omega). \qquad (8.22)$$

Thus we obtain the complex conjugate of $GH(j\omega)$, and the plot for the portion of the polar plot from $\omega = -\infty$ to $\omega = 0^-$ is symmetrical to the polar plot from $\omega = +\infty$ to $\omega = 0^+$. This symmetrical polar plot is shown on the $GH(s)$-plane in Fig. 8.9(b).

Now, in order to investigate the stability of this second-order system, we first note that the number of poles, P, within the right-hand s-plane is zero. Therefore, for this sytem to be stable, we require $N = Z = 0$, and the contour Γ_{GH} must not encircle the -1 point in the GH-plane. Examining Fig. 8.9(b), we find that irrespective of the value of the gain K and the time constant τ, the contour does not encircle the -1 point, and the system is always stable. As in Chapter 6, we are considering positive values of gain K. If negative values of gain are to be considered, one should use $-K$, where $K \geqslant 0$.

We may draw two general conclusions from this example as follows:

(1) The plot of the contour Γ_{GH} for the range $-\infty < \omega < 0^-$ will be the complex conjugate of the plot for the range $0^+ < \omega < +\infty$ and the polar plot of $GH(s)$ will be symmetrical in the $GH(s)$-plane about the u-axis. Therefore *it is sufficient to construct the contour Γ_{GH} for the frequency range $0^+ < \omega < +\infty$ in order to investigate the stability.*

(2) The magnitude of $GH(s)$ as $s = re^{j\phi}$ and $r \to \infty$ will normally approach zero or a constant.

Example 8.2. Let us again consider the single-loop system shown in Fig. 8.8 when

$$GH(s) = \frac{K}{s(\tau_1 s + 1)(\tau_2 s + 1)}. \tag{8.23}$$

The Nyquist contour Γ_s is shown in Fig. 8.9(a). Again this mapping is symmetrical for $GH(j\omega)$ and $GH(-j\omega)$ so that it is sufficient to investigate the $GH(j\omega)$-locus. The origin of the s-plane maps into a semicircle of infinite radius as in the last example. Also, the semicircle $re^{j\phi}$ in the s-plane maps into the point $GH(s) = 0$ as we expect. Therefore, in order to investigate the stability of the system, it is sufficient to plot the portion of the contour Γ_{GH} which is the real frequency polar plot $GH(j\omega)$ for $0^+ < \omega < +\infty$. Therefore, when $s = +j\omega$, we have

$$
\begin{aligned}
GH(j\omega) &= \frac{K}{j\omega(j\omega\tau_1 + 1)(j\omega\tau_2 + 1)} \\
&= \frac{-K(\tau_1 + \tau_2) - jK(1/\omega)(1 - \omega^2\tau_1\tau_2)}{1 + \omega^2(\tau_1^2 + \tau_2^2) + \omega^4\tau_1^2\tau_2^2} \\
&= \frac{K}{[\omega^4(\tau_1 + \tau_2)^2 + \omega^2(1 - \omega^2\tau_1\tau_2)^2]^{1/2}} \\
&\quad \times \underline{/-\tan^{-1}\omega\tau_1 - \tan^{-1}\omega\tau_2 - (\pi/2)}.
\end{aligned}
\tag{8.24}
$$

When $\omega = 0^+$, the magnitude of the locus is infinite at an angle of $-90°$ in the $GH(s)$-plane. When ω approaches $+\infty$, we have

$$
\begin{aligned}
\lim_{\omega \to \infty} GH(j\omega) &= \lim_{\omega \to \infty} \left|\frac{1}{\omega^3}\right| \underline{/-(\pi/2) - \tan^{-1}\omega\tau_1 - \tan^{-1}\omega\tau_2} \\
&= \left(\lim_{\omega \to \infty} \left|\frac{1}{\omega^3}\right|\right) \underline{/-(3\pi/2)}.
\end{aligned}
\tag{8.25}
$$

Therefore $GH(j\omega)$ approaches a magnitude of zero at an angle of $-270°$. In order for the locus to approach at an angle of $-270°$, the locus must cross the u-axis in the $GH(s)$-plane as shown in Fig. 8.10. Thus it is possible to encircle the -1 point as is shown in Fig. 8.10. The number of encirclements, when the -1 point lies within the locus as shown in Fig. 8.10, is equal to two and the system is unstable with two roots in the right-hand s-plane. The point where the $GH(s)$-locus intersects the real axis can be found by setting the imaginary part of $GH(j\omega) = u + jv$ equal to zero. Then we have from Eq. (8.24)

$$v = \frac{-K(1/\omega)(1 - \omega^2\tau_1\tau_2)}{1 + \omega^2(\tau_1^2 + \tau_2^2) + \omega^4\tau_1^2\tau_2^2} = 0. \tag{8.26}$$

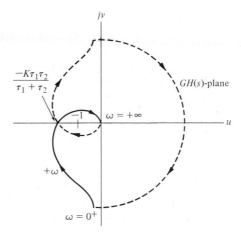

Fig. 8.10. Nyquist diagram for $GH(s) = K/s(\tau_1 s + 1)(\tau_2 s + 1)$.

Thus, $v = 0$ when $1 - \omega^2 \tau_1 \tau_2 = 0$ or $\omega = 1/\sqrt{\tau_1 \tau_2}$. The magnitude of the real part, u, of $GH(j\omega)$ at this frequency is

$$
\begin{aligned}
u &= \left. \frac{-K(\tau_1 + \tau_2)}{1 + \omega^2(\tau_1^2 + \tau_2^2) + \omega^4 \tau_1^2 \tau_2^2} \right|_{\omega^2 = 1/\tau_1 \tau_2} \\
&= \frac{-K(\tau_1 + \tau_2)\tau_1 \tau_2}{\tau_1 \tau_2 + (\tau_1^2 + \tau_2^2) + \tau_1 \tau_2} = \frac{-K\tau_1 \tau_2}{\tau_1 + \tau_2}.
\end{aligned}
\tag{8.27}
$$

Therefore, the system is stable when

$$
\frac{-K\tau_1 \tau_2}{\tau_1 + \tau_2} > -1
$$

or

$$
K < \frac{\tau_1 + \tau_2}{\tau_1 \tau_2}.
\tag{8.28}
$$

Example 8.3. Again let us determine the stability of the single-loop system shown in Fig. 8.8 when

$$
GH(s) = \frac{K}{s^2(\tau s + 1)}.
\tag{8.29}
$$

The real frequency polar plot is obtained when $s = j\omega$, and we have

$$
\begin{aligned}
GH(j\omega) &= \frac{K}{-\omega^2(j\omega\tau + 1)} \\
&= \frac{K}{[\omega^4 + \tau^2 \omega^6]^{1/2}} \underline{/-\pi - \tan^{-1} \omega\tau}.
\end{aligned}
\tag{8.30}
$$

We note that the angle of $GH(j\omega)$ is always $-180°$ or greater, and the locus of $GH(j\omega)$ is above the u-axis for all values of ω. As ω approaches 0^+, we have

$$\lim_{\omega \to 0^+} GH(j\omega) = \left(\lim_{\omega \to 0^+} \left| \frac{K}{\omega^2} \right| \right) \underline{/-\pi}. \tag{8.31}$$

As ω approaches $+\infty$, we have

$$\lim_{\omega \to +\infty} GH(j\omega) = \left(\lim_{\omega \to +\infty} \frac{K}{\omega^3} \right) \underline{/-3\pi/2}. \tag{8.32}$$

At the small semicircular detour at the origin of the s-plane where $s = \epsilon e^{j\phi}$, we have

$$\lim_{\epsilon \to 0} GH(s) = \lim_{\epsilon \to 0} \frac{K}{\epsilon^2} e^{-2j\phi}, \tag{8.33}$$

where $-\pi/2 \le \phi \le \pi/2$. Thus the contour Γ_{GH} ranges from an angle of $+\pi$ at $\omega = 0^+$ to $-\pi$ at $\omega = 0^+$ and passes through a full circle of 2π rad as ω changes from $\omega = 0^-$ to $\omega = 0^+$. The complete contour plot of Γ_{GH} is shown in Fig. 8.11. Since the contour encircles the -1 point twice, there are two roots of the closed-loop system in the right-hand plane and the system, irrespective of the gain K, is unstable.

Example 8.4.   Let us consider the control system shown in Fig. 8.12 and determine the stability of the system. First, let us consider the system without derivative feedback so that $K_2 = 0$. Then we have the open-loop transfer function

$$GH(s) = \frac{K_1}{s(s-1)}. \tag{8.34}$$

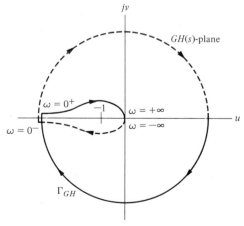

Fig. 8.11. Nyquist contour plot for $GH(s) = K/s^2(\tau s + 1)$.

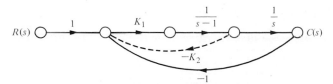

Fig. 8.12. Second-order feedback control system.

Thus the open-loop transfer function has one pole in the right-hand plane, and therefore $P = 1$. In order for this system to be stable, we require $N = -P = -1$, one counterclockwise encirclement of the -1 point. At the semicircular detour at the origin of the s-plane, we let $s = \epsilon e^{j\phi}$ when $-\pi/2 \leqslant \phi \leqslant \pi/2$. Then we have, when $s = \epsilon e^{j\phi}$,

$$\lim_{\epsilon \to 0} GH(s) = \lim_{\epsilon \to 0} \frac{K_1}{-\epsilon e^{j\phi}} = \left(\lim_{\epsilon \to 0} \left|\frac{K_1}{\epsilon}\right|\right) \underline{/-180° - \phi}. \tag{8.35}$$

Therefore, this portion of the contour Γ_{GH} is a semicircle of infinite magnitude in the left-hand GH-plane as shown in Fig. 8.13. When $s = j\omega$, we have

$$GH(j\omega) = \frac{K_1}{j\omega(j\omega - 1)} = \frac{K_1}{(\omega^2 + \omega^4)^{1/2}} \underline{/(-\pi/2) - \tan^{-1}(-\omega)}$$

$$= \frac{K_1}{(\omega^2 + \omega^4)^{1/2}} \underline{/(+\pi/2) + \tan^{-1}\omega}. \tag{8.36}$$

Finally, for the semicircle of radius r as r approaches infinity, we have

$$\lim_{r \to \infty} GH(s)\big|_{s=re^{j\phi}} = \left(\lim_{r \to \infty} \left|\frac{K_1}{r^2}\right|\right) e^{-2j\phi}, \tag{8.37}$$

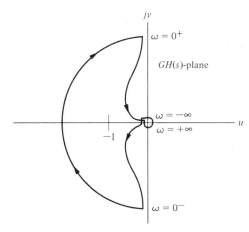

Fig. 8.13. Nyquist diagram for $GH(s) = K_1/s(s - 1)$.

Table 8.2

s	$j0^-$	$j0^+$	$j1$	$+j\infty$	$-j\infty$
$\|GH\|/K_1$	∞	∞	$1/\sqrt{2}$	0	0
$\underline{/GH}$	$-90°$	$+90°$	$+135°$	$+180°$	$-180°$

where ϕ varies from $\pi/2$ to $-\pi/2$ in a clockwise direction. Therefore the contour Γ_{GH}, at the origin of the GH-plane, varies 2π rad in a counterclockwise direction as shown in Fig. 8.13. Several important values of the $GH(s)$-locus are given in Table 8.2. The contour Γ_{GH} in the $GH(s)$-plane encircles the -1 point once in the clockwise direction and $N = +1$. Therefore

$$Z = N + P = 2, \tag{8.38}$$

and the system is unstable since two zeros of the characteristic equation, irrespective of the value of the gain K, lie in the right half of the s-plane.

Let us now reconsider the system when the derivative feedback is included in the system shown in Fig. 8.12. Then the open-loop transfer function is

$$GH(s) = \frac{K_1(1 + K_2 s)}{s(s - 1)}. \tag{8.39}$$

The portion of the contour Γ_{GH}, when $s = \epsilon e^{j\phi}$, is the same as the system without derivative feedback as is shown in Fig. 8.14. However, when $s = re^{j\phi}$ as r approaches infinity, we have

$$\lim_{r \to \infty} GH(s)\big|_{s=r\,e^{j\phi}} = \lim_{r \to \infty} \left|\frac{K_1 K_2}{r}\right| e^{-j\phi}, \tag{8.40}$$

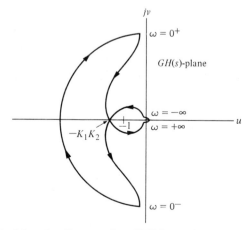

Fig. 8.14. Nyquist diagram for $GH(s) = K_1(1 + K_2 s)/s(s - 1)$.

and the Γ_{GH}-contour at the origin of the GH-plane varies π rad in a counterclockwise direction as shown in Fig. 8.14. The frequency locus $GH(j\omega)$ crosses the u-axis and is determined by considering the real frequency transfer function

$$
\begin{aligned}
GH(j\omega) &= \frac{K_1(1 + K_2j\omega)}{-\omega^2 - j\omega} \\
&= \frac{-K_1(\omega^2 + \omega^2 K_2) + j(\omega - K_2\omega^3)K_1}{\omega^2 + \omega^4}.
\end{aligned} \tag{8.41}
$$

The $GH(j\omega)$-locus intersects the u-axis at a point where the imaginary part of $GH(j\omega)$ is zero. Therefore,

$$
\omega - K_2\omega^3 = 0
$$

at this point, or $\omega^2 = 1/K_2$. The value of the real part of $GH(j\omega)$ at the intersection is then

$$
u\big|_{\omega^2 = 1/K_2} = \frac{-\omega^2 K_1(1 + K_2)}{\omega^2 + \omega^4}\bigg|_{\omega^2 = 1/K_2} = -K_1K_2. \tag{8.42}
$$

Therefore, when $-K_1K_2 < -1$ or $K_1K_2 > 1$, the contour Γ_{GH} encircles the -1 point once in a counterclockwise direction, and therefore $N = -1$. Then Z, the number of zeros of the system in the right-hand plane, is

$$
Z = N + P = -1 + 1 = 0. \tag{8.43}
$$

Thus the system is stable when $K_1K_2 > 1$. Often, it may be useful to utilize a computer or calculator program to calculate the Nyquist diagram [5].

8.4 RELATIVE STABILITY AND THE NYQUIST CRITERION

We discussed the relative stability of a system in terms of the s-plane in Section 5.3. For the s-plane, we defined the relative stability of a system as the property measured by the relative settling time of each root or pair of roots. We would like to determine a similar measure of relative stability useful for the frequency-response method. The Nyquist criterion provides us with suitable information concerning the absolute stability, and, furthermore, can be utilized to define and ascertain the relative stability of a system.

The Nyquist stability criterion is defined in terms of the $(-1, 0)$ point on the polar plot or the 0 db, 180° point on the Bode diagram or log magnitude-phase diagram. Clearly, the proximity of the $GH(j\omega)$-locus to this stability point is a measure of the relative stability of a system. The polar plot for $GH(j\omega)$ for several values of K and

$$
GH(j\omega) = \frac{K}{j\omega(j\omega\tau_1 + 1)(j\omega\tau_2 + 1)} \tag{8.44}
$$

is shown in Fig. 8.15. As K increases, the polar plot approaches the -1 point and eventually encircles the -1 point for a gain $K = K_3$. We determined in Section 8.3 that the locus intersects the u-axis at a point

$$u = \frac{-K\tau_1\tau_2}{\tau_1 + \tau_2}. \tag{8.45}$$

Therefore the system has roots on the $j\omega$-axis when

$$u = -1 \quad \text{or} \quad K = \left(\frac{\tau_1 + \tau_2}{\tau_1\tau_2}\right).$$

As K is decreased below this marginal value, the stability is increased and the margin between the gain $K = (\tau_1 + \tau_2)/\tau_1\tau_2$ and a gain $K = K_2$ is a measure of the relative stability. This measure of relative stability is called the *gain margin* and is defined as *the reciprocal of the gain* $|GH(j\omega)|$ *at the frequency at which the phase angle reaches 180°* (i.e., $v = 0$). The gain margin is a measure of the factor by which the system gain would have to be increased for the $GH(j\omega)$ locus to pass through the $u = -1$ point. Thus, for a gain $K = K_2$ in Fig. 8.15, the gain margin is equal to the reciprocal of $GH(j\omega)$ when $v = 0$. Since $\omega = 1/\sqrt{\tau_1\tau_2}$ when the phase shift is 180°, we have a gain margin equal to

$$\frac{1}{|GH(j\omega)|} = \left[\frac{K_2\tau_1\tau_2}{\tau_1 + \tau_2}\right]^{-1} = \frac{1}{d}. \tag{8.46}$$

The gain margin may be defined in terms of a logarithmic (decibel) measure as

$$20 \log\left(\frac{1}{d}\right) = -20 \log d \text{ db}. \tag{8.47}$$

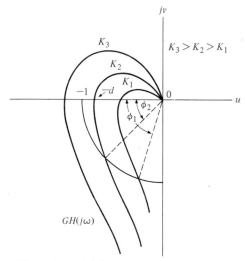

Fig. 8.15. The polar plot for $GH(j\omega)$ for three values of gain.

Therefore, for example, when $\tau_1 = \tau_2 = 1$, the system is stable when $K \leq 2$. Thus when $K = K_2 = 0.5$, the gain margin is equal to

$$\frac{1}{d} = \left[\frac{K_2 \tau_1 \tau_2}{\tau_1 + \tau_2} \right]^{-1} = 4, \tag{8.48}$$

or, in logarithmic measure,

$$20 \log 4 = 12 \text{ db.} \tag{8.49}$$

Therefore the gain margin indicates that the system gain may be increased by a factor of four (12 db) before the stability boundary is reached.

An alternative measure of relative stability can be defined in terms of the phase angle margin between a specific system and a system which is marginally stable. Several roots of the characteristic equation lie on the $j\omega$-axis when the $GH(j\omega)$-locus intersects the $u = -1$, $v = 0$ point in the GH-plane. Therefore, a measure of relative stability, *the phase margin,* is defined as *the phase angle through which the $GH(j\omega)$ locus must be rotated in order that the unity magnitude $|GH(j\omega)| = 1$ point passes through the $(-1, 0)$ point in the $GH(j\omega)$ plane.* This measure of relative stability is called the phase margin and is equal to the additional phase lag required before the system becomes unstable. This information can be determined from the Nyquist diagram as shown in Fig. 8.15. For a gain $K = K_2$, an additional phase angle, ϕ_2, may be added to the system before the system becomes unstable. Furthermore, for the gain K_1, the phase margin is equal to ϕ_1 as shown in Fig. 8.15.

The gain and phase margins are easily evaluated from the Bode diagram, and since it is preferable to draw the Bode diagram in contrast to the polar plot, it is worthwhile to illustrate the relative stability measures for the Bode diagram. The critical point for stability is $u = -1$, $v = 0$ in the $GH(j\omega)$ plane which is equivalent to a logarithmic magnitude of 0 db and a phase angle of 180° in the Bode diagram.

The gain margin and phase margin may be readily calculated by utilizing a computer program [6]. A computer program for accomplishing this calculation is given in Table 8.3 in the computer language BASIC [7]. This program can readily be converted to FORTRAN. The symbols used are: $W = \omega$; $G2 = |G(s)|^2$, $P = $ phase of $G(s)$; $PM = $ phase margin. The program is shown for the case where $GH(j\omega)$ is as given in Eq. (8.50). The calculations commence at $\omega = 0.1$ and increase by two percent at each iteration at line 50.

The Bode diagram of

$$GH(j\omega) = \frac{1}{j\omega(j\omega + 1)(0.2j\omega + 1)} \tag{8.50}$$

is shown in Fig. 8.16. The phase angle when the logarithmic magnitude is zero decibels is equal to 137°. Thus the phase margin is $180° - 137° = 43°$ as shown in Fig. 8.16. The logarithmic magnitude when the phase angle is $-180°$ is -15 db, and therefore the gain margin is equal to 15 db as shown in Fig. 8.16.

The frequency response of a system can be graphically portrayed on the logarithmic magnitude-phase angle diagram. For the log magnitude-phase diagram, the

Table 8.3. A Computer Program in BASIC Computer Language for Calculating the Gain Margin and Phase Margin for the Third Order System $GH(j\omega) = 1/j\omega(j\omega + 1)(0.2j\omega + 1)$

```
10  LET W = 0.1
20  GOSUB 100
30  IF G2 < =1 THEN 60
35  IF G2 < =100 THEN 50
40  LET W = 2*W
45  GO TO 20
50  LET W = 1.02*W
55  GO TO 20
60  IF P> =180 THEN 140
65  PRINT "UNITY GAIN", "W =" W, "P=" P
70  LET W=1.02*W
75  GOSUB 100
80  IF P> =180 THEN 90
85  GO TO 70
90  PRINT "W=" W, "GAIN MARGIN=" 4.343*LOG(1/G2)
95  GO TO 200
100 LET P = 57.3*(ATN(W) + ATN(0.2*W) + 1.571)
110 LET X = W*W
120 LET G2 = 1/((1 + X)*(1 + 0.04*X)*X)
130 RETURN
140 PRINT "W =" W, "SYSTEM UNSTABLE"
200 END
```

critical stability point is the 0 db, $-180°$ point, and the gain margin and phase margin may be easily determined and indicated on the diagram. The log magnitude-phase locus of

$$GH_1(j\omega) = \frac{1}{j\omega(j\omega + 1)(0.2j\omega + 1)} \tag{8.51}$$

is shown in Fig. 8.17. The indicated phase margin is $43°$ and the gain margin is 15 db. For comparison, the locus for

$$GH_2(j\omega) = \frac{1}{j\omega(j\omega + 1)^2} \tag{8.52}$$

is also shown in Fig. 8.17. The gain margin for GH_2 is equal to 5.7 db, and the phase margin for GH_2 is equal to $20°$. Clearly, the feedback system $GH_2(j\omega)$ is relatively less stable than the system $GH_1(j\omega)$. However, the question still remains: How much less stable is the system $GH_2(j\omega)$ in comparison to the system $GH_1(j\omega)$? In the following paragraph we shall answer this question for a second-order system, and the usefulness of the relation which we develop will depend upon the presence of dominant roots.

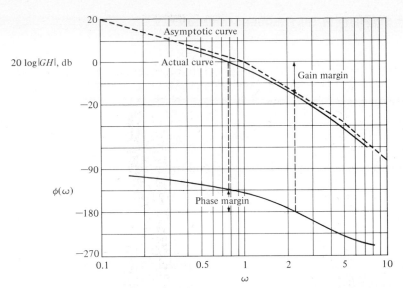

Fig. 8.16. Bode diagram for $GH_1(j\omega) = 1/j\omega(j\omega + 1)(0.2j\omega + 1)$.

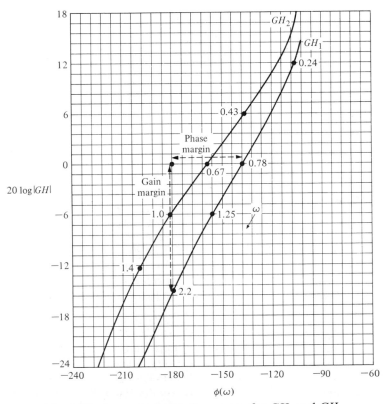

Fig. 8.17. Log magnitude-phase curve for GH_1 and GH_2.

Let us now determine the phase margin of a second-order system and relate the phase margin to the damping ratio ζ of an underdamped system. Consider the loop-transfer function

$$GH(j\omega) = \frac{\omega_n^2}{j\omega(j\omega + 2\zeta\omega_n)}. \tag{8.53}$$

The characteristic equation for this second-order system is

$$s^2 + 2\zeta\omega_n s + \omega_n^2 = 0.$$

Therefore the closed-loop roots are

$$s = -\zeta\omega_n \pm j\omega_n\sqrt{1 - \zeta^2}.$$

The magnitude of the frequency response is equal to 1 at a frequency ω_c and thus

$$\frac{\omega_n^2}{\omega_c(\omega_c^2 + 4\zeta^2\omega_n^2)^{1/2}} = 1. \tag{8.54}$$

Rearranging Eq. (8.54), we obtain

$$(\omega_c^2)^2 + 4\zeta^2\omega_n^2(\omega_c^2) - \omega_n^4 = 0. \tag{8.55}$$

Solving for ω_c, we find that

$$\frac{\omega_c^2}{\omega_n^2} = (4\zeta^4 + 1)^{1/2} - 2\zeta^2. \tag{8.56}$$

The phase margin for this system is

$$\begin{aligned}
\phi_{pm} &= 180° - 90° - \tan^{-1}\left(\frac{\omega_c}{2\zeta\omega_n}\right) \\
&= 90° - \tan^{-1}\left(\frac{1}{2\zeta}[(4\zeta^4 + 1)^{1/2} - 2\zeta^2]^{1/2}\right) \\
&= \tan^{-1}\left(2\zeta\left[\frac{1}{(4\zeta^4 + 1)^{1/2} - 2\zeta^2}\right]^{1/2}\right).
\end{aligned} \tag{8.57}$$

Equation (8.57) is the relationship between the damping ratio ζ and the phase margin ϕ_{pm} that provides a correlation between the frequency response and the time response. A plot of ζ versus ϕ_{pm} is shown in Fig. 8.18. The actual curve of ζ versus ϕ_{pm} may be approximated by the dotted line shown in Fig. 8.18. The slope of the linear approximation is equal to 0.01, and therefore an approximate linear relationship between the damping ratio and the phase margin is

$$\zeta = 0.01\phi_{pm}, \tag{8.58}$$

where the phase margin is measured in degrees. This approximation is reasonably accurate for $\zeta \leqslant 0.7$, and is a useful index for correlating the frequency response with the transient performance of a system. Equation (8.58) is a suitable approxi-

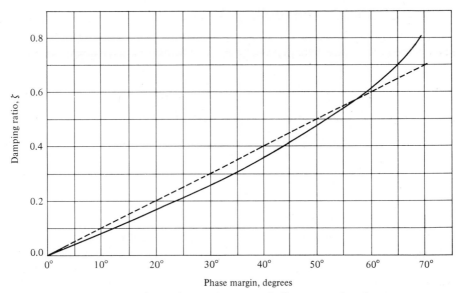

Fig. 8.18. Damping ratio vs. phase margin for a second-order system.

mation for a second-order system and may be used for higher-order systems if one can assume that the transient response of the system is primarily due to a pair of dominant underdamped roots. The approximation of a higher-order system by a dominant second-order system is a useful approximation indeed! While it must be used with care, control engineers find this approach to be a simple, yet fairly accurate, technique of setting the specifications of a control system.

Therefore, for the system with a loop-transfer function

$$GH(j\omega) = \frac{1}{j\omega(j\omega + 1)(0.2j\omega + 1)}, \qquad (8.59)$$

we found that the phase margin was 43° as shown in Fig. 8.16. Thus the damping ratio is approximately

$$\zeta \approx 0.01\phi_{pm} = 0.43. \qquad (8.60)$$

Then the peak response to a step input for this system is approximately

$$M_{pt} = 1.22 \qquad (8.61)$$

as obtained from Fig. 4.8 for $\zeta = 0.43$.

The phase margin of a system is a quite suitable frequency response measure for indicating the expected transient performance of a system. Another useful index of performance in the frequency domain is M_{p_ω}, the maximum magnitude of the closed-loop frequency response, and we shall now consider this practical index.

8.5 THE CLOSED-LOOP FREQUENCY RESPONSE

The transient performance of a feedback system can be estimated from the closed-loop frequency response. The open- and closed-loop frequency responses for a single-loop system are related as follows:

$$\frac{C(j\omega)}{R(j\omega)} = T(j\omega) = \frac{G(j\omega)}{1 + GH(j\omega)}. \tag{8.62}$$

The Nyquist criterion and the phase margin index are defined for the open-loop transfer function $GH(j\omega)$. However, as we found in Section 7.2, the maximum magnitude of the closed-loop frequency response can be related to the damping ratio of a second-order system as

$$M_{p_\omega} = |G(\omega_r)| = (2\zeta\sqrt{1 - \zeta^2})^{-1}, \qquad \zeta < 0.707. \tag{8.63}$$

This relation is graphically portrayed in Fig. 7.10. Since this relationship between the closed-loop frequency response and the transient response is a useful relationship, we would like to be able to determine M_{p_ω} from the plots completed for the investigation of the Nyquist criterion. That is, it is desirable to be able to obtain the closed-loop frequency response (Eq. 8.62) from the open-loop frequency response. Of course, one could determine the closed-loop roots of $1 + GH(s)$ and plot the closed-loop frequency response. However, once we have invested all the effort necessary to find the closed-loop roots of a characteristic equation, then a closed-loop frequency response is not necessary.

The relation between the closed-loop and open-loop frequency response is easily obtained by considering Eq. (8.62) when $H(j\omega) = 1$. If the system is not in fact a unity feedback system where $H(j\omega) = 1$, we will simply redefine the system output to be equal to the output of $H(j\omega)$. Then Eq. (8.62) becomes

$$T(j\omega) = M(\omega)e^{j\phi(\omega)} = \frac{G(j\omega)}{1 + G(j\omega)}. \tag{8.64}$$

The relationship between $T(j\omega)$ and $G(j\omega)$ is readily obtained in terms of complex variables utilizing the $G(j\omega)$-plane. The coordinates of the $G(j\omega)$-plane are u and v, and we have

$$G(j\omega) = u + jv. \tag{8.65}$$

Therefore, the magnitude of the closed-loop response $M(\omega)$ is

$$M = \left|\frac{G(j\omega)}{1 + G(j\omega)}\right| = \left|\frac{u + jv}{1 + u + jv}\right| = \frac{(u^2 + v^2)^{1/2}}{((1 + u)^2 + v^2)^{1/2}}. \tag{8.66}$$

Squaring Eq. (8.66) and rearranging, we obtain

$$(1 - M^2)u^2 + (1 - M^2)v^2 - 2M^2u = M^2. \tag{8.67}$$

Dividing Eq. (8.67) by $(1 - M^2)$ and adding the term $(M^2/(1 - M^2))^2$ to both sides of Eq. (8.67), we have

$$u^2 + v^2 - \frac{2M^2 u}{1 - M^2} + \left(\frac{M^2}{1 - M^2}\right)^2 = \left(\frac{M^2}{1 - M^2}\right) + \left(\frac{M^2}{1 - M^2}\right)^2. \qquad (8.68)$$

Rearranging, we obtain

$$\left(u - \frac{M^2}{1 - M^2}\right)^2 + v^2 = \left(\frac{M}{1 - M^2}\right)^2, \qquad (8.69)$$

which is the equation of a circle on the u, v-plane with the center at

$$u = \frac{M^2}{1 - M^2}, \qquad v = 0.$$

The radius of the circle is equal to $|M/(1 - M^2)|$. Therefore we may plot several circles of constant magnitude M in the $G(j\omega) = u + jv$ plane. Several constant M circles are shown in Fig. 8.19. The circles to the left of $u = -\tfrac{1}{2}$ are for $M > 1$, and the circles to the right of $u = -\tfrac{1}{2}$ are for $M < 1$. When $M = 1$, the circle becomes the straight line $u = -\tfrac{1}{2}$, which is evident from inspection of Eq. (8.67).

The open-loop frequency response for a system is shown in Fig. 8.20 for two gain values where $K_2 > K_1$. The frequency response curve for the system with gain K_1 is tangent to magnitude circle M_1 at a frequency ω_{r_1}. Similarly, the frequency response curve for gain K_s is tangent to magnitude circle M_2 at the frequency ω_{r_2}. Therefore the closed-loop frequency response magnitude curves are estimated as shown in Fig. 8.21. Clearly, one can obtain the closed-loop frequency response of a system from the $(u + jv)$ plane. If the maximum magnitude, M_{p_ω}, is the only information desired, then it is sufficient to read this value directly from the polar plot. The maximum magnitude of the closed-loop frequency response, M_{p_ω}, is the

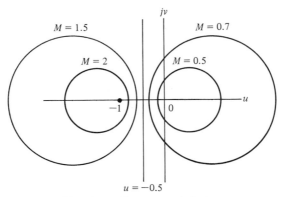

Fig. 8.19. Constant M circles.

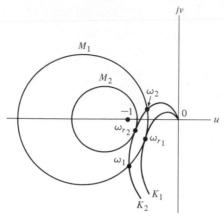

Fig. 8.20. Polar plot of $G(j\omega)$ for two values of a gain.

value of the M circle which is tangent to the $G(j\omega)$-locus. The point of tangency occurs at the frequency ω_r, the resonant frequency. The complete closed-loop frequency response of a system can be obtained by reading the magnitude M of the circles which the $G(j\omega)$-locus intersects at several frequencies. Therefore the system with a gain $K = K_2$ has a closed-loop magnitude M_1 at the frequencies ω_1 and ω_2. This magnitude is read from Fig. 8.20 and is shown on the closed-loop frequency response in Fig. 8.21.

In a similar manner, one may obtain circles of constant closed-loop phase angles. Thus, for Eq. (8.64), the angle relation is

$$\phi = \underline{/T(j\omega)} = \underline{/(u + jv)/(1 + u + jv)}$$

$$= \tan^{-1}\left(\frac{v}{u}\right) - \tan^{-1}\left(\frac{v}{1 + u}\right). \tag{8.70}$$

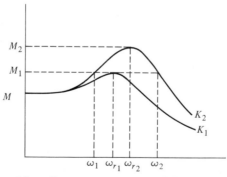

Fig. 8.21. Closed-loop frequency response of $T(j\omega) = G(j\omega)/1 + G(j\omega)$.

Taking the tangent of both sides and rearranging, we have

$$u^2 + v^2 + u - \frac{v}{N} = 0, \qquad (8.71)$$

where $N = \tan \phi = $ constant. Adding the term $\frac{1}{4}(1 + (1/N^2))$ to both sides of the equation and simplifying, we obtain

$$(u + 0.5)^2 + \left(v - \frac{1}{2N}\right)^2 = \frac{1}{4}\left(1 + \frac{1}{N^2}\right), \qquad (8.72)$$

which is the equation of a circle with its center at $u = -0.5$ and $v = +(1/2N)$. The radius of the circle is equal to $\frac{1}{2}(1 + (1/N^2))^{1/2}$. Therefore, the constant phase angle curves can be obtained for various values of N in a manner similar to the M circles.

The constant M and N circles can be used for analysis and design in the polar plane. However, it is much easier to obtain the Bode diagram for a system, and it would be preferable if the constant M and N circles were translated to a logarithmic gain phase. N. B. Nichols transformed the constant M and N circles to the log magnitude-phase diagram, and the resulting chart is called the "Nichols chart" [8]. The M and N circles appear as contours on the Nichols chart as shown in Fig. 8.22. The coordinates of the log magnitude-phase diagram are the same as those used in Section 7.5. However, superimposed on the log magnitude-phase plane we find constant M and N lines. The constant M lines are given in decibels and the N lines in degrees. An example will illustrate the use of the Nichols chart to determine the closed-loop frequency response.

Example 8.5. Consider a feedback system with a loop transfer function

$$G(j\omega) = \frac{1}{j\omega(j\omega + 1)(0.2j\omega + 1)}. \qquad (8.73)$$

The $G(j\omega)$-locus is plotted on the Nichols chart and is shown in Fig. 8.23. The maximum magnitude, M_{p_ω}, is equal to $+2.5$ db and occurs at a frequency $\omega_r = 0.8$. The closed-loop phase angle at ω_r is equal to $-72°$. The 3-db closed-loop bandwidth is equal to $\omega_B = 1.35$ as shown in Fig. 8.23. The closed-loop phase angle at ω_B is equal to $-145°$.

Example 8.6. Let us consider a system with an open-loop transfer function

$$G(j\omega) = \frac{0.64}{j\omega((j\omega)^2 + j\omega + 1)}, \qquad (8.74)$$

where $\zeta = 0.5$ for the complex poles and $H(j\omega) = 1$. The Nichols diagram for this system is shown in Fig. 8.24. The phase margin for this system as it is determined from the Nichols chart is $30°$. On the basis of the phase, we estimate the system

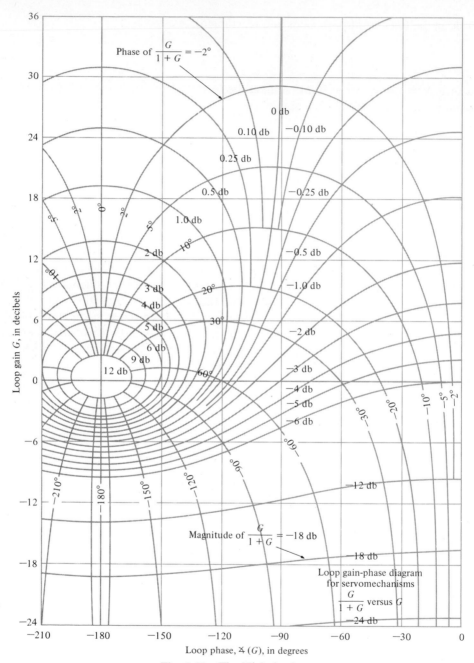

Fig. 8.22. The Nichols chart.

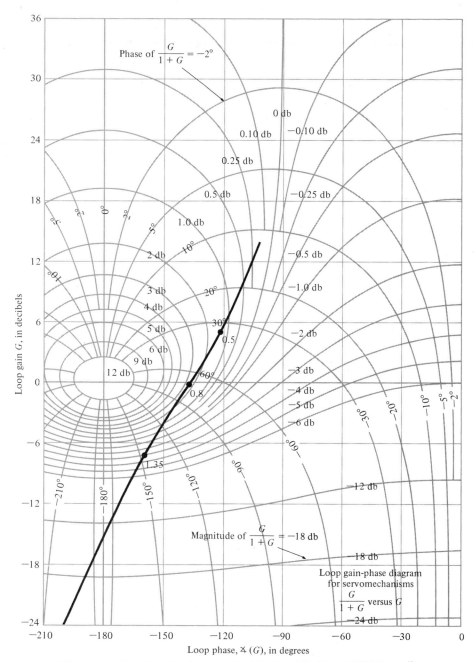

Fig. 8.23. The Nichols diagram for $G(j\omega) = 1/j\omega(j\omega + 1)(0.2j\omega + 1)$.

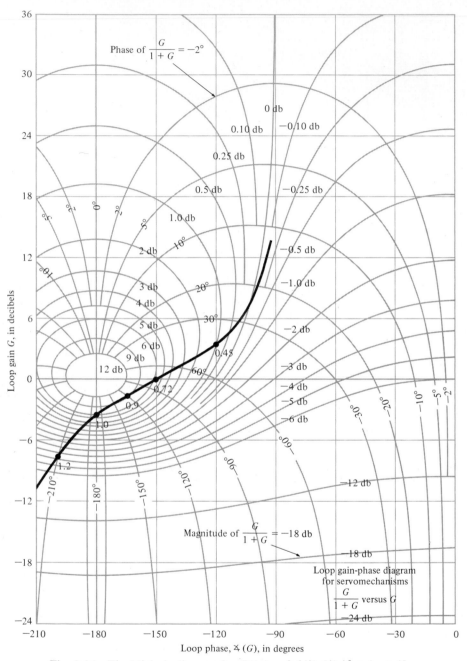

Fig. 8.24. The Nichols diagram for $G(j\omega) = 0.64/j\omega[(j\omega)^2 + j\omega + 1]$.

damping ratio as $\zeta = 0.30$. The maximum magnitude is equal to $+9$ db occurring at a frequency $\omega_r = 0.88$. Therefore,

$$20 \log M_{p_\omega} = 9 \text{ db}$$

or

$$M_{p_\omega} = 2.8.$$

Utilizing Fig. 7.10 to estimate the damping ratio, we find that $\zeta \simeq 0.175$.

We are confronted with two conflicting damping ratios, where one is obtained from a phase margin measure and another from a peak frequency-response measure. In this case, we have discovered an example where the correlation between the frequency domain and the time domain is unclear and uncertain. This apparent conflict is caused by the nature of the $G(j\omega)$ locus which slopes rapidly toward the 180° line from the 0-db axis. If we determine the roots of the characteristic equation for $1 + GH(s)$, we obtain

$$q(s) = (s + 0.77)(s^2 + 0.225s + 0.826) = 0. \tag{8.75}$$

The damping ratio of the complex conjugate roots is equal to 0.124, where the complex roots do not dominate the response of the system. Therefore the real root will add some damping to the system and one might estimate the damping ratio as being approximately the value determined from the M_{p_ω} index; that is, $\zeta = 0.175$. A designer must use the frequency-domain to time-domain correlations with caution. However, one is usually safe if the lower value of the damping ratio resulting from the phase margin and the M_{p_ω} relation is utilized for analysis and design purposes.

The Nichols chart may be used for design purposes by altering the $G(j\omega)$-locus in a suitable manner in order to obtain a desirable phase margin and M_{p_ω}. The system gain K is readily adjusted in order to provide a suitable phase margin and M_{p_ω} by inspecting the Nichols chart. For example, let us reconsider the previous example, where

$$G(j\omega) = \frac{K}{j\omega((j\omega)^2 + j\omega + 1)}. \tag{8.76}$$

The $G(j\omega)$-locus on the Nichols chart for $K = 0.64$ is shown in Fig. 8.24. Let us determine a suitable value for K so that the system damping is greater than 0.40. Examining Fig. 7.10, we find that it is required that M_{p_ω} be less than 1.75 (4.9 db). From Fig. 8.24, we find that the $G(j\omega)$-locus will be tangent to the 4.9 db curve if the $G(j\omega)$ locus is lowered by a factor of 2.2 db. Therefore K should be reduced by 2.2 db or the factor antilog $(2.2/20) = 1.28$. Thus the gain K must be less than 0.64/1.28 = 0.50 if the system damping ratio is to be greater than 0.40.

8.6 THE STABILITY OF CONTROL SYSTEMS WITH TIME DELAYS

The Nyquist stability criterion has been discussed and illustrated in the previous sections for control systems whose transfer functions are rational polynomials of $j\omega$. There are many control systems which have a time delay within the closed loop of the system which affects the stability of the system. Fortunately, the Nyquist criterion can be utilized to determine the effect of the time delay on the relative stability of the feedback system. A pure time delay, without attenuation, is represented by the transfer function

$$G_d(s) = e^{-sT}, \tag{8.77}$$

where T is the delay time. The Nyquist criterion remains valid for a system with a time delay since the factor e^{-sT} does not introduce any additional poles or zeros within the contour. The factor adds a phase shift to the frequency response without altering the magnitude curve.

This type of time delay occurs in systems which have a movement of a material that requires a finite time to pass from an input or control point to an output or measured point [10, 11].

For example, a steel rolling mill control system is shown in Fig. 8.25. The motor adjusts the separation of the rolls so that the thickness error is minimized. If the steel is traveling at a velocity v, then the time delay between the roll adjustment and the measurement is

$$T = \frac{d}{v}. \tag{8.78}$$

Therefore in order to have a negligible time delay, one must decrease the distance to the measurement and increase the velocity of the flow of steel. Usually, one cannot eliminate the effect of time delay and thus the loop transfer function is

$$G(s)G_c(s)e^{-sT}. \tag{8.79}$$

However, one notes that the frequency response of this system is obtained from the loop-transfer function

$$GH(j\omega) = GG_c(j\omega)e^{-j\omega T}. \tag{8.80}$$

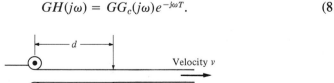

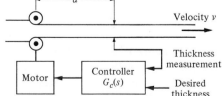

Fig. 8.25. A steel rolling mill control system.

The usual loop-transfer function is plotted on the $GH(j\omega)$-plane and the stability ascertained relative to the -1 point. Alternatively, one can plot the Bode diagram, including the delay factor, and investigate the stability relative to the 0 db, $-180°$ point. The delay factor $e^{-j\omega T}$ results in a phase shift

$$\phi(\omega) = -\omega T \tag{8.81}$$

and is readily added to the phase shift resulting from $GG_c(j\omega)$. An illustration will show the simplicity of this approach on the Bode diagram.

Example 8.6.   A level control system is shown in Fig. 8.26(a), and the block diagram in Fig. 8.26(b). The time delay between the valve adjustment and the fluid output is $T = d/v$. Therefore if the flow rate is 5 m³/sec, the cross-sectional area of the pipe is 1 m², and the distance is equal to 5 m, we have a time delay $T = 1$ sec. The loop-transfer function is then

$$GH(s) = G_A(s)G(s)G_f(s)e^{-sT}$$

$$= \frac{31.5}{(s+1)(30s+1)((s^2/9)+(s/3)+1)} e^{-sT}. \tag{8.82}$$

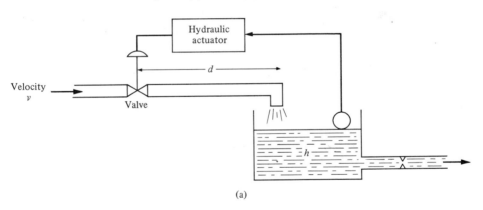

(a)

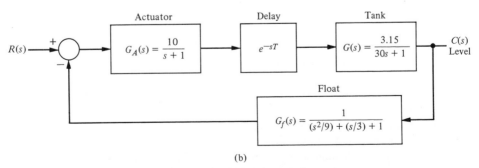

(b)

Fig. 8.26. A liquid level control system.

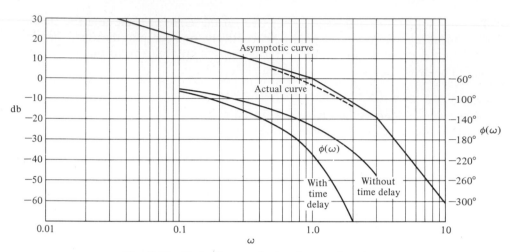

Fig. 8.27. Bode diagram for level control system.

The Bode diagram for this system is shown in Fig. 8.27. The phase angle is shown both for the denominator factors alone and with the additional phase lag due to the time delay. The logarithmic gain curve crosses the 0-db line at $\omega = 0.8$. Therefore the phase margin of the system without the pure time delay would be 40°. However, with the time delay added, we find that the phase margin is equal to $-3°$, and the system is unstable. Therefore the system gain must be reduced in order to provide a reasonable phase margin. In order to provide a phase margin of 30°, the gain would have to be decreased by a factor of 5 db to $K = 31.5/1.78 = 17.7$.

A time delay, e^{-sT}, in a feedback system introduces an additional phase lag and results in a less stable system. Therefore as pure time delays are unavoidable in many systems, it is often necessary to reduce the loop gain in order to obtain a stable response. However, the cost of stability is the resulting increase in the steady-state error of the system as the loop gain is reduced.

8.7 SUMMARY

The stability of a feedback control system can be determined in the frequency domain by utilizing Nyquist's criterion. Furthermore, Nyquist's criterion provides us with two relative stability measures: (1) gain margin and (2) phase margin. These relative stability measures may be utilized as indices of the transient performance on the basis of correlations established between the frequency domain and the transient response. The magnitude and phase of the closed-loop system may be determined from the frequency response of the open-loop transfer function by utilizing constant magnitude and phase circles on the polar plot. Alternatively, one may utilize a log magnitude-phase diagram with closed-loop magnitude and phase curves superimposed (called the Nichols chart) to obtain the closed-loop frequency response. A measure of relative stability, the maximum magnitude of the closed-loop frequency response, M_{p_ω}, is available from the Nichols chart. The frequency measure, M_{p_ω}, may be correlated with the damping ratio of the time response and is a useful index of performance. Finally, it was found that a control system with a pure time delay can be investigated in a similar manner to systems without time delay. A summary of the Nyquist criterion, the relative stability measures, and the Nichols diagram is given in Table 8.4 for several transfer functions.

PROBLEMS

8.1. For the polar plots of Problem 7.1, use the Nyquist criterion to ascertain the stability of the various systems. In each case specify the values of N, P, and Z.

8.2. Sketch the polar plots of the following loop transfer functions $GH(s)$ and determine whether the system is stable by utilizing the Nyquist criterion.

(a) $GH(s) = \dfrac{K}{s(s^2 + s + 4)}$

(b) $GH(s) = \dfrac{K(s + 1)}{s^2(s + 2)}$

If the system is stable, find the maximum value for K by determining the point where the polar plot crosses the u-axis.

8.3. The polar plot of a conditionally stable system is shown in Fig. P8.3 for a specific gain K. Determine whether the system is stable and find the number of roots (if any) in the right-hand s-plane. The system has no poles of $GH(s)$ in the right-half plane.

8.4. (a) Find a suitable contour Γ_s in the s-plane that can be used to determine whether all the roots of the characteristic equation have damping ratios greater than ζ_1. (b) Find a suitable contour Γ_s in the s-plane that can be used to determine whether all the roots of the characteristic equation have real parts less than $s = -\sigma_1$. (c) Using the contour of part (b) and Cauchy's theorem, determine whether the following characteristic equation has roots with real parts less than $s = -1$:

$$q(s) = s^3 + 8s^2 + 30s + 36.$$

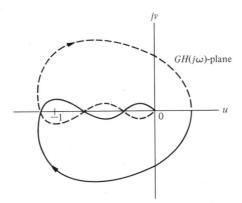

Figure P8.3

Table 8.4 Transfer-function Plots for Typical Transfer Function

$G(s)$	Polar plot	Bode diagram

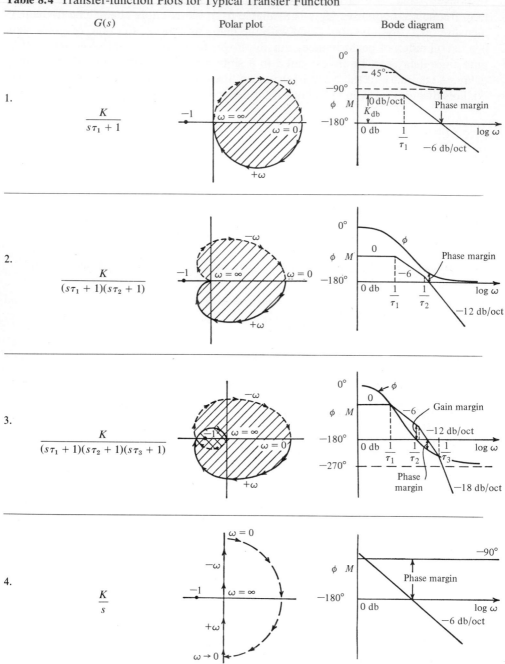

1. $\dfrac{K}{s\tau_1 + 1}$

2. $\dfrac{K}{(s\tau_1 + 1)(s\tau_2 + 1)}$

3. $\dfrac{K}{(s\tau_1 + 1)(s\tau_2 + 1)(s\tau_3 + 1)}$

4. $\dfrac{K}{s}$

Nichols diagram	Root locus	Comments
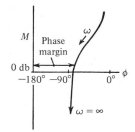	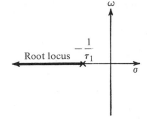	Stable; gain margin = ∞
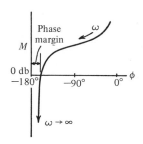	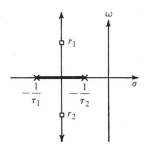	Elementary regulator; stable; gain margin = ∞
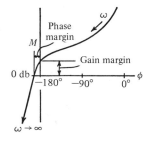	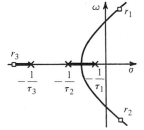	Regulator with additional energy-storage component; unstable, but can be made stable by reducing gain
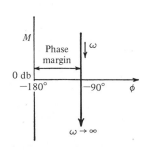	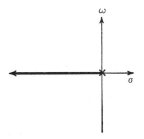	Ideal integrator; stable

Table 8.4 Transfer-function Plots for Typical Transfer Function (*cont.*)

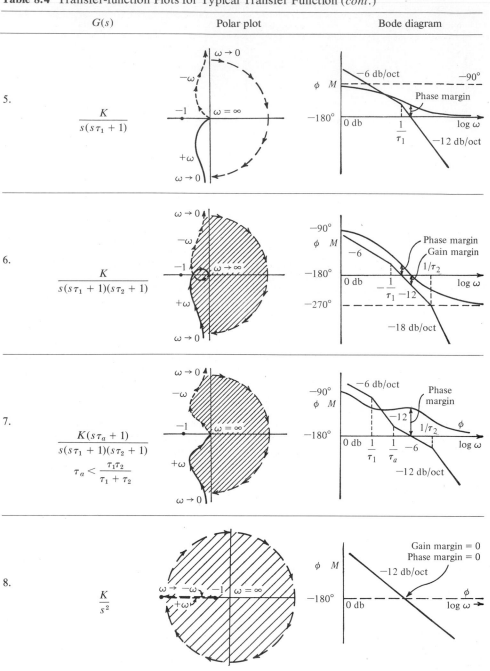

	$G(s)$	Polar plot	Bode diagram
5.	$\dfrac{K}{s(s\tau_1 + 1)}$		
6.	$\dfrac{K}{s(s\tau_1 + 1)(s\tau_2 + 1)}$		
7.	$\dfrac{K(s\tau_a + 1)}{s(s\tau_1 + 1)(s\tau_2 + 1)}$ $\tau_a < \dfrac{\tau_1 \tau_2}{\tau_1 + \tau_2}$		
8.	$\dfrac{K}{s^2}$		

Nichols diagram	Root locus	Comments
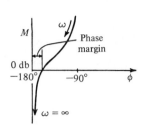	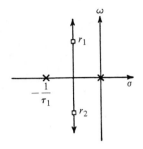	Elementary instrument servo; inherently stable; gain margin = ∞
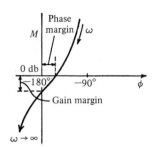	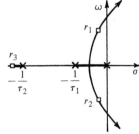	Instrument servo with field-control motor or power servo with elementary Ward-Leonard drive; stable as shown, but may become unstable with increased gain
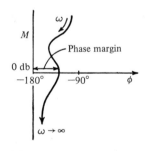	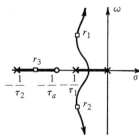	Elementary instrument servo with phase-lead (derivative) compensator; stable
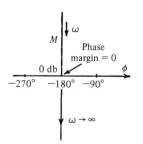	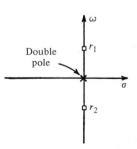	Inherently unstable; must be compensated

Table 8.4 Transfer-function Plots for Typical Transfer Function (*cont.*)

	$G(s)$	Polar plot	Bode diagram

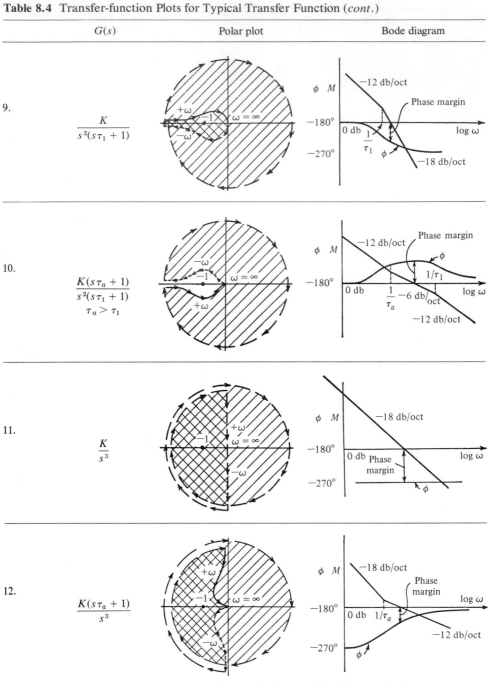

9. $\dfrac{K}{s^2(s\tau_1 + 1)}$

10. $\dfrac{K(s\tau_a + 1)}{s^2(s\tau_1 + 1)}$
     $\tau_a > \tau_1$

11. $\dfrac{K}{s^3}$

12. $\dfrac{K(s\tau_a + 1)}{s^3}$

Nichols diagram	Root locus	Comments

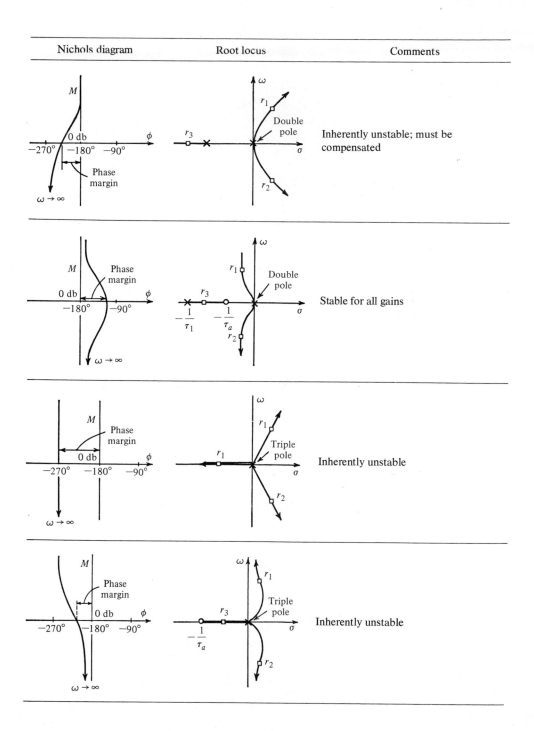

Inherently unstable; must be compensated

Stable for all gains

Inherently unstable

Inherently unstable

Table 8.4 Transfer-function Plots for Typical Transfer Function (*cont.*)

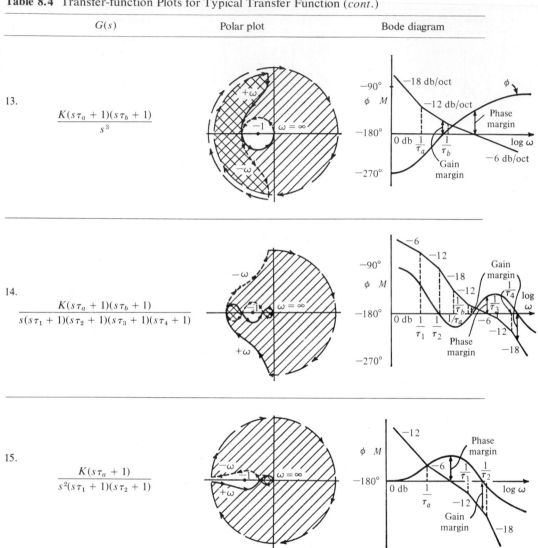

$G(s)$	Polar plot	Bode diagram
13. $\dfrac{K(s\tau_a + 1)(s\tau_b + 1)}{s^3}$		
14. $\dfrac{K(s\tau_a + 1)(s\tau_b + 1)}{s(s\tau_1 + 1)(s\tau_2 + 1)(s\tau_3 + 1)(s\tau_4 + 1)}$		
15. $\dfrac{K(s\tau_a + 1)}{s^2(s\tau_1 + 1)(s\tau_2 + 1)}$		

Nichols diagram	Root locus	Comments
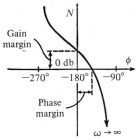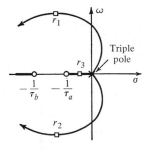		Conditionally stable; becomes unstable if gain is too low
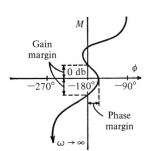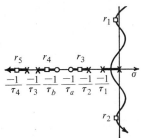		Conditionally stable; stable at low gain, becomes unstable as gain is raised, again becomes stable as gain is further increased, and becomes unstable for very high gains
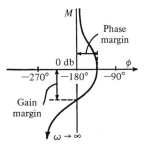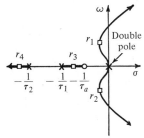		Conditionally stable; becomes unstable at high gain

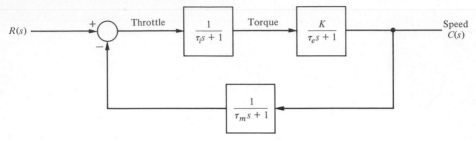

R(s)

Throttle

Torque   K

Speed
C(s)

Figure P8.5

8.5. A speed control for a gasoline engine is shown in Fig. P8.5. Because of the restriction at the carburetor intake and the capacitance of the reduction manifold, the lag τ_t occurs and is equal to 1 sec. The engine time constant τ_e is equal to $J/f = 4$ sec. The speed measurement time constant is $\tau_m = 0.5$ sec. (a) Determine the necessary gain K if the steadystate speed error is required to be less than 7% of the speed reference setting. (b) With the gain determined from (a), utilize the Nyquist criterion to investigate the stability of the system. (c) Determine the phase and gain margins of the system.

8.6. A two-tank level control system is shown in Fig. P8.6. The linearized equation representing one tank is

$$G(s) = \frac{h(s)}{q_i(s)} = \frac{1}{As + R},$$

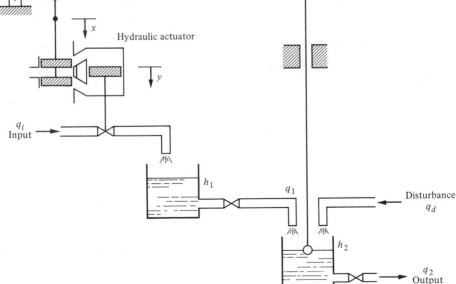

Figure P8.6

where $1/R$ is the orifice constant and A equals the cross-sectional area of the tank. Assume that each tank is the same and that $A = 100$ and $R = 1$. Since the inertia of the control valve is small, the hydraulic actuator transfer function (see Table 2.4) may be represented by

$$G_A(s) = \frac{Y(s)}{X(s)} = \frac{K_A}{s}.$$

The ratio of the lever arms is a/b and is equal to 0.10. The actuator gain is $K_A = 0.10$. The value gain, $q_i(s)/Y(s)$, is -1. (a) Sketch the Bode diagram for the level control system and estimate the gain and phase margins. (b) Sketch the Nichols plot and estimate M_{p_ω} and ω_r. (c) For a step disturbance, $q_d(s) = 1/s$, estimate the settling time for the disturbance to complete the transient. Also, calculate the resulting steady-state error in the outputs. (d) Compare the estimate of the damping ratio, ζ, obtained from the phase margin and from M_{p_ω}.

8.7. A vertical takeoff (VTOL) aircraft is an inherently unstable vehicle and requires an automatic stabilization system. An attitude stabilization system for the K-16B U.S. Army VTOL aircraft has been designed and is shown in block diagram form in Fig. P8.7 [23]. At 40 knots, the dynamics of the vehicle are approximately represented by the transfer function

$$G(s) = \frac{10}{(s^2 + 0.25)}.$$

The actuator and filter is represented by the transfer function

$$G_1(s) = \frac{K_1(s + 7)}{(s + 2)}.$$

(a) Draw the Bode diagram of the loop transfer function $G_1(s)G(s)H(s)$ when the gain is $K_1 = 6$. (b) Determine the gain and phase margins of this system. (c) Determine the steady-state error for a wind disturbance of $T(s) = 1/s$. (d) Determine the maximum amplitude of the resonant peak of the closed-loop frequency response and the frequency of the resonance. (e) Estimate the damping ratio of the system from M_{p_ω} and the phase margin.

8.8. Electrohydraulic servomechanisms are utilized in control systems requiring a rapid response for a large mass. An electrohydraulic servomechanism can provide an output of 100kW or greater. A photo of a servo valve and actuator is shown in Figure P8.8(a). The output

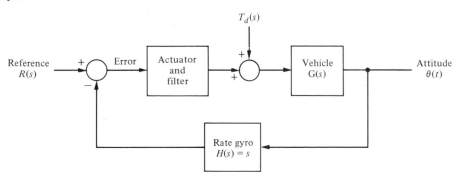

Figure P8.7

Figure P8.8(a) (Courtesy of Moog, Inc., Industrial Division)

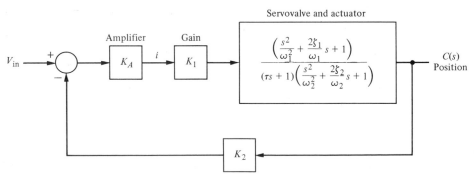

Figure P8.8(b)

sensor yields a measurement of actuator position which is compared with V_{in}. The error is amplified and controls the hydraulic valve position thus controlling the hydraulic fluid flow to the actuator. The block diagram of a closed-loop electrohydraulic servomechanism using pressure feedback to obtain damping is shown in Fig. P8.8(b) [12,13]. Typical values for this system are $\tau = 0.02$ sec, and for the hydraulic system are $\omega_2 = 7(2\pi)$ and $\zeta_2 = 0.05$. The structural resonance ω_1 is equal to $10(2\pi)$ and the damping is $\zeta_1 = 0.05$. The loop gain is $K_A K_1 K_2 = 1.0$. (a) Sketch the Bode diagram and determine the phase margin of the system. (b) The damping of the system may be increased by drilling a small hole in the piston so that $\zeta_2 = 0.25$. Sketch the Bode diagram and determine the phase margin of this system.

8.9. The key to future exploration and use of space is the reusable earth-to-orbit transport system, popularly known as the space shuttle. The first unpowered test of the space shuttle occurred in October 1977 as shown in Fig. 8.9(a). The shuttle will carry large payloads into space and return them to earth for reuse and should be fully operational by 1981 [14,15]. The

(a)

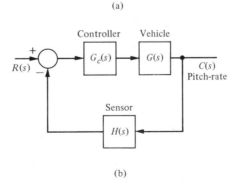

(b)

Figure P8.9 (a) The space shuttle "Enterprise" glides toward its first landing in October 1977 (NASA Photo). (b) Pitch rate control system.

shuttle, roughly the size of a DC-9 with an empty weight of 75,000 Kg, uses elevons at the trailing edge of the wing and a brake on the tail to control the flight. The block diagram of a pitch rate control system is shown in Fig. P8.9(b). The sensor is represented by a gain, $H(s)$ = 0.5, and the vehicle by the transfer function

$$G(s) = \frac{0.30(s + 0.05)(s^2 + 1600)}{(s^2 + 0.05s + 16)(s + 70)}.$$

The controller $G_c(s)$ can be a gain or any suitable transfer function. (a) Draw the Bode diagram of the system when $G_c(s)$ = 2 and determine the stability margin. (b) Draw the Bode diagram of the system when

$$G_c(s) = K_1 + K_2/s \quad \text{and} \quad K_2/K_1 = 0.5.$$

The gain k_1 should be selected so that the gain margin is 10 db.

8.10. Machine tools are often automatically controlled by a punched tape reader as shown in Fig. P8.10 [16,17]. These automatic systems are often called numerical machine controls. Considering one axis, the desired position of the machine tool is compared with the actual position and is used to actuate a solenoid coil and the shaft of a hydraulic actuator. The transfer function of the actuator (see Table 2.4) is

$$G_a(s) = \frac{X(s)}{Y(s)} = \frac{K_a}{s(\tau_a s + 1)},$$

where K_a = 1 and τ_a = 0.4 sec. The output voltage of the difference amplifier is

$$E_0(s) = K_1(X(s) - X_d(s)),$$

where $x_d(t)$ is the desired position input from the tape reader. The force on the shaft is proportional to the current i so that $F = K_2 i(t)$, where K_2 = 3.0. The spring constant K_s is equal to 1.5 and R = 0.1 and L = 0.2. (a) Determine the gain K_1 which results in a system

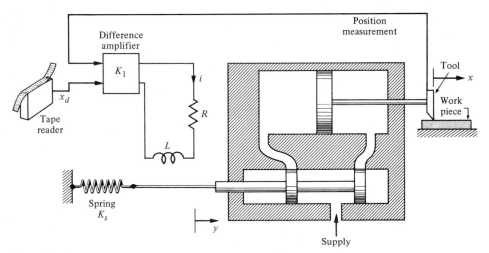

Figure P8.10

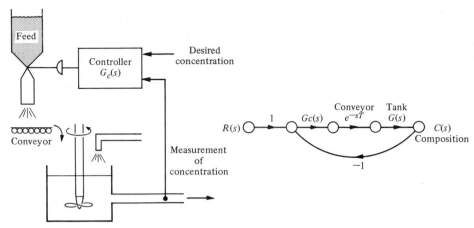

Figure P8.11

with a phase margin of 30°. (b) For the gain K_1 of part (a), determine M_{p_ω}, ω_r, and the closed-loop system bandwidth. (c) Estimate the percent overshoot of the transient response for a step input, $X_d(s) = 1/s$, and the settling time.

8.11. A control system for a chemical concentration control system is shown in Fig. P8.11 [10]. The system receives a granular feed of varying composition, and it is desired to maintain a constant composition of the output mixture by adjusting the feed-flow valve. The transfer function of the tank and output valve is

$$G(s) = \frac{5}{5s + 1} \quad \text{and the controller is} \quad G_c(s) = K_1 + \frac{K_2}{s}.$$

The transport of the feed along the conveyor requires a transport (or delay) time, $T = 2$ sec. (a) Sketch the Nyquist diagram when $K_1 = K_2 = 1$, and investigate the stability of the system. (b) Sketch the Nyquist diagram when $K_1 = 0.1$ and $K_2 = 0.05$, and investigate the stability of the system. (c) When $K_1 = 0.2$, use the Nyquist criterion to calculate the maximum allowable gain K_2 for the system to remain stable.

8.12. A simplified model of the control system for regulating the pupillary aperture in the human eye is shown in Fig. P8.12 [18]. The gain K represents the pupillary gain and τ is the pupil time constant which is 0.5 sec. The time delay T is equal to 0.2 sec. The pupillary gain is equal to 4.0. (a) Assuming the time delay is negligible, draw the Bode diagram for the system. Determine the phase margin of the system. (b) Include the effect of the time delay by adding the phase shift due to the delay. Determine the phase margin of the system with the time delay included.

8.13. A controller is used to control the temperature of a mold for plastic part fabrication, as shown in Fig. P8.13 [19]. The value of the delay time is estimated as 6 seconds. (a) Utilizing the Nyquist criterion, determine the stability of the system for $K = 1$. (b) Determine a suitable value for K for a stable system.

8.14. The closed-loop performance of a high-frequency operational amplifier can be pre-

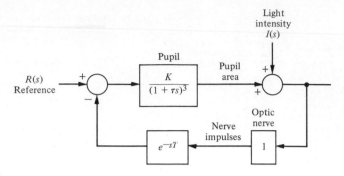

Figure P8.12

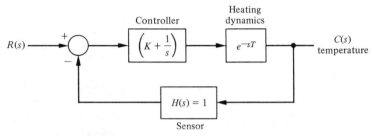

Figure P8.13

dicted using the Nichols chart.[20] One inverting operational amplifier has an open-loop response as given below.

f(MHz)	GH (dB)	$\underline{/GH}$ (degrees)
1	33	-132
2	26	-120
4	21	-114
6	17	-114
10	12	-120
15	6	-128
17	5	-132
20	3	-137
25	0	-142
30	-2	-148
38	-5	-155
42	-7	-160

Obtain the Nichols chart and show that the M_{p_ω} is +4.0db when $\omega = 25$Mhz.

8.15. Electronics and computers are being used to control automobiles. The trip computer is shown in Fig. P8.15(a) in a 1978 Cadillac. This device tells the driver the amount of fuel

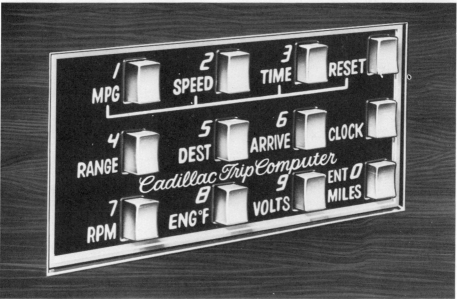

Figure P8.15 (a) The Trip Computer for Cadillac automobiles. (Photo courtesy of General Motors Corp., Cadillac Motor Car Division) (b—next page) The Unicontrol control stick for General Motors Firebird III. (Photo courtesy of General Motors Corp.)

(b)

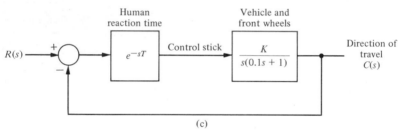

(c)

Figure P8.15 (cont.)

remaining, speed, average and instantaneous fuel consumption, and driving range [24]. Another example of an automobile control system, the steering control for the General Motors Firebird III research automobile, is a control stick as shown in Fig. P8.15(b). The control stick, called a Unicontrol, is used for steering and controlling the throttle. A typical driver has a reaction time of $T = 0.3$ sec. (a) Using the Nichols chart, determine the magnitude of the gain K which will result in a system with a peak magnitude of the closed-loop frequency response M_{p_ω} less than or equal to 2 db. (b) Estimate the damping ratio of the system based on (1) M_{p_ω} and (2) the phase margin. Compare the results and explain the difference, if any. (c) Determine the closed-loop 3-db bandwidth of the system.

8.16. Consider the automatic ship steering system discussed in Problem 7.13. The frequency response of the open-loop portion of the ship steering control system is shown in Fig. P7.13. The deviation of the tanker from the straight track is measured by radar and is used to generate the error signal as shown in Fig. P8.16. This error signal is used to control the rudder angle $\delta(s)$. (a) Is this system stable? Discuss what an unstable ship steering system indicates in terms of the transient response of the system. Recall that the system under consideration

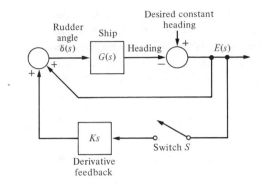

Figure P8.16

is a ship attempting to follow a straight track. (b) Is it possible to stabilize this system by lowering the gain of the transfer function $G(s)$? (c) Is it possible to stabilize this system? Can you suggest a suitable feedback compensator? (d) Repeat parts (a), (b), and (c) when the switch S is closed.

8.17. The increasing demand for short-haul air travel has contributed to the growth of the number of intermediate sized aircraft. One electromechanical nose wheel steering system supplies general aviation aircraft with the needed maneuverability and durability. During takeoff from an unimproved landing strip, the task of keeping the aircraft on the proper heading is facilitated by nose wheel power steering. The block diagram of one nose wheel steering system is shown in Fig. P8.17 [21]. The system uses a magnetic particle clutch to actuate the rotation of the wheel heading. Draw the Bode diagram and the Nichols chart for the system. What gain is required for K in order to have a system with a phase margin of 20 degrees? Let $K_1 = 1$, $L/R = 0.05$, $J = 1$, $f = 1$ and $k = 4$.

8.18. The primary objective of many control systems is to maintain the output variable at the desired or reference condition when the system is subjected to a disturbance [10]. A typical chemical reactor control scheme is shown in Fig. P8.18. The disturbance is represented by $U(s)$ and the chemical process by G_3 and G_4. The controller is represented by G_1 and the valve by G_2. The feedback sensor is $H(s)$ and will be assumed to be equal to one. We will assume that G_2, G_3 and G_4 are all of the form

$$G_i(s) = \frac{K_i}{1 + \tau_i s},$$

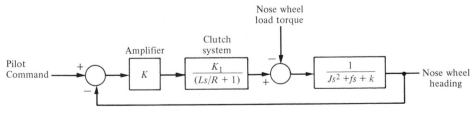

Figure P8.17

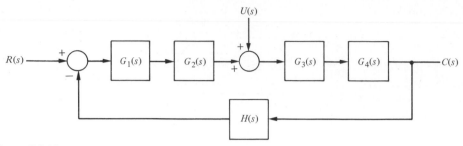

Figure P8.18

where $\tau_3 = \tau_4 = 4$ seconds and $K_3 = K_4 = 0.1$. The valve constants are $K_2 = 20$ and $\tau_2 = 0.5$ seconds. It is desired to maintain a steady-state error less than 5 percent of the desired reference position. (a) When $G_1(s) = K_1$, find the necessary gain to satisfy the error constant requirement. For this condition, determine the expected overshoot to a step change in the reference signal $r(t)$. (b) If the controller has a proportional term plus an integral term so that $G_1(s) = K_1(1 + 1/s)$, determine a suitable gain to yield a system with an overshoot less than 30 percent but greater than 5 percent. For parts (a) and (b) use the approximation of damping ratio as a function of phase margin that yields $\zeta = 0.01 \, \phi_{pm}$. For these calculations, assume that $U(s) = 0$. (c) Estimate the settling time of the step response of the system for the controller of parts (a) and (b). (d) The system is expected to be subjected to a step disturbance $U(s) = A/s$. For ease, assume that the desired reference is $r(t) = 0$ when the system has settled. Determine the response of the system of part (b) to the disturbance.

8.19. In the U.S. over 1.6 billion dollars are spent annually for solid waste collection and disposal. One system which uses a remote control pick-up arm for collecting waste bags is shown in Fig. P8.19 [22]. The open-loop transfer of the remote pick-up arm is

$$GH(s) = \frac{0.3}{s(1 + 3s)(1 + s)}$$

(a) Plot the Nichols chart and show that the gain margin is approximately 11.5 db. (b) Determine the phase margin and the $M_{p\omega}$ for the closed loop. Also determine the closed-loop bandwidth.

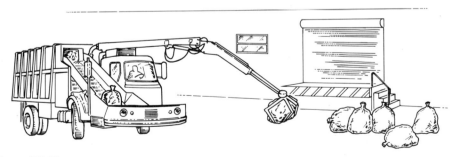

Figure P8.19

REFERENCES

1. H. Nyquist, "Regneration Theory," *Bell System Technical Journal,* January 1932; pp. 126–147. Also in *Automatic Control: Classical Linear Theory,* G. J. Thaler, ed., Dowden, Hutchinson and Ross, Inc., Stroudsburg, Pa.; pp. 105–126.

2. M. D. Fagen, *A History of Engineering and Science in the Bell System,* Bell Telephone Laboratories, Inc., Murray Hill, N.J., 1978; Ch. 5.

3. C. R. Wylie, Jr., *Advanced Engineering Mathematics,* 4th ed., McGraw-Hill, New York, 1975.

4. G. Polya and G. Latta, *Complex Variables,* Wiley, New York, 1974.

5. F. Elices, "Nyquist-plot Routine Predicts Closed-Loop Stability Limits," *Electronics,* July 20, 1978; pp. 145–147.

6. R. C. Dorf, *Introduction to Computers and Computer Science,* 2nd ed., Boyd and Fraser Publishing Co., San Francisco, 1977; Ch. 7.

7. S. M. Shinners, *Modern Control System Theory and Application,* 2nd ed., Addison-Wesley, Reading, Mass., 1978; pp. 220–223.

8. H. M. James. N. B. Nichols, and R. S. Phillips, *Theory of Servomechanisms,* McGraw-Hill, New York, 1947.

9. E. C. Hind, "Closed-loop Transient Response From the Open-Loop Frequency Response," *Measurement and Control,* August 1978; pp. 302–308.

10. D. Y. Etchart, "Forecasting and Compensating the Effects of Deadtime on a Commonly Applied Chemical Pacing Control Loop," *ISA Transactions,* **16,** 4, 1977; pp. 59–67.

11. Z. J. Palmor and R. Shinnar, "Design and Tuning of Dead-Time Compensators," *Proceedings of the JACC,* 1978; pp. 59–70.

12. A. Kandelman and D. J. Nelson, "Simplified Model Eases Hydraulic System Simulation," *Control Engineering,* April 1978; pp. 65–66.

13. "Servovalves," *Machine Design,* September 28, 1978; pp. 67–69.

14. M. S. Malkin, "The Space Shuttle," *American Scientist,* December 1978; pp. 718–723.

15. C. T. Sheridan, "Space Shuttle Software," *Datamation,* July 1978; pp. 128–140.

16. "Mike It As You Cut It," *Production Engineering,* May 1978; p. 28.

17. K. Srinivasan and C. L. Nachtigal, "Analysis and design of Machine Tool Chatter Control Systems Using the Regeneration Spectrum," *J. of Dynamic Systems, Measurement and Control,* September 1978; pp. 191–200.

18. A. T. Bahill and L. Stark, "The Trajectories of Saccadic Eye Movements," *Scientific American,* January 1979; pp. 108–117.

19. R. Floersch, "A Digital Controller for Cyclic Temperature Control," *Control Engineering,* October 1978; pp. 58–61.

20. E. Thibodeaux, "Predict Wideband Amplifier Response," *Electronic Design,* December 6, 1974; pp. 68–71.

21. J. Camp and M. J. Campbell, "Aircraft Power Steering," *Sperry Rand Engineering Review,* **24,** 2, 1971; pp. 37–40.

22. P. H. Har-Oz, "A Remote Waste Bags Collector," *Proceedings of JACC,* 1976, *American Society of Mechanical Engineers;* pp. 279–286.

23. J. Van Train, "Automatic Flight Control for Army VTOL/STOL Aircraft," *Report No. G-168,* Kamen Aircraft Corp., May 1964.

24. C. E. Wise, "Cars and Computers Come Together," *Machine Design,* November 23, 1978; pp. 24–30.

9 / Time-Domain Analysis of Control Systems

9.1 INTRODUCTION

In the preceding chapters, we have developed and studied several useful approaches to the analysis and design of feedback systems. The Laplace transform was utilized to transform the differential equations representing the system into an algebraic equation expressed in terms of the complex variable, s. Utilizing this algebraic equation, we were able to obtain a transfer function representation of the input-output relationship. Then the root locus and s-plane methods were developed on the basis of the Laplace transform representation. Furthermore, the steady-state representation of the system in terms of the real frequency variable, ω, was developed, and several useful techniques for analysis were studied. The frequency-domain approach, in terms of the complex variable s or the real frequency variable ω, is extremely useful; it is and will remain one of the primary tools of the control engineer. However, the limitations of the frequency-domain techniques and the recently acquired attractiveness of the time-domain approach require a reconsideration of the time-domain formulation of the equations representing control systems.

The frequency-domain techniques are limited in applicability to linear, time-invariant systems. Furthermore, they are particularly limited in their usefulness for multivariable control systems due to the emphasis on the input-output relationship of transfer functions. By contrast, the time-domain techniques may be readily utilized for nonlinear, time-varying, and multivariable systems. *A time-varying control system is a system for which one or more of the parameters of the system may vary as a function of time.* For example, the mass of a missile varies as a function of time as the fuel is expended during flight. A multivariable system, as discussed in Section 2.6, is a system with several input and output signals. The solution of a time-domain formulation of a control system problem is facilitated by the availability and ease of use of digital and analog computers. Therefore, we are interested in reconsidering the time-domain description of dynamic systems as they are represented by the system differential equation.

The time-domain representation of control systems is an essential basis for modern control theory and system optimization. In Chapter 10, we shall have an opportunity to design an optimum control system by utilizing time-domain methods. In this chapter, we shall develop the time-domain representation of control systems, investigate the stability of these systems, and illustrate several methods for the solution of the system time response.

9.2 THE STATE VARIABLES OF A DYNAMIC SYSTEM

The time-domain analysis and design of control systems utilize the concept of the state of a system [1, 2, 3, 4]. *The state of a system is a set of numbers such that the knowledge of these numbers and the input functions will, with the equations describing the dynamics, provide the future state and output of the system.* For a dynamic system, the state of a system is described in terms of a set of *state variables* $(x_1(t), x_2(t), \ldots, x_n(t))$. The state variables are those variables which determine the future behavior of a system when the present state of the system and the excitation signals are known. Consider the system shown in Fig. 9.1, where $c_1(t)$ and $c_2(t)$ are the output signals and $u_1(t)$ and $u_2(t)$ are the input signals. A set of state variables $(x_1, x_2, \ldots, x_n)$ for the system shown in Fig. 9.1 is a set such that knowledge of the initial values of the state variables $(x_1(t_0), x_2(t_0), \ldots, x_n(t_0))$ at the initial time t_0, and of the input signals $u_1(t)$ and $u_2(t)$ for $t \geq t_0$, suffices to determine the future values of the outputs and state variables [2].

A simple example of a state variable is the state of an On-Off light switch. The switch can be in either the On or Off position and thus the state of the switch can assume one of two possible values. Thus, if we know the present state (position) of the switch at t_0 and if an input is applied, we are able to determine the future value of the state of the element.

The concept of a set of state variables which represent a dynamic system can be illustrated in terms of the spring-mass-damper system shown in Fig. 9.2. The number of state variables chosen to represent this system should be as few as possible in order to avoid redundant state variables. A set of state variables sufficient to describe this system is the position and the velocity of the mass. Therefore, we will define a set of state variables as (x_1, x_2) where

$$x_1(t) = y(t) \quad \text{and} \quad x_2(t) = \frac{dy(t)}{dt}.$$

Fig. 9.1. System block diagram.

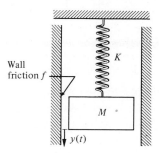

Fig. 9.2. A spring-mass-damper system.

The differential equation describes the behavior of the system and is usually written as

$$M \frac{d^2 y}{dt^2} + f \frac{dy}{dt} + Ky = u(t). \tag{9.1}$$

In order to write Eq. (9.1) in terms of the state variables, we substitute the definition of the state variables and obtain

$$M \frac{dx_2}{dt} + fx_2 + Kx_1 = u(t). \tag{9.2}$$

Therefore we may write the differential equations which describe the behavior of the spring-mass-damper system as a set of two first-order differential equations as follows:

$$\frac{dx_1}{dt} = x_2, \tag{9.3}$$

$$\frac{dx_2}{dt} = \frac{-f}{M} x_2 - \frac{K}{M} x_1 + \frac{1}{M} u. \tag{9.4}$$

This set of differential equations describes the behavior of the state of the system in terms of the rate of change of each state variable.

As another example of the state variable characterization of a system, let us consider the *RLC* circuit shown in Fig. 9.3. The state of this system may be described in terms of a set of state variables (x_1, x_2), where x_1 is the capacitor

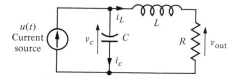

Fig. 9.3. An *RLC* circuit.

voltage $v_c(t)$ and x_2 is equal to the inductor current $i_L(t)$. This choice of state variables is intuitively satisfactory, since the stored energy of the network can be described in terms of these variables as

$$\mathcal{E} = 1/2Li_L^2 + 1/2Cv_c^2. \tag{9.5}$$

Therefore, $x_1(t_0)$ and $x_2(t_0)$ represent the total initial energy of the network and thus the state of the system at $t = t_0$. For a passive RLC network, the number of state variables required is equal to the number of independent energy storage elements. Utilizing Kirchhoff's current law at the junction, we obtain a first-order differential equation by describing the rate of change of capacitor voltage as

$$i_c = C\frac{dv_c}{dt} = +u(t) - i_L. \tag{9.6}$$

Kirchhoff's voltage law for the right-hand loop provides the equation describing the rate of change of inductor current as

$$L\frac{di_L}{dt} = -Ri_L + v_c. \tag{9.7}$$

The output of this system is represented by the linear algebraic equation

$$v_{out} = Ri_L(t).$$

We may rewrite Eqs. (9.6) and (9.7) as a set of two first-order differential equations in terms of the state variables x_1 and x_2 as follows:

$$\frac{dx_1}{dt} = -\frac{1}{C}x_2 + \frac{1}{C}u(t), \tag{9.8}$$

$$\frac{dx_2}{dt} = +\frac{1}{L}x_1 - \frac{R}{L}x_2. \tag{9.9}$$

The output signal is then

$$c_1(t) = v_{out}(t) = Rx_2. \tag{9.10}$$

Utilizing Eqs. (9.8) and (9.9) and the intial conditions of the network represented by $(x_1(t_0), x_2(t_0))$, we may determine the system's future behavior and its output.

The state variables which describe a system are not a unique set, and several alternative sets of state variables may be chosen. For example, for a second-order system, such as the mass-spring-damper or RLC circuit, the state variables may be any two independent linear combinations of $x_1(t)$ and $x_2(t)$. Therefore, for the RLC circuit, one might choose the set of state variables as the two voltages, $v_c(t)$ and $v_L(t)$, where v_L is the voltage drop across the inductor. Then the new state variables, x_1^* and x_2^*, are related to the old state variables, x_1 and x_2, as

$$x_1^* = v_c = x_1, \tag{9.11}$$

$$x_2^* = v_L = v_c - Ri_L = x_1 - Rx_2. \tag{9.12}$$

Equation (9.12) represents the relation between the inductor voltage and the former state variables v_c and i_L. Thus, in an actual system, there are several choices of a set of state variables which specify the energy stored in a system and therefore adequately describe the dynamics of the system. A widely used choice is a set of state variables which can be readily measured.

The state variables of a system characterize the dynamic behavior of a system. The engineer's interest is primarily in physical systems, where the variables are voltages, currents, velocities, positions, pressures, temperatures, and similar physical variables. However, the concept of system state is not limited to the analysis of physical systems and is particularly useful for analyzing biological, social, economic, and physical systems. For these systems, the concept of state is extended beyond the concept of energy of a physical system to the broader viewpoint of variables which describe the future behavior of the system.

9.3 THE STATE VECTOR DIFFERENTIAL EQUATION

The state of a system is described by the set of first-order differential equations written in terms of the state variables $(x_1, x_2, \ldots, x_n)$. These first-order differential equations may be written in general form as

$$
\begin{aligned}
\dot{x}_1 &= a_{11}x_1 + a_{12}x_2 + \cdots + a_{1n}x_n + b_{11}u_1 + \cdots + b_{1m}u_m, \\
\dot{x}_2 &= a_{21}x_1 + a_{22}x_2 + \cdots + a_{2n}x_n + b_{21}u_1 + \cdots + b_{2m}u_m, \\
&\vdots \\
\dot{x}_n &= a_{n1}x_1 + a_{n2}x_2 + \cdots + a_{nn}x_n + b_{n1}u_1 + \cdots + b_{nm}u_m,
\end{aligned}
\tag{9.13}
$$

where $\dot{x} = dx/dt$. Thus this set of simultaneous differential equations may be written in matrix form as follows [5, 6]:

$$
\frac{d}{dt}\begin{bmatrix} x_1 \\ x_2 \\ \vdots \\ x_n \end{bmatrix} = \begin{bmatrix} a_{11} & a_{12} & \cdots & a_{1n} \\ a_{21} & a_{22} & \cdots & a_{2n} \\ \vdots & & \cdots & \vdots \\ a_{n1} & a_{n2} & \cdots & a_{nn} \end{bmatrix}\begin{bmatrix} x_1 \\ x_2 \\ \vdots \\ x_n \end{bmatrix} + \begin{bmatrix} b_{11} \\ \vdots \\ b_{n1} \end{bmatrix}\begin{bmatrix} \cdots \\ \cdots \end{bmatrix}\begin{bmatrix} b_{1m} \\ \vdots \\ b_{nm} \end{bmatrix}\begin{bmatrix} u_1 \\ \vdots \\ u_m \end{bmatrix}.
\tag{9.14}
$$

The column matrix consisting of the state variables is called the *state vector* and is written as

$$
\mathbf{x} = \begin{bmatrix} x_1 \\ x_2 \\ \vdots \\ x_n \end{bmatrix},
\tag{9.15}
$$

where the boldface indicates a matrix. The matrix of input signals is defined as $\mathbf{u}$. Then the system may be represented by the compact notation of the *system vector differential equation* as

$$
\dot{\mathbf{x}} = \mathbf{A}\mathbf{x} + \mathbf{B}\mathbf{u}.
\tag{9.16}
$$

The matrix **A** is an $n \times n$ square matrix and **B** is an $n \times m$ matrix.* The vector matrix differential equation relates the rate of change of the state of the system to the state of the system and the input signals. In general, the outputs of a linear system may be related to the state variables and the input signals by the vector matrix equation

$$\mathbf{c} = \mathbf{Dx} + \mathbf{Hu}, \tag{9.17}$$

where **c** is the set of output signals expressed in column vector form.

The solution of the state vector differential equation (Eq. 9.16) may be obtained in a manner similar to the approach we utilize for solving a first-order differential equation. Consider the first-order differential equation

$$\dot{x} = ax + bu, \tag{9.18}$$

where $x(t)$ and $u(t)$ are scalar functions of time. We expect an exponential solution of the form e^{at}. Taking the Laplace transform of Eq. (9.18), we have

$$sX(s) - x(0) = aX(s) + bU(s),$$

and therefore

$$X(s) = \frac{x(0)}{s - a} + \frac{b}{s - a} U(s). \tag{9.19}$$

The inverse Laplace transform of Eq. (9.19) results in the solution

$$x(t) = e^{at}x(0) + \int_0^t e^{+a(t-\tau)}bu(\tau)\,d\tau. \tag{9.20}$$

We expect the solution of the vector differential equation to be similar to Eq. (9.20) and of exponential form. The matrix exponential function is defined as

$$e^{\mathbf{A}t} = \exp(\mathbf{A}t) = \mathbf{I} + \mathbf{A}t + \frac{\mathbf{A}^2 t^2}{2!} + \cdots + \frac{\mathbf{A}^k t^k}{k!} + \cdots, \tag{9.21}$$

which converges for all finite t and any **A** [6]. Then the solution of the vector differential equation is found to be [1]

$$\mathbf{x}(t) = \exp(\mathbf{A}t)\mathbf{x}(0) + \int_0^t \exp[\mathbf{A}(t - \tau)]\mathbf{Bu}(\tau)\,d\tau. \tag{9.22}$$

Equation (9.22) may be obtained by taking the Laplace transform of Eq. (9.16) and rearranging to obtain

$$\mathbf{X}(s = [s\mathbf{I} - \mathbf{A}]^{-1}\mathbf{x}(0) + [s\mathbf{I} - \mathbf{A}]^{-1}\mathbf{BU}(s), \tag{9.23}$$

---

* Bold-faced lower case letters denote vector quantities and bold-faced upper case letters denote matrices. For an introduction to matrices and elementary matrix operations, the reader is referred to Appendix C and references [5] and [6].

where we note that $[s\mathbf{I} - \mathbf{A}]^{-1} = \boldsymbol{\phi}(s)$, which is the Laplace transform of $\boldsymbol{\phi}(t) = \exp{(\mathbf{A}t)}$. Taking the inverse Laplace transform of Eq. (9.23) and noting that the second term on the right-hand side involves the product $\boldsymbol{\phi}(s)\mathbf{B}U(s)$, we obtain Eq. (9.22). The matrix exponential function describes the unforced response of the system and is called the *fundamental* or *transition matrix* $\boldsymbol{\phi}(t)$. Therefore Eq. (9.22) may be written as

$$\mathbf{x}(t) = \boldsymbol{\phi}(t)\mathbf{x}(0) + \int_0^t \boldsymbol{\phi}(t - \tau)\mathbf{B}\mathbf{u}(\tau)\,d\tau. \tag{9.24}$$

The solution to the unforced system (that is, when $\mathbf{u} = 0$) is simply

$$\begin{bmatrix} x_1(t) \\ x_2(t) \\ \vdots \\ x_n(t) \end{bmatrix} = \begin{bmatrix} \phi_{11}(t) & \cdots & \phi_{1n}(t) \\ \phi_{21}(t) & \cdots & \phi_{2n}(t) \\ \vdots & & \vdots \\ \phi_{n1}(t) & \cdots & \phi_{nn}(t) \end{bmatrix} \begin{bmatrix} x_1(0) \\ x_2(0) \\ \vdots \\ x_n(0) \end{bmatrix}. \tag{9.25}$$

Hence one notes that in order to determine the transition matrix, all initial conditions are set to zero except for one state variable, and the output of each state variable is evaluated. That is, the term $\phi_{ij}(t)$ is the response of the ith state variable due to an initial condition on the jth state variable when there are zero initial conditions on all the other states. We shall utilize this relationship between the initial conditions and the state variables to evaluate the coefficients of the transition matrix in a later section. However, first we shall develop several suitable signal-flow state models of systems and investigate the stability of the systems by utilizing these flow graphs.

9.4 SIGNAL FLOW GRAPH STATE MODELS

The state of a system describes a system's dynamic behavior where the dynamics of the system are represented by a series of first-order differential equations. Alternatively, the dynamics of the system may be represented by a vector differential equation as in Eq. (9.16). In either case, it is useful to develop a state flow graph model of the system and use this model to relate the state variable concept to the familiar transfer function representation.

As we have learned in previous chapters, a system can be meaningfully described by an input-output relationship, the transfer function $G(s)$. For example, if we are interested in the relation between the output voltage and the input voltage of the network of Fig. 9.3, we may obtain the transfer function

$$G(s) = \frac{V_0(s)}{U(s)}.$$

The transfer function for the *RLC* network is of the form

$$G(s) = \frac{V_0(s)}{U(s)} = \frac{\alpha}{s^2 + \beta s + \gamma}, \tag{9.26}$$

where α, β, and γ are functions of the circuit parameters R, L, and C. The values of α, β, and γ may be determined from the flow graph representing the differential equations which describe the circuit. For the RLC circuit (see Eqs. 9.8 and 9.9), we have

$$\dot{x}_1 = -\frac{1}{C}x_2 + \frac{1}{C}u(t), \tag{9.27}$$

$$\dot{x}_2 = \frac{1}{L}x_1 - \frac{R}{L}x_2, \tag{9.28}$$

$$v_{\text{out}} = Rx_2. \tag{9.29}$$

The flow graph representing these simultaneous equations is shown in Fig. 9.4, where $1/s$ indicates an integration. Using Mason's signal flow gain formula, we obtain the transfer function

$$\frac{V_{\text{out}}(s)}{U(s)} = \frac{+R/LCs^2}{1 + (R/Ls) + (1/LCs^2)} = \frac{+R/LC}{s^2 + (R/L)s + (1/LC)}. \tag{9.30}$$

Unfortunately, many elecric circuits, electromechanical systems, and other control systems are not so simple as the RLC circuit of Fig. 9.3, and it is often a difficult task to determine a series of first-order differential equations describing the system. Therefore, it is often simpler to derive the transfer function of the system by the techniques of Chapter 2 and then derive the state model from the transfer function.

The signal flow graph state model can be readily derived from the transfer function of a system. However, as we noted in the previous section, there is more than one alternative set of state variables, and therefore there is more than one possible form for the signal flow graph state model. In general, we may represent a transfer function as

$$G(s) = \frac{C(s)}{U(s)} = \frac{s^m + b_{m-1}s^{m-1} + \cdots + b_1s + b_0}{s^n + a_{n-1}s^{n-1} + \cdots + a_1s + a_0}, \tag{9.31}$$

where $n \geq m$ and all the a coefficients are real positive numbers. If we multiply the numerator and denominator by s^{-n}, we obtain

$$G(s) = \frac{s^{-(n-m)} + b_{m-1}s^{-(n-m+1)} + \cdots + b_1s^{-(n-1)} + b_0s^{-n}}{1 + a_{n-1}s^{-1} + \cdots + a_1s^{-(n-1)} + a_0s^{-n}}. \tag{9.32}$$

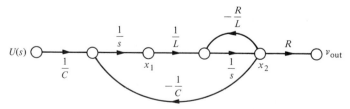

Fig. 9.4. Flow graph for the RLC network.

Our familiarity with Mason's flow graph gain formula causes us to recognize the familiar feedback factors in the denominator and the forward-path factors in the numerator. Mason's flow graph formula was discussed in Section 2.7 and is written as

$$G(s) = \frac{C(s)}{U(s)} = \frac{\Sigma_k P_k \, \Delta_k}{\Delta}. \tag{9.33}$$

When all the feedback loops are touching and all the forward paths touch the feedback loops, Eq. (9.33) reduces to

$$G(s) = \frac{\Sigma_k P_k}{1 - \Sigma_{q=1}^N L_q} = \frac{\text{Sum of the forward-path factors}}{1 - \text{sum of the feedback loop factors}}. \tag{9.34}$$

There are several flow graphs which could represent the transfer function. Two flow graph configurations are of particular interest, and we will consider these in greater detail.

In order to illustrate the derivation of the signal flow graph state model, let us initially consider the fourth-order transfer function

$$G(s) = \frac{C(s)}{U(s)} = \frac{b_0}{s^4 + a_3 s^3 + a_2 s^2 + a_1 s + a_0}$$

$$= \frac{b_0 s^{-4}}{1 + a_3 s^{-1} + a_2 s^{-2} + a_1 s^{-3} + a_0 s^{-4}}. \tag{9.35}$$

First we note that the system is fourth-order, and hence we identify four state variables (x_1, x_2, x_3, x_4). Recalling Mason's gain formula, we note that the denominator may be considered to be one minus the sum of the loop gains. Furthermore, the numerator of the transfer function is equal to the forward-path factor of the flow graph. The flow graph must utilize a minimum number of integrators equal to the order of the system. Therefore, we use four integrators to represent this system. The necessary flow graph nodes and the four integrators are shown in Fig. 9.5. Considering the simplest series interconnection of integrators, we may represent the transfer function by the flow graph of Fig. 9.6. Examining Fig. 9.6, we note that all the loops are touching and that the transfer function of this flow graph is indeed Eq. (9.35). This can be readily verified by the reader by noting that the forward-path factor of the flow graph is b_0/s^4 and the denominator is equal to 1 minus the sum of the loop gains.

Now, consider the fourth-order transfer function when the numerator polyno-

Fig. 9.5. Flow graph nodes and integrators for fourth-order system.

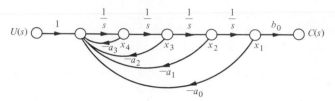

Fig. 9.6. A flow graph state model for $G(s)$ of Eq. (9.35).

mial is a polynomial in s so that we have

$$G(s) = \frac{b_3 s^3 + b_2 s^2 + b_1 s + b_0}{s^4 + a_3 s^3 + a_2 s^2 + a_1 s + a_0}$$

$$= \frac{b_3 s^{-1} + b_2 s^{-2} + b_1 s^{-3} + b_0 s^{-4}}{1 + a_3 s^{-1} + a_2 s^{-2} + a_1 s^{-3} + a_0 s^{-4}}. \tag{9.36}$$

The numerator terms represent forward-path factors in Mason's gain formula. The forward paths will touch all the loops, and a suitable signal flow graph realization of Eq. (9.36) is shown in Fig. 9.7. The forward-path factors are b_3/s, b_2/s^2, b_1/s^3, and b_0/s^4 as required to provide the numerator of the transfer function. Recall that Mason's flow graph gain formula indicates that the numerator of the transfer function is simply the sum of the forward-path factors. This general form of a signal flow graph can represent the general transfer function of Eq. (9.36) by utilizing n feedback loops involving the a_n coefficients and m forward-path factors involving the b_m coefficients.

The state variables are identified in Fig. 9.7 as the output of each energy storage element; that is, the output of each integrator. In order to obtain the set of first-order differential equations representing the state model of Fig. 9.7, we will introduce a new set of flow graph nodes immediately preceding each integrator of Fig. 9.7 [1, 8]. Since the nodes are placed before each integrator, they represent the derivative of the output of each integrator. The signal flow graph, including the added nodes, is shown in Fig. 9.8. Using the flow graph of Fig. 9.8, we are able to

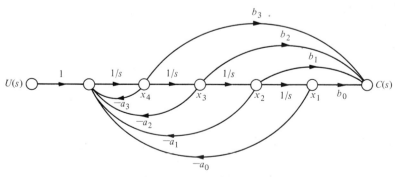

Fig. 9.7. A flow graph state model for $G(s)$ of Eq. (9.36).

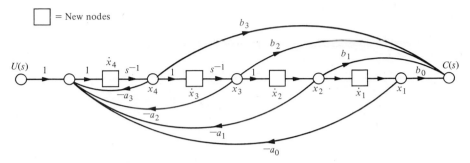

Fig. 9.8. Flow graph of Fig. 9.7 with nodes inserted.

obtain the following set of first-order differential equations describing the state of the model:

$$\dot{x}_1 = x_2,$$
$$\dot{x}_2 = x_3,$$
$$\dot{x}_3 = x_4,$$
$$\dot{x}_4 = -a_0 x_1 - a_1 x_2 - a_2 x_3 - a_3 x_4 + u. \tag{9.37}$$

Furthermore, the output is simply

$$c(t) = b_0 x_1 + b_1 x_2 + b_2 x_3 + b_3 x_4. \tag{9.38}$$

Then, in matrix form, we have

$$\dot{\mathbf{x}} = \mathbf{Ax} + \mathbf{b}u \tag{9.39}$$

or

$$\frac{d}{dt} \begin{bmatrix} x_1 \\ x_2 \\ x_3 \\ x_4 \end{bmatrix} = \begin{bmatrix} 0 & 1 & 0 & 0 \\ 0 & 0 & 1 & 0 \\ 0 & 0 & 0 & 1 \\ -a_0 & -a_1 & -a_2 & -a_3 \end{bmatrix} \begin{bmatrix} x_1 \\ x_2 \\ x_3 \\ x_4 \end{bmatrix} + \begin{bmatrix} 0 \\ 0 \\ 0 \\ 1 \end{bmatrix} u(t), \tag{9.40}$$

and the output is

$$c(t) = \mathbf{Dx} = [b_0, b_1, b_2, b_3] \begin{bmatrix} x_1 \\ x_2 \\ x_3 \\ x_4 \end{bmatrix}. \tag{9.41}$$

The flow graph structure of Fig. 9.7 is not a unique representation of Eq. (9.36), and another equally useful structure may be obtained. A flow graph which represents Eq. (9.36) equally well is shown in Fig. 9.9. In this case, the forward-path factors are obtained by feeding forward the signal $U(s)$.

Then the output signal $c(t)$ is equal to the first state variable $x_1(t)$. This flow

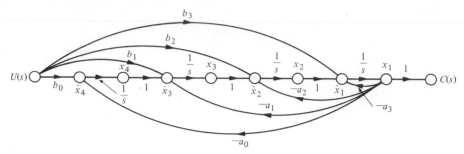

Fig. 9.9. An alternative flow graph state model for Eq. (9.36).

graph structure has the forward-path factors b_0/s^4, b_1/s^3, b_2/s^2, b_3/s, and all the forward paths touch the feedback loops. Therefore, the resulting transfer function is indeed equal to Eq. (9.36).

Using the flow graph of Fig. 9.9 to obtain the set of first-order differential equations, we obtain

$$
\begin{aligned}
\dot{x}_1 &= -a_3 x_1 + x_2 + b_3 u, \\
\dot{x}_2 &= -a_2 x_1 + x_3 + b_2 u, \\
\dot{x}_3 &= -a_1 x_1 + x_4 + b_1 u, \\
\dot{x}_4 &= -a_0 x_1 + b_0 u.
\end{aligned}
\tag{9.42}
$$

Thus, in matrix form, we have

$$
\frac{d\mathbf{x}}{dt} =
\begin{bmatrix}
-a_3 & 1 & 0 & 0 \\
-a_2 & 0 & 1 & 0 \\
-a_1 & 0 & 0 & 1 \\
-a_0 & 0 & 0 & 0
\end{bmatrix}
\mathbf{x} +
\begin{bmatrix}
b_3 \\
b_2 \\
b_1 \\
b_0
\end{bmatrix}
u(t).
\tag{9.43}
$$

While the flow graph of Fig. 9.9 represents the same transfer function as the flow graph of Fig. 9.7, the state variables of each graph are not equal since the structure of each flow graph is different. The signal flow graphs may be recognized as being equivalent to an analog computer diagram. Furthermore, we recognize that the initial conditions of the system may be represented by the initial conditions of the integrators, $x_1(0), x_2(0), \ldots, x_n(0)$. Let us consider a control system and determine the state vector differential equation by utilizing the two forms of flow graph state models.

Example 9.1. A single-loop control system is shown in Fig. 9.10. The closed-loop transfer function of the system is

$$
T(s) = \frac{C(s)}{R(s)} = \frac{2s^2 + 8s + 6}{s^3 + 8s^2 + 16s + 6}.
\tag{9.44}
$$

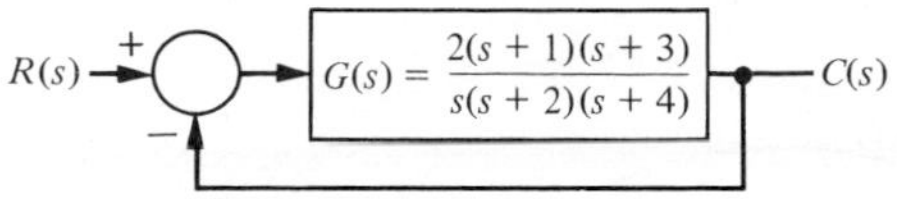

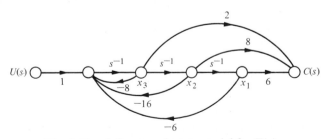

$$R(s) \xrightarrow{+} \bigcirc \longrightarrow \boxed{G(s) = \frac{2(s + 1)(s + 3)}{s(s + 2)(s + 4)}} \longrightarrow C(s)$$

Fig. 9.10. Single-loop control system.

Multiplying the numerator and denominator by s^{-3}, we have

$$T(s) = \frac{C(s)}{R(s)} = \frac{2s^{-1} + 8s^{-2} + 6s^{-3}}{1 + 8s^{-1} + 16s^{-2} + 6s^{-3}}. \tag{9.45}$$

The signal flow graph state model using the feedforward of the state variables to provide the output signal is shown in Fig. 9.11. The vector differential equation for this flow graph is

$$\dot{\mathbf{x}} = \begin{bmatrix} 0 & 1 & 0 \\ 0 & 0 & 1 \\ -6 & -16 & -8 \end{bmatrix} \mathbf{x} + \begin{bmatrix} 0 \\ 0 \\ 1 \end{bmatrix} u(t), \tag{9.46}$$

and the output is

$$c(t) = [6, 8, 2] \begin{bmatrix} x_1 \\ x_2 \\ x_3 \end{bmatrix}. \tag{9.47}$$

The flow graph state model using the feedforward of the input variable is shown in Fig. 9.12. The vector differential equation for this flow graph is

$$\dot{\mathbf{x}} = \begin{bmatrix} -8 & 1 & 0 \\ -16 & 0 & 1 \\ -6 & 0 & 0 \end{bmatrix} \mathbf{x} + \begin{bmatrix} 2 \\ 8 \\ 6 \end{bmatrix} u(t), \tag{9.48}$$

and the output is $c(t) = x_1(t)$.

We note that both signal flow graph representations of the transfer function $T(s)$ are readily obtained. Furthermore, it was not necessary to factor the numerator or denominator polynomial in order to obtain the state differential equations. Avoiding the factoring of polynomials permits us to avoid the tedious effort involved. Each

Fig. 9.11. A flow graph state model for $T(s)$.

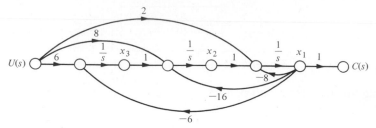

Fig. 9.12. An alternative flow graph state model for $T(s)$.

of the two signal flow graph state models represents an analog computer simulation of the transfer function. Both models require three integrators, since the system is third-order. However, it is important to emphasize that the state variables of the state model of Fig. 9.11 are not identical to the state variables of the state model of Fig. 9.12. Of course, one set of state variables is related to the other set of state variables by a suitable linear transformation of variables. A linear matrix transformation is represented by $\mathbf{y} = \mathbf{Bx}$ which transforms the x-vector into the y-vector by means of the $\mathbf{B}$-matrix (see Appendix C, especially Section C-3, for an introduction to matrix algebra). Finally, we note that the transfer function of Eq. (9.31) represents a single output linear constant coefficient system, and thus the transfer function may represent an nth-order differential equation

$$\frac{d^n c}{dt^n} + a_{n-1}\frac{d^{n-1}c}{dt^{n-1}} + \cdots + a_0 c(t) = \frac{d^m u}{dt^m} + b_{m-1}\frac{d^{m-1}u}{dt^{m-1}} + \cdots + b_0 u(t). \quad (9.49)$$

Thus one may obtain the n first-order equations for the nth-order differential equation by utilizing the signal flow graph state models of this section.

9.5 THE STABILITY OF SYSTEMS IN THE TIME DOMAIN

The stability of a system modeled by a state variable flow graph model may be readily ascertained. The stability of a system with an input-output transfer function $T(s)$ may be determined by examining the denominator polynomial of $T(s)$. Therefore, if the transfer function is written as

$$T(s) = \frac{p(s)}{q(s)},$$

where $p(s)$ and $q(s)$ are polynomials in s, the stability of the system is represented by the roots of $q(s)$. The polynomial $q(s)$, when set equal to zero, is called the characteristic equation and was discussed in Section 2.4. The roots of the characteristic equation must lie in the left-hand s-plane for the system to exhibit a stable time response. Therefore, in order to ascertain the stability of a system represented by a transfer function, we investigate the characteristic equation and utilize the

Routh-Hurwitz criterion. If the system we are investigating is represented by a signal flow graph state model, we may obtain the characteristic equation by evaluating the flow graph determinant. As an illustration of this method, let us investigate the stability of the system of Example 9.1.

Example 9.2. The transfer function $T(s)$ examined in Example 9.1 is

$$T(s) = \frac{2s^2 + 8s + 6}{s^3 + 8s^2 + 16s + 6}. \tag{9.50}$$

Clearly, the characteristic equation for this system is

$$q(s) = s^3 + 8s^2 + 16s + 6. \tag{9.51}$$

Of course, this characteristic equation is also readily obtained from either flow graph model shown in Fig. 9.7 or Fig. 9.9. Using the Routh-Hurwitz criterion, we find that the system is stable and all the roots of $q(s)$ lie in the left-hand s-plane.

   Often we determine the flow graph state model directly from a set of state differential equations. In this case, one may use the flow graph to directly determine the stability of the system by obtaining the characteristic equation from the flow graph determinant $\Delta(s)$. An illustration of this approach will aid in comprehending this method.

Example 9.3. The spread of an epidemic disease can be described by a set of differential equations [9]. The population under study is made up of three groups, x_1, x_2, and x_3, such that the group x_1 is susceptible to the epidemic disease, group x_2 is infected with the disease, and group x_3 has been removed from x_1 and x_2. The removal of x_3 will be due to immunization, death, or isolation from x_1. The feedback system can be represented by the following equations:

$$\frac{dx_1}{dt} = -\alpha x_1 - \beta x_2 + u_1(t), \tag{9.52}$$

$$\frac{dx_2}{dt} = \beta x_1 - \gamma x_2 + u_2(t), \tag{9.53}$$

$$\frac{dx_3}{dt} = \alpha x_1 + \gamma x_2. \tag{9.54}$$

The rate at which new susceptibles are added to the population is equal to $u_1(t)$ and the rate at which new infectives are added to the population is equal to $u_2(t)$. For a closed population, we have $u_1(t) = u_2(t) = 0$. It is interesting to note that these equations could equally well represent the spread of information of a new idea through a populace.

   The state variables for this system are x_1, x_2, and x_3. The signal flow diagram that represents this set of differential equations is shown in Fig. 9.13. The vector

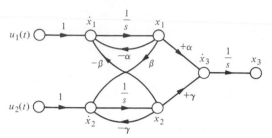

Fig. 9.13. State model flow graph for the spread of an epidemic disease.

differential equation is equal to

$$\frac{d}{dt}\begin{bmatrix} x_1 \\ x_2 \\ x_3 \end{bmatrix} = \begin{bmatrix} -\alpha & -\beta & 0 \\ \beta & -\gamma & 0 \\ \alpha & \gamma & 0 \end{bmatrix} \begin{bmatrix} x_1 \\ x_2 \\ x_3 \end{bmatrix} + \begin{bmatrix} 1 & 0 \\ 0 & 1 \\ 0 & 0 \end{bmatrix} \begin{bmatrix} u_1(t) \\ u_2(t) \end{bmatrix}. \tag{9.55}$$

By examining Eq. (9.55) and the signal flow graph, we find that the state variable x_3 is dependent upon x_1 and x_2 and does not affect the variables x_1 and x_2.

Let us consider a closed population so that $u_1(t) = u_2(t) = 0$. The equilibrium point in the state space for this system is obtained by setting $dx/dt = 0$. The equilibrium point in the state space is the point to which the system settles in the equilibrium or rest condition. Examining Eq. (9.55), we find that the equilibrium point for this system is $x_1 = x_2 = 0$. Thus, in order to determine if the system is stable and the epidemic disease is eliminated from the population, we must obtain the characteristic equation of the system. From the signal flow graph shown in Fig. 9.13, we obtain the flow graph determinant

$$\Delta(s) = 1 - (-\alpha s^{-1} - \gamma s^{-1} - \beta^2 s^{-2}) + (\alpha\gamma s^{-2}), \tag{9.56}$$

where there are three loops, two of which are nontouching. Thus, the characteristic equation is

$$q(s) = s^2 \, \Delta(s) = s^2 + (\alpha + \gamma)s + (\alpha\gamma + \beta^2) = 0. \tag{9.57}$$

Examining Eq. (9.57), we find that this system is stable when $(\alpha + \gamma) > 0$ and $(\alpha\gamma + \beta^2) > 0$.

A method of obtaining the characteristic equation directly from the vector differential equation is based on the fact that the solution to the unforced system is an exponential function. The vector differential equation without input signals is

$$\dot{x} = Ax, \tag{9.58}$$

where x is the state vector. The solution is of exponential form and one may define a constant λ such that the solution of the system for one state might be $x_i(t) = k_i e^{\lambda_i t}$. The λ_i are called the characteristic roots of the system, which are simply the roots of the characteristic equation. If we let $x = c e^{\lambda t}$ and substitute into Eq.(9.58), we have

$$\lambda \mathbf{c} e^{\lambda t} = \mathbf{A} \mathbf{c} e^{\lambda t} \tag{9.59}$$

or

$$\lambda \mathbf{x} = \mathbf{A} \mathbf{x}. \tag{9.60}$$

Equation (9.60) may be rewritten as

$$(\lambda \mathbf{I} - \mathbf{A})\mathbf{x} = \mathbf{0}, \tag{9.61}$$

where $\mathbf{I}$ equals the identity matrix and $\mathbf{0}$ equals the null matrix. The solution of this set of simultaneous equations has a nontrivial solution if and only if the determinant vanishes, that is, only if

$$\det (\lambda \mathbf{I} - \mathbf{A}) = 0. \tag{9.62}$$

The nth-order equation in λ resulting from the evaluation of this determinant is the characteristic equation and the stability of the system may be readily ascertained. Let us reconsider the previous example in order to illustrate this approach.

Example 9.4.   The vector differential equation of the epidemic system is given in Eq. (9.55). The characteristic equation is then

$$
\det (\lambda \mathbf{I} - \mathbf{A}) = \det \left\{ \begin{bmatrix} \lambda & 0 & 0 \\ 0 & \lambda & 0 \\ 0 & 0 & \lambda \end{bmatrix} - \begin{bmatrix} -\alpha & -\beta & 0 \\ \beta & -\gamma & 0 \\ \alpha & \gamma & 0 \end{bmatrix} \right\}
$$

$$
= \det \begin{bmatrix} (\lambda + \alpha) & \beta & 0 \\ -\beta & (\lambda + \gamma) & 0 \\ -\alpha & -\gamma & \lambda \end{bmatrix}
$$

$$
= \lambda [(\lambda + \alpha)(\lambda + \gamma) + \beta^2]
$$

$$
= \lambda [\lambda^2 + (\alpha + \gamma)\lambda + (\alpha\gamma + \beta^2)] = 0. \tag{9.63}
$$

Thus we obtain the characteristic equation of the system, and it is similar to that obtained in Eq. (9.57) by flow graph methods. The additional root $\lambda = 0$ results from the definition of x_3 as the integral of $(\alpha x_1 + \gamma x_2)$, and x_3 does not affect the other state variables. Thus the root $\lambda = 0$ indicates the integration connected with x_3. The characteristic equation indicates that the system is stable when $(\alpha + \gamma) > 0$ and $(\alpha\gamma + \beta^2) > 0$.

Example 9.5.   The problem of balancing a broomstick on the end of one's finger is not unlike the problem of controlling the attitude of a missile during the intitial stages of launch. This problem is the classic and intriguing problem of the inverted pendulum mounted on a cart as shown in Fig. 9.14. The cart must be moved so that mass m is always in an upright position. The state variables must be expressed in

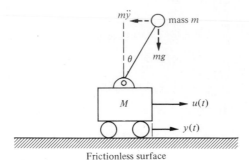

Fig. 9.14. A cart and inverted pendulum.

terms of the angular rotation $\theta(t)$ and the position of the cart $y(t)$. The differential equations describing the motion of the system may be obtained by writing the sum of the forces in the horizontal direction and the sum of the moments about the pivot point [10, 11, 27]. We will assume that $M \gg m$ and the angle of rotation θ is small so that the equations are linear. The sum of the forces in the horizontal direction is

$$M\ddot{y} + ml\ddot{\theta} - u(t) = 0, \tag{9.64}$$

where $u(t)$ equals the force on the cart and l is the distance from the mass m to the pivot point. The sum of the torques about the pivot point is

$$ml\ddot{y} + ml^2\ddot{\theta} - mlg\theta = 0. \tag{9.65}$$

The state variables for the two second-order equations are chosen as (x_1, x_2, x_3, x_4) $= (y, \dot{y}, \theta, \dot{\theta})$. Then Eqs. (9.64) and (9.65) are written in terms of the state variables as

$$M\dot{x}_2 + ml\dot{x}_4 - u(t) = 0 \tag{9.66}$$

and

$$\dot{x}_2 + l\dot{x}_4 - gx_3 = 0. \tag{9.67}$$

In order to obtain the necessary first-order differential equations, we solve for $l\dot{x}_4$ in Eq. (9.67) and substitute into Eq. (9.66) to obtain

$$m\dot{x}_2 + mgx_3 = u(t), \tag{9.68}$$

since $M \gg m$. Substituting $\dot{x}_2$ from Eq. (9.66) into Eq. (9.67), we have

$$Ml\dot{x}_4 + Mgx_3 + u(t) = 0. \tag{9.69}$$

Therefore the four first-order differential equations may be written as

$$\dot{x}_1 = x_2,$$

$$\dot{x}_2 = \frac{-mg}{M}x_3 + \frac{1}{M}u(t),$$

$$\dot{x}_3 = x_4,$$

$$\dot{x}_4 = \frac{g}{l}x_3 - \frac{1}{Ml}u(t). \tag{9.70}$$

Thus the system matrix is

$$\mathbf{A} = \begin{bmatrix} 0 & 1 & 0 & 0 \\ 0 & 0 & -(m/Mg) & 0 \\ 0 & 0 & 0 & 1 \\ 0 & 0 & g/l & 0 \end{bmatrix}. \tag{9.71}$$

The characteristic equation can be obtained from the determinant of $(\lambda \mathbf{I} - \mathbf{A})$ as follows

$$\det \begin{bmatrix} \lambda & -1 & 0 & 0 \\ 0 & \lambda & m/Mg & 0 \\ 0 & 0 & \lambda & -1 \\ 0 & 0 & -(g/l) & \lambda \end{bmatrix} = \lambda\left(\lambda\left(\lambda^2 - \frac{g}{l}\right)\right) = \lambda^2\left(\lambda^2 - \frac{g}{l}\right) = 0. \tag{9.72}$$

The characteristic equation indicates that there are two roots at $\lambda = 0$, a root at $\lambda = +\sqrt{g/l}$, and a root at $\lambda = -\sqrt{g/l}$. Clearly, the system is unstable, since there is a root in the right-hand plane at $\lambda = +\sqrt{g/l}$.

A control system may be designed so that if $u(t)$ is a function of the state variables, a stable system will result. The design of a stable feedback control system is based on a suitable selection of a feedback system structure. Therefore, considering the control of the cart and the unstable inverted pendulum shown in Fig. 9.14, one must measure and utilize the state variables of the system in order to control the cart. Thus if one desires to measure the state variable $x_3 = \theta$, a potentiometer connected to the shaft of the pendulum hinge could be used. Similarly, the rate of change of the angle, $x_4 = \dot{\theta}$, could be measured by using a tachometer generator. The state variables, x_1 and x_2, which are the position and velocity of the cart, may also be measured by suitable sensors. If the state variables are all measured, then they may be utilized in a feedback controller so that $u = \mathbf{hx}$, where $\mathbf{h}$ is the feedback matrix. Since the state vector $\mathbf{x}$ represents the state of the system, knowledge of $\mathbf{x}(t)$ and the equations describing the system dynamics provide sufficient information for control and stabilization of a system. This design approach is called *state variable feedback* [3, 4]. In order to illustrate the utilization of state variable feedback, let us reconsider the unstable portion of the inverted pendulum system and design a suitable state variable feedback control system.

Example 9.6. In order to investigate the unstable portion of the inverted pendulum system, let us consider a reduced system. If we assume that the control signal is an acceleration signal and the mass of the cart is negligible, we may focus on the unstable dynamics of the pendulum. When $u(t)$ is an acceleration signal, Eq. (9.67) becomes

$$gx_3 - l\dot{x}_4 = \dot{x}_2 = \ddot{y} = u(t). \tag{9.73}$$

For the reduced system, where the control signal is an acceleration signal, the position and velocity of the cart are integral functions of $u(t)$. The portion of the state vector under consideration is $[x_3, x_4] = [\theta, \dot{\theta}]$. Thus the state vector differential equation reduces to

$$\frac{d}{dt}\begin{bmatrix} x_3 \\ x_4 \end{bmatrix} = \begin{bmatrix} 0 & 1 \\ g/l & 0 \end{bmatrix}\begin{bmatrix} x_3 \\ x_4 \end{bmatrix} + \begin{bmatrix} 0 \\ -(1/l) \end{bmatrix}u(t). \tag{9.74}$$

Clearly, the **A** matrix of Eq. (9.74) is simply the lower right-hand portion of the **A** matrix of Eq. (9.71) and the system has the characteristic equation $(\lambda^2 - (g/l))$ with one root in the right-hand s-plane. In order to stabilize the system, we generate a control signal which is a function of the two state variables x_3 and x_4. Then we have

$$u(t) = \mathbf{hx}$$
$$= [h_1, h_2]\begin{bmatrix} x_3 \\ x_4 \end{bmatrix}$$
$$= h_1 x_3 + h_2 x_4. \tag{9.75}$$

Substituting this control signal relationship into Eq. (9.74), we have

$$\begin{bmatrix} \dot{x}_3 \\ \dot{x}_4 \end{bmatrix} = \begin{bmatrix} 0 & 1 \\ g/l & 0 \end{bmatrix}\begin{bmatrix} x_3 \\ x_4 \end{bmatrix} + \begin{bmatrix} 0 \\ -(1/l)(h_1 x_3 + h_2 x_4) \end{bmatrix}. \tag{9.76}$$

Combining the two additive terms on the right side of the equation, we find

$$\begin{bmatrix} \dot{x}_3 \\ \dot{x}_4 \end{bmatrix} = \begin{bmatrix} 0 & 1 \\ 1/l\,(g - h_1) & -(h_2/l) \end{bmatrix}\begin{bmatrix} x_3 \\ x_4 \end{bmatrix}. \tag{9.77}$$

Therefore, obtaining the characteristic equation, we have

$$\det\begin{bmatrix} +\lambda & -1 \\ -(1/l)(g - h_1) & \lambda + h_2/l \end{bmatrix} = \lambda\left(\lambda + \frac{h_2}{l}\right) - \frac{1}{l}(g - h_1)$$
$$= \lambda^2 + \left(\frac{h_2}{l}\right)\lambda + \frac{1}{l}(h_1 - g). \tag{9.78}$$

Thus for the system to be stable, we require that $(h_2/l) > 0$ and $h_1 > g$. Hence we have stabilized an unstable system by measuring the state variables x_3 and x_4 and using the control function $u = h_1 x_3 + h_2 x_4$ to obtain a stable system.

In this section we have developed a useful method for investigating the stability of a system represented by a state vector differential equation. Furthermore, we have established an approach for the design of a feedback control system by using the state variables as the feedback variables in order to increase the stability of the system. In the next two sections, we will be concerned with developing two methods for obtaining the time response of the state variables.

9.6 THE TIME RESPONSE AND THE TRANSITION MATRIX

It is often desirable to obtain the time response of the state variables of a control system and thus examine the performance of the system. The transient response of a system can be readily obtained by evaluating the solution to the state vector differential equation. We found in Section 9.3 that the solution for the state vector differential equation (Eq. 9.24) was

$$\mathbf{x}(t) = \boldsymbol{\phi}(t)\mathbf{x}(0) + \int_0^t \boldsymbol{\phi}(t - \tau)\mathbf{Bu}(\tau) \, d\tau. \tag{9.79}$$

Clearly, if the initial conditions $\mathbf{x}(0)$, the input $\mathbf{u}(\tau)$, and the transition matrix $\boldsymbol{\phi}(t)$ are known, the time response of $\mathbf{x}(t)$ can be numerically evaluated. Thus the problem focuses on the evaluation of $\boldsymbol{\phi}(t)$, the transition matrix which represents the response of the system. Fortunately, the transition matrix may be readily evaluated by using the signal flow graph techniques with which we are already familiar.

However, before proceeding to the evaluation of the transition matrix using signal flow graphs, we should note that several other methods exist for evaluating the transition matrix such as the evaluation of the exponential series

$$\boldsymbol{\phi}(t) = e^{At} = \sum_{k=0}^{\infty} \frac{\mathbf{A}^k t^k}{k!}$$

in a truncated form [12,13]. A digital computer program titled Program Transition which utilizes a truncated series to evaluate $\boldsymbol{\phi}(t)$ and the state variables of a system is given in Appendix E. Also, there exist several efficient methods for the evaluation of $\boldsymbol{\phi}(t)$ by means of a computer algorithm; these methods are discussed in Appendix D [13,14].

A series of computer programs to assist in the calculation of the state variable response of systems is often available in local computer centers. Such a series of programs is available written in Fortran IV. This series provides the transition matrix as well as the time response of a system [15].

In addition, we found in Eq. 9.23 that $\boldsymbol{\phi}(s) = [s\mathbf{I} - \mathbf{A}]^{-1}$. Therefore, if $\boldsymbol{\phi}(s)$ is obtained by completing the matrix inversion, we can obtain $\boldsymbol{\phi}(t)$ by noting that $\boldsymbol{\phi}(t) = \mathcal{L}^{-1}\{\boldsymbol{\phi}(s)\}$. However, the matrix inversion process is unwieldy for higher-order systems.

The usefulness of the signal flow graph state model for obtaining the transition matrix becomes clear upon consideration of the Laplace transformation version of

Eq. (9.79) when the input is zero. Taking the Laplace transformation of Eq. (9.79), when $\mathbf{u}(\tau) = 0$, we have

$$\mathbf{X}(s) = \boldsymbol{\phi}(s)\mathbf{x}(0). \tag{9.80}$$

Therefore one may evaluate the Laplace transform of the transition matrix from the signal flow graph by determining the relation between a state variable $X_i(s)$ and the state initial conditions $[x_1(0), x_2(0), \ldots, x_n(0)]$. Then the state transition matrix is simply the inverse transform of $\boldsymbol{\phi}(s)$; that is,

$$\boldsymbol{\phi}(t) = \mathcal{L}^{-1} \{\boldsymbol{\phi}(s)\}. \tag{9.81}$$

The relationship between a state variable $X_i(s)$ and the initial conditions $\mathbf{x}(0)$ is obtained by using Mason's gain formula. Thus for a second-order system, we would have

$$X_1(s) = \phi_{11}(s)x_1(0) + \phi_{12}(s)x_2(0),$$
$$X_2(s) = \phi_{21}(s)x_1(0) + \phi_{22}(s)x_2(0), \tag{9.82}$$

and the relation between $X_1(s)$ as an output and $x_1(0)$ as an input may be evaluated by Mason's formula. All the elements of the transition matrix, $\phi_{ij}(s)$, may be obtained by evaluating the individual relationships between $X_i(s)$ and $x_j(0)$ from the state model flow graph. An example will illustrate this approach to the determination of the transition matrix.

Example 9.7.   The signal flow graph state model of the *RLC* network of Fig. 9.3 is shown in Fig. 9.4. This *RLC* network, which was discussed in Sections 9.3 and 9.4, may be represented by the state variables $x_1 = v_c$ and $x_2 = i_L$. The initial conditions, $x_1(0)$ and $x_2(0)$, represent the initial capacitor voltage and inductor current, respectively. The flow graph, including the initial conditions of each state variable, is shown in Fig. 9.15. The initial conditions appear as the initial value of the state variable at the output of each integrator.

In order to obtain $\boldsymbol{\phi}(s)$, we set $U(s) = 0$. When $R = 3$, $L = 1$, and $C = \frac{1}{2}$, we obtain the signal flow graph shown in Fig. 9.16, where the output and input nodes are deleted since they are not involved in the evaluation of $\boldsymbol{\phi}(s)$. Then, using Mason's gain formula, we obtain $X_1(s)$ in terms of $x_1(0)$ as

$$X_1(s) = \frac{1 \cdot \Delta_1(s) \cdot (x_1(0)/s)}{\Delta(s)}, \tag{9.83}$$

where $\Delta(s)$ is the graph determinant and $\Delta_1(s)$ is the path cofactor. The graph determinant is

$$\Delta(s) = 1 + 3s^{-1} + 2s^{-2}.$$

The path cofactor is $\Delta_1 = 1 + 3s^{-1}$ since the path between $x_1(0)$ and $X_1(s)$ does not touch the loop with the factor $-3s^{-1}$. Therefore the first element of the transition

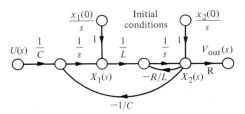

Fig. 9.15. Flow graph of the RLC network.

matrix is

$$\phi_{11}(s) = \frac{(1 + 3s^{-1})(1/s)}{1 + 3s^{-1} + 2s^{-2}} = \frac{(s + 3)}{(s^2 + 3s + 2)}. \tag{9.84}$$

The element $\phi_{12}(s)$ is obtained by evaluating the relationship between $X_1(s)$ and $x_2(0)$ as

$$X_1(s) = \frac{(-2s^{-1})(x_2(0)/s)}{1 + 3s^{-1} + 2s^{-2}}. \tag{9.85}$$

Therefore we obtain

$$\phi_{12}(s) = \frac{-2}{s^2 + 3s + 2}. \tag{9.86}$$

Similarly, for $\phi_{21}(s)$ we have

$$\phi_{21}(s) = \frac{(s^{-1})(1/s)}{1 + 3s^{-1} + 2s^{-2}} = \frac{1}{s^2 + 3s + 2}. \tag{9.87}$$

Finally, for $\phi_{22}(s)$ we obtain

$$\phi_{22}(s) = \frac{1(1/s)}{1 + 3s^{-1} + 2s^{-2}} = \frac{s}{s^2 + 3s + 2}. \tag{9.88}$$

Therefore, the transition matrix in Laplace transformation form is

$$\phi(s) = \begin{bmatrix} (s + 3)/(s^2 + 3s + 2) & -2/(s^2 + 3s + 2) \\ 1/(s^2 + 3s + 2) & s/(s^2 + 3s + 2) \end{bmatrix} \tag{9.89}$$

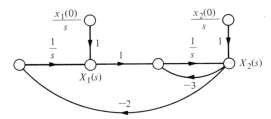

Fig. 9.16. Flow graph of the RLC network with $U(s) = 0$.

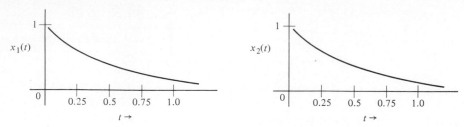

Fig. 9.17. Time response of the state variables of the *RLC* network for $x_1(0) = x_2(0) = 1$.

The factors of the characteristic equation are $(s + 1)$ and $(s + 2)$ so that

$$(s + 1)(s + 2) = s^2 + 3s + 2.$$

Then the transition matrix is

$$\boldsymbol{\phi}(t) = \mathcal{L}^{-1}\{\boldsymbol{\phi}(s)\} = \begin{bmatrix} (2e^{-t} - e^{-2t}) & (-2e^{-t} + 2e^{-2t}) \\ (e^{-t} - e^{-2t}) & (-e^{-t} + 2e^{-2t}) \end{bmatrix}. \tag{9.90}$$

The evaluation of the time response of the *RLC* network to various initial conditions and input signals may now be evaluated by utilizing Eq. (9.79). For example, when $x_1(0) = x_2(0) = 1$ and $u(t) = 0$, we have

$$\begin{bmatrix} x_1(t) \\ x_2(t) \end{bmatrix} = \boldsymbol{\phi}(t) \begin{bmatrix} 1 \\ 1 \end{bmatrix} = \begin{bmatrix} e^{-2t} \\ e^{-2t} \end{bmatrix}. \tag{9.91}$$

The response of the system for these initial conditions is shown in Fig. 9.17. The trajectory of the state vector $(x_1(t), x_2(t))$ on the $(x_1$ vs. $x_2)$-plane is shown in Fig. 9.18.

The evaluation of the time response is facilitated by the determination of the transition matrix. While this approach is limited to linear systems, it is a powerful method and utilizes the familiar signal flow graph to evaluate the transition matrix.

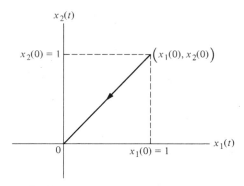

Fig. 9.18. Trajectory of the state vector in the x_1-x_2 plane.

9.7 A DISCRETE-TIME EVALUATION OF THE TIME RESPONSE

The response of a system represented by a state vector differential equation may be obtained by utilizing a *discrete-time approximation*. The discrete-time approximation is based on the division of the time axis into sufficiently small time increments. Then the values of the state variables are evaluated at the successive time intervals; that is, $t = 0, T, 2T, 3T, \ldots$, where T is the increment of time $\Delta t = T$. This approach is a familiar method utilized in numerical analysis and digital computer numerical methods. If the time increment, T, is sufficiently small compared with the time constants of the system, the response evaluated by discrete-time methods will be reasonably accurate.

The linear state vector differential equation is written as

$$\dot{\mathbf{x}} = \mathbf{A}\mathbf{x} + \mathbf{B}\mathbf{u}. \tag{9.92}$$

The basic definition of a derivative is

$$\dot{\mathbf{x}}(t) = \lim_{\Delta t \to 0} \frac{\mathbf{x}(t + \Delta t) - \mathbf{x}(t)}{\Delta t}. \tag{9.93}$$

Therefore we may utilize this definition of the derivative and determine the value of $\mathbf{x}(t)$ when t is divided in small intervals $\Delta t = T$. Thus, approximating the derivative as

$$\dot{\mathbf{x}} = \frac{\mathbf{x}(t + T) - \mathbf{x}(t)}{T}, \tag{9.94}$$

we substitute into Eq. (9.92) to obtain

$$\frac{\mathbf{x}(t + T) - \mathbf{x}(t)}{T} = \mathbf{A}\mathbf{x}(t) + \mathbf{B}\mathbf{u}(t). \tag{9.95}$$

Solving for $\mathbf{x}(t + T)$, we have

$$\begin{aligned}\mathbf{x}(t + T) &= T\mathbf{A}\mathbf{x}(t) + \mathbf{x}(t) + T\mathbf{B}\mathbf{u}(t) \\ &= (T\mathbf{A} + \mathbf{I})\mathbf{x}(t) + T\mathbf{B}\mathbf{u}(t),\end{aligned} \tag{9.96}$$

where t is divided into intervals of width T. Therefore the time t is written as $t = kT$, where k is an integer index so that $k = 0, 1, 2, 3, \ldots$. Then Eq. (9.96) is written as

$$\mathbf{x}[(k + 1)T] = (T\mathbf{A} + \mathbf{I})\mathbf{x}(kT) + T\mathbf{B}\mathbf{u}(kT). \tag{9.97}$$

Therefore the value of the state vector at the $(k + 1)$st time instant is evaluated in terms of the value of $\mathbf{x}$ and $\mathbf{u}$ at the kth time instant. Equation (9.97) may be rewritten as

$$\mathbf{x}(k + 1) = \boldsymbol{\psi}(T)\mathbf{x}(k) + T\mathbf{B}\mathbf{u}(k), \tag{9.98}$$

where $\psi(T) = (T\mathbf{A} + \mathbf{I})$ and the symbol T is omitted from the arguments of the variables. Equation (9.98) clearly relates the resulting operation for obtaining $\mathbf{x}(t)$ by evaluating the discrete-time approximation $\mathbf{x}(k + 1)$ in terms of the previous value $\mathbf{x}(k)$. This recurrence operation is a sequential series of calculations and is very suitable for digital computer calculation. In order to illustrate this approximate approach, let us reconsider the evaluation of the response of the *RLC* network of Fig. 9.3.

Example 9.8.   We shall evaluate the time response of the *RLC* network without determining the transition matrix by using the discrete-time approximation. Therefore, as in Example 9.7, we will let $R = 3$, $L = 1$, and $C = \frac{1}{2}$. Then, as we found in Eqs. (9.27) and (9.28), the state vector differential equation is

$$\dot{\mathbf{x}} = \begin{bmatrix} 0 & -(1/C) \\ 1/L & -(R/L) \end{bmatrix} \mathbf{x} + \begin{bmatrix} +(1/C) \\ 0 \end{bmatrix} u(t) = \begin{bmatrix} 0 & -2 \\ 1 & -3 \end{bmatrix} \mathbf{x} + \begin{bmatrix} +2 \\ 0 \end{bmatrix} u(t). \quad (9.99)$$

Now we must choose a sufficiently small timer interval T so that the approximation of the derivative (Eq. 9.94) is reasonably accurate. A suitable interval T must be chosen so that the solution to Eq. (9.97) is stable. Usually, one chooses T to be less than one-half of the smallest time constant of the system. Therefore, since the shortest time constant of this system is 0.5 sec [recalling that the characteristic equation is $(s + 1)(s + 2)$], we might choose $T = 0.2$. Alternatively, if a digital computer is used for the calculations and the number of calculations is not important, we would choose $T = 0.05$ in order to obtain greater accuracy. However, we note that as we decrease the increment size, the number of calculations increases proportionally if we wish to evaluate the output from 0 to 10 sec, for example. Using $T = 0.2$ sec, Eq. (9.97) is

$$\mathbf{x}(k + 1) = (0.2\mathbf{A} + \mathbf{I})\mathbf{x}(k) + 0.2\mathbf{B}\mathbf{u}(k). \quad (9.100)$$

Therefore

$$\psi(T) = \begin{bmatrix} 1 & -0.4 \\ 0.2 & 0.4 \end{bmatrix} \quad (9.101)$$

and

$$T\mathbf{B} = \begin{bmatrix} +0.4 \\ 0 \end{bmatrix}. \quad (9.102)$$

Now let us evaluate the response of the system when $x_1(0) = x_2(0) = 1$ and $u(t) = 0$ as in Example 9.7. The response at the first instant, when $t = T$, or $k = 0$, is

$$\mathbf{x}(1) = \begin{bmatrix} 1 & -0.4 \\ 0.2 & 0.4 \end{bmatrix} \mathbf{x}(0) = \begin{bmatrix} 0.6 \\ 0.6 \end{bmatrix}. \quad (9.103)$$

Table 9.1

Time t	0	0.2	0.4	0.6	0.8
Exact $x_1(t)$	1	0.67	0.448	0.30	0.20
Approximate $x_1(t)$, $T = 0.1$	1	0.64	0.41	0.262	0.168
Approximate $x_1(t)$, $T = 0.2$	1	0.60	0.36	0.216	0.130

Then the response at the time $t = 2T = 0.4$ sec, or $k = 1$, is

$$\mathbf{x}(2) = \begin{bmatrix} 1 & -0.4 \\ 0.2 & 0.4 \end{bmatrix} \mathbf{x}(1) = \begin{bmatrix} 0.36 \\ 0.36 \end{bmatrix} \tag{9.104}$$

The value of the response as $k = 2, 3, 4, \ldots$ is then evaluated in a similar manner.

Now let us compare the actual response of the system evaluated in the previous section using the transition matrix with the approximate response determined by the discrete-time approximation. We found in Example 9.7 that the exact value of the state variables, when $x_1(0) = x_2(0) = 1$, is $x_1(t) = x_2(t) = e^{-2t}$. Therefore the exact values may be readily calculated and compared with the approximate values of the time response as in Table 9.1. The approximate time response values for $T = 0.1$ sec are also given in Table 9.1. The error, when $T = 0.2$, is approximately a constant equal to 0.07, and thus the percentage error compared to the initial value is 7%. When T is equal to 0.1 sec, the percentage error compared to the initial value is approximately 3.5%. If we use $T = 0.05$, the value of the approximation, when time $t = 0.2$ sec, is $x_1(t) = 0.655$, and the error has been reduced to 1.5% of the initial value.

Therefore, if one is using a digital computer to evaluate the transient response by evaluating the discrete-time response equations, a value of T equal to one-tenth of the smallest time constant of the system would be selected. The value of this method merits another illustration of the evaluation of the time response of a system.

Example 9.9. Let us reconsider the state variable representation of the spread of an epidemic disease presented in Example 9.3. The state vector differential equation was given in Eq. (9.55). When the constants are $\alpha = \beta = \gamma = 1$, we have

$$\dot{\mathbf{x}} = \begin{bmatrix} -1 & -1 & 0 \\ 1 & -1 & 0 \\ 1 & 1 & 0 \end{bmatrix} \mathbf{x} + \begin{bmatrix} 1 & 0 \\ 0 & 1 \\ 0 & 0 \end{bmatrix} \mathbf{u}. \tag{9.105}$$

The characteristic equation of this system, as determined in Eq. (9.57), is $s(s^2 + s + 2) = 0$, and thus the system has complex roots. Let us determine the transient response of the spread of disease when the rate of new susceptibles is zero; that is, $u_1 = 0$. The rate of adding new infectives is represented by $u_2(0) = 1$ and $u_2(k) =$

0 for $k \geq 1$; that is, one new infective is added at the initial time only (this is equivalent to a pulse input). Since the time constant of the complex roots is $1/\zeta\omega_n$ = 2 sec, we will use $T = 0.2$ sec. (Note that the actual time units might be months and the units of the input in thousands.)

Then the discrete-time equation is

$$\mathbf{x}(k + 1) = \begin{bmatrix} 0.8 & -0.2 & 0 \\ 0.2 & 0.8 & 0 \\ 0.2 & 0.2 & 1 \end{bmatrix} \mathbf{x}(k) + \begin{bmatrix} 0 \\ 0.2 \\ 0 \end{bmatrix} u_2(k). \tag{9.106}$$

Therefore the response at the first instant, $t = T$, is obtained when $k = 0$ as

$$\mathbf{x}(1) = \begin{bmatrix} 0 \\ 0.2 \\ 0 \end{bmatrix}, \tag{9.107}$$

when $x_1(0) = x_2(0) = x_3(0) = 0$. Then the input $u_2(k)$ is zero for $k \geq 1$ and the response at $t = 2T$ is

$$\mathbf{x}(2) = \begin{bmatrix} 0.8 & -0.2 & 0 \\ 0.2 & 0.8 & 0 \\ 0.2 & 0.2 & 1 \end{bmatrix} \begin{bmatrix} 0 \\ 0.2 \\ 0 \end{bmatrix} = \begin{bmatrix} -0.04 \\ 0.16 \\ 0.04 \end{bmatrix}. \tag{9.108}$$

The response at $t = 3T$ is then

$$\mathbf{x}(3) = \begin{bmatrix} 0.8 & -0.2 & 0 \\ 0.2 & 0.8 & 0 \\ 0.2 & 0.2 & 1 \end{bmatrix} \begin{bmatrix} -0.04 \\ 0.16 \\ 0.04 \end{bmatrix} = \begin{bmatrix} -0.064 \\ 0.120 \\ 0.064 \end{bmatrix}, \tag{9.109}$$

and the ensuing values may then be readily evaluated.

The discrete-time approximate method is particularly useful for evaluating the time response of nonlinear systems. The transition matrix approach is limited to linear systems, but the discrete-time approximation is not limited to linear systems and may be readily applied to nonlinear and time-varying systems. The basic state vector differential equation may be written as

$$\dot{\mathbf{x}} = \mathbf{f}(\mathbf{x}, \mathbf{u}, t), \tag{9.110}$$

where $\mathbf{f}$ is a function, not necessarily linear, of the state vector $\mathbf{x}$ and the input vector $\mathbf{u}$. The column vector $\mathbf{f}$ is the column matrix of functions of $\mathbf{x}$ and $\mathbf{u}$. If the system is a linear function of the control signals, Eq. (9.110) becomes

$$\dot{\mathbf{x}} = \mathbf{f}(\mathbf{x}, t) + \mathbf{Bu}. \tag{9.111}$$

If the system is not time-varying, that is, if the coefficients of the differential equation are constants, Eq. (9.111) is then

$$\dot{\mathbf{x}} = \mathbf{f}(\mathbf{x}) + \mathbf{Bu}. \tag{9.112}$$

Let us consider Eq. (9.112) for a nonlinear system and determine the discrete-time approximation. Using Eq. (9.94) as the approximation to the derivative, we have

$$\frac{\mathbf{x}(t + T) - \mathbf{x}(t)}{T} = \mathbf{f}(\mathbf{x}(t)) + \mathbf{B}\mathbf{u}(t). \tag{9.113}$$

Therefore, solving for $\mathbf{x}(k + 1)$ when $t = kT$, we obtain

$$\mathbf{x}(k + 1) = \mathbf{x}(k) + T[\mathbf{f}(\mathbf{x}(k)) + \mathbf{B}\mathbf{u}(k)]. \tag{9.114}$$

Similarly, the general discrete-time approximation to Eq. (9.110) is

$$\mathbf{x}(k + 1) = \mathbf{x}(k) + T\mathbf{f}(\mathbf{x}(k), \mathbf{u}(k), k). \tag{9.115}$$

Now let us reconsider the previous example when the system is nonlinear.

Example 9.10.   The spread of an epidemic disease is actually best represented by a set of nonlinear equations as

$$\begin{aligned}
\dot{x}_1 &= -\alpha x_1 - \beta x_1 x_2 + u_1(t), \\
\dot{x}_2 &= \beta x_1 x_2 - \gamma x_2 + u_2(t), \\
\dot{x}_3 &= \alpha x_1 + \gamma x_2,
\end{aligned} \tag{9.116}$$

where the interaction between the groups is represented by the nonlinear term $x_1 x_2$. Now, the transition matrix approach and the characteristic equation are not applicable since the system is nonlinear. As in the previous example, we will let $\alpha = \beta = \gamma = 1$ and $u_1(t) = 0$. Also, $u_2(0) = 1$ and $u_2(k) = 0$ for $k \geq 1$. We will again select the time increment as $T = 0.2$ sec and the initial conditions as $\mathbf{x}^T(0) = [1, 0, 0]$. Then, substituting the $t = kT$ and

$$\dot{x}_i(k) = \frac{x_i(k + 1) - x_i(k)}{T} \tag{9.117}$$

into Eq. (9.116), we obtain

$$\frac{x_1(k + 1) - x_1(k)}{T} = -x_1(k) - x_1(k)x_2(k),$$

$$\frac{x_2(k + 1) - x_2(k)}{T} = +x_1(k)x_2(k) - x_2(k) + u_2(k),$$

$$\frac{x_3(k + 1) - x_3(k)}{T} = x_1(k) + x_2(k). \tag{9.118}$$

Solving these equations for $x_i(k + 1)$ and recalling that $T = 0.2$, we have

$$\begin{aligned}
x_1(k + 1) &= 0.8x_1(k) - 0.2x_1(k)x_2(k), \\
x_2(k + 1) &= 0.8x_2(k) + 0.2x_1(k)x_2(k) + 0.2u_2(k), \\
x_3(k + 1) &= x_3(k) + 0.2x_1(k) + 0.2x_2(k).
\end{aligned} \tag{9.119}$$

Then the response of the first instant, $t = T$, is

$$
\begin{aligned}
x_1(1) &= 0.8x_1(0) = 0.8, \\
x_2(1) &= 0.2u_2(k) = 0.2, \\
x_3(1) &= 0.2x_1(0) = 0.2.
\end{aligned}
\tag{9.120}
$$

Again, using Eq. (9.119) and noting that $u_2(1) = 0$, we have

$$
\begin{aligned}
x_1(2) &= 0.8x_1(1) - 0.2x_1(1)x_2(1) = 0.608, \\
x_2(2) &= 0.8x_2(1) + 0.2x_1(1)x_2(1) = 0.192, \\
x_3(2) &= x_3(1) + 0.2x_1(1) + 0.2x_2(1) = 0.40.
\end{aligned}
\tag{9.121}
$$

At the third instant, when $t = 3T$, we obtain

$$
\begin{aligned}
x_1(3) &= 0.463, \\
x_2(3) &= 0.177, \\
x_3(3) &= 0.56.
\end{aligned}
\tag{9.122}
$$

The evaluation of the ensuing values follows in a similar manner. We note that the response of the nonlinear system differs considerably from the response of the linear model considered in the previous example.

Finally, in order to illustrate the utility of the time-domain approach, let us consider a system which is both nonlinear and time-varying.

Example 9.11.   Let us again consider the spread of an epidemic disease which is represented by the nonlinear differential equations of Eq. (9.116) when the coefficient β is time-varying. The time variation, $\beta(t)$, might represent the cyclic seasonal variation of interaction between the susceptible group and the infected group; that is, the interaction between $x_1(t)$ and $x_2(t)$ is greatest in the winter when people are together indoors and is least during the summer. Therefore, we represent the time variation of β as

$$
\beta(t) = \beta_0 + \sin \omega_0 t
$$

$$
= 1 + \sin \left(\frac{\pi}{2} \right) t.
\tag{9.123}
$$

Then, the nonlinear time-varying differential equations are

$$
\begin{aligned}
\dot{x}_1(t) &= -\alpha_1 x_1(t) - \beta(t)x_1(t)x_2(t) + u_1(t), \\
\dot{x}_2(t) &= \beta(t)x_1(t)x_2(t) - \gamma x_2(t) + u_2(t), \\
\dot{x}_3(t) &= \alpha x_1(t) + \gamma x_2(t).
\end{aligned}
\tag{9.124}
$$

As in Example 9.10, we will let $\alpha = \gamma = 1$, $u_1(t) = 0$, $u_2(0) = 1$ and $u_2(k) = 0$ for $k \geq 1$. Also, we will again select the time increment as $T = 0.2$ sec and the initial conditions as $\mathbf{x}^T(0) = [1, 0, 0]$. Then substituting $t = kT$ and

$$
\dot{x}_i(k) = \frac{x_i(k + 1) - x_i(k)}{T}
\tag{9.125}
$$

into Eq. (9.124), we obtain

$$\frac{x_1(k+1) - x_1(k)}{T} = -x_1(k) - \beta(k)x_1(k)x_2(k),$$

$$\frac{x_2(k+1) - x_2(k)}{T} = \beta(k)x_1(k)x_2(k) - x_2(k) + u_2(k),$$

$$\frac{x_3(k+1) - x_3(k)}{T} = x_1(k) + x_2(k). \tag{9.126}$$

Solving these equations for $x_i(k+1)$ and again recalling that $T = 0.2$, we obtain

$$x_1(k+1) = 0.8x_1(k) - 0.2\beta(k)x_1(k)x_2(k),$$
$$x_2(k+1) = 0.8x_2(k) + 0.2\beta(k)x_1(k)x_2(k) + 0.2u_2(k),$$
$$x_3(k+1) = x_3(k) + 0.2x_1(k) + 0.2x_2(k). \tag{9.127}$$

Then the response at the first instant, $t = T$, is

$$x_1(1) = 0.8x_1(0) = 0.8,$$
$$x_2(1) = 0.2u_2(0) = 0.2,$$
$$x_3(1) = 0.2x_1(0) = 0.2. \tag{9.128}$$

Noting that

$$\beta(k) = 1 + \sin(\pi/2)kT = 1 + \sin(0.314k), \tag{9.129}$$

we evaluate the response at the second instant as

$$x_1(2) = 0.8x_1(1) - 0.2\beta(1)x_1(1)x_2(1) = 0.598,$$
$$x_2(2) = 0.8x_2(1) + 0.2\beta(1)x_1(1)x_2(1) = 0.202,$$
$$x_3(2) = x_3(1) + 0.2x_1(1) + 0.2x_2(1) = 0.40. \tag{9.130}$$

The evaluation of the ensuing values of the time response follows in a similar manner by using Eq. (6.129) to account for the time-varying parameter.

The evaluation of the time response of the state variables of linear systems is readily accomplished by using either (1) the transition matrix approach or (2) the discrete-line approximation. The transition matrix of linear systems is readily obtained from the signal flow graph state model. For a nonlinear system, the discrete-time approximation provides a suitable approach, and the discrete-time approximation method is particularly useful if a digital computer is used for numerical calculations.

9.8 SUMMARY

In this chapter we have considered the description and analysis of systems in the time-domain. The concept of the state of a system and the definition of the state variables of a system were discussed. The selection of a set of state variables in terms of the variables which describe the stored energy of a system was examined, and the nonuniqueness of a set of state variables was noted. The state vector differential equation and the solution for x(t) were discussed. Two alternative signal flow graph model structures were considered for representing the transfer function (or differential equation) of a system. Using Mason's gain formula, we noted the ease of obtaining the flow graph model. The vector differential equation representing these flow graph models was also examined. Then the stability of a system represented by a state variable formulation was considered in terms of the vector differential equation. The time response of a linear system and its associated transition matrix were discussed and the utility of Mason's gain formula for obtaining the transition matrix was illustrated. Finally, a discrete-time evaluation of the time response of a nonlinear system and a time-varying system were considered. It was noted that the flow graph state model is equivalent to an analog computer diagram where the output of each integrator is a state variable. Also, we found that the discrete-time approximation for a time response, as well as the transition matrix formulation for linear systems, is readily applicable for programming and solution by using a digital computer. The time-domain approach is applicable to biological, chemical, sociological, business, and physiological systems as well as physical systems and thus appears to be an approach of general interest.

PROBLEMS

9.1. An *RLC* circuit is shown in Fig. P9.1. (a) Identify a suitable set of state variables. (b) Obtain the set of first-order differential equations in terms of the state variables. (c) Write the state equations in matrix form. (d) Draw the state variable flow graph.

9.2. A *balanced* bridge network is shown in Fig. P9.2. (a) Show that the **A** and **B** matrices for this circuit are

$$\mathbf{A} = \begin{bmatrix} -(2/(R_1 + R_2)C & 0 \\ 0 & 2R_1R_2/(R_1 + R_2)L \end{bmatrix}$$

$$\mathbf{B} = 1/(R_1 + R_2) \begin{bmatrix} 1/C & 1/C \\ R_2/L & -R_2/L \end{bmatrix}$$

(b) Draw the state model flow graph. The state variables are $(x_1, x_2) = (v_c, i_L)$.

9.3. An *RLC* network is shown in Fig. P9.3. Define the state variables as $x_1 = i_L$ and $x_2 = v_c$. (a) Obtain the vector differential equation. (b) Draw the state model flow graph.

Partial Answer:

$$\mathbf{A} = \begin{bmatrix} 0 & 1/L \\ -(1/C) & -(1/RC) \end{bmatrix}$$

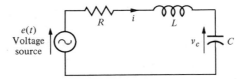

Figure P9.1

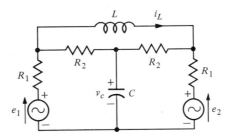

Figure P9.2

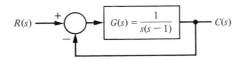

Figure P9.3

9.4. The transfer function of a system is

$$T(s) = \frac{C(s)}{R(s)} = \frac{s^2 + 3s + 3}{s^3 + 2s^2 + 3s + 1}.$$

(a) Draw the flow graph state model where all the state variables are fed back to the input node and the state variables are fed forward to the output signal as in Fig. 9.7. (b) Determine the vector differential equation for the flow graph of (a). (c) Draw the flow graph state model where the input signal is fed forward to the state variables as in Fig. 9.9. (d) Determine the vector differential equation for the flow graph of (c).

9.5. A closed-loop control system is shown in Fig. P9.5. (a) Determine the closed-loop transfer function $T(s) = C(s)/R(s)$. (b) Draw the state model flow graph for the system using the form of Fig. 9.7, where the state variables are fed back to the input node. (c) Determine the state vector differential equation.

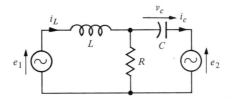

Figure P9.5

9.6. Consider the case of the rabbits and foxes in Australia. The number of rabbits is x_1 and if left alone would grow indefinitely (until the food supply was exhausted) so that [9,16]

$$\dot{x}_1 = kx_1.$$

However, with foxes present on the continent, we have

$$\dot{x}_1 = kx_1 - ax_2,$$

where x_2 is the number of foxes, Now, if the foxes must have rabbits to exist, we have

$$\dot{x}_2 = -hx_2 + bx_1.$$

Determine if this system is stable and thus decays to the condition $x_1(t) = x_2(t) = 0$ at $t = \infty$. What are the requirements on a, b, h, and k for a stable system? What is the result when k is greater than h?

9.7. An automatic depth-control system for a robot submarine is shown in Fig. 9.7 [17]. The depth is measured by a pressure transducer. The gain of the stern plane actuator is $K = 10^{-4}$ when the velocity is 25 ft/sec. The submarine has the approximate transfer function.

$$G(s) = \frac{(s + 0.2)^2}{(s^2 + 0.01)},$$

and the feedback transducer is $H(s) = 1$. (a) Obtain a flow graph state model. (b) Determine the vector differential equation for the system. (c) Determine whether the system is stable.

9.8. Recent attempts to model some cybernetic aspects of sociology and economics are interesting and illustrative of the state model concept. The state variable representing the number of underdeveloped, unindustrialized nations is x_1, which is equal to n_u/n_d, where n_u = number of underdeveloped nations and n_d = number of developed (industrial) nations. This variable, x_1, is an index of the development and industrialization of the nations of the world. The second state variable, x_2, represents the tendency toward underdevelopment or the lack of industrialization, which presumably can be reduced by means of foreign aid, technical development assistance, and education. However, it has been noted that the gap between the developed and underdeveloped nations is growing, for underdeveloped nations tend to remain underdeveloped compared, on a relative basis, to the highly industrialized nations [18,30]. Therefore, we might use the set of equations

$$\dot{x}_1 = -c_1x_1 + c_2x_2,$$
$$\dot{x}_2 = -c_3x_2 + c_4x_1.$$

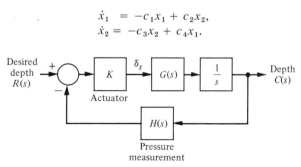

Figure P9.7

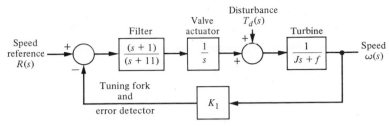

Figure P9.9

Determine whether the system is stable. That is, in this case, do the variables $x_1(t)$ and $x_2(t)$ eventually decrease to zero? For what relationship between the coefficients is this system stable?

9.9. A speed-control system utilizing fluid flow components may be designed [19]. The system is a pure fluid-control system, since the system does not have any moving mechanical parts. The fluid may be a gas or liquid. A system is desired which maintains the speed within 0.5% of the desired speed by using a tuning fork reference and a valve actuator. Fluid-control systems are insensitive and reliable over a wide range of temperature, electromagnetic and nuclear radiation, acceleration, and vibration. The amplification within the system is achieved by using a fluid jet deflection amplifier. The system may be designed for a 500-kw steam turbine with a speed of 12,000 rpm. The block diagram of the system is shown in Fig. P9.9. The friction of the large inertia turbine is negligible and thus $f = 0$. The closed-loop gain is $K_1/J = 1$, where $K_1 = J = 10^4$. (a) Determine the closed-loop transfer function

$$T(s) = \frac{\omega(s)}{R(s)},$$

and draw the state model flow graph for the form of Fig. 9.7, where all the state variables are fed back to the input node. (b) Determine the state vector differential equation. (c) Determine whether the system is stable by investigating the characteristic equation obtained from the **A** matrix.

9.10. Many control systems must operate in two dimensions, for example, the x- and y-axes. A two-axis control system is shown in Fig. P9.10, where a set of state variables is identified. The gain of each axis is K_1 and K_2 respectively. (a) Obtain the state vector differ-

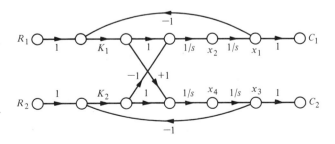

Figure P9.10

ential equation. (b) Find the characteristic equation from the **A** matrix, and determine whether the system is stable for suitable values of K_1 and K_2.

9.11. Consider the problem of rabbits and foxes in Australia as discussed in Problem 9.6. The values of the constants are $k = 1$, $h = 3$, and $a = b = 2$. Is this system stable? (a) Determine the transition matrix for this system. (b) Using the transition matrix, determine the response of the system when $x_1(0) = x_2(0) = 10$.

9.12. Consider the model of the development of unindustrialized nations discussed in Problem 9.8. (a) Determine the system's transition matrix when $c_1 = c_3 = 1$, and $c_2 = c_4 = 2$. (b) Determine the response of the system when $x_1(0) = 0.5$ and $x_2(0) = 0.8$.

9.13. Reconsider the *RLC* circuit of Problem 9.1 when $R = 6.0$, $L = \frac{1}{4}$, and $C = \frac{1}{2}$. (a) Determine whether the system is stable by finding the characteristic equation with the aid of the **A** matrix. (b) Determine the transition matrix of the network. (c) When the initial inductor current is 0.1 ampere, $v_c(0) = 0$, and $e(t) = 0$, determine the response of the system. (d) Repeat part (b) when the initial conditions are zero and $e(t) = E$, for $t > 0$, where E is a constant.

9.14. Consider the system discussed in Problem 9.5. (a) Determine whether the system is stable by obtaining the characteristic equation from the **A** matrix. (b) An additional feedback signal is to be added in order to stabilize the system. If $R(s) = -K_s C(s)$, the state variable x_2 is fedback so that $r(t) = -Kx_2(t)$. Determine the minimum value of K so that the system is stable.

9.15. Reconsider the rabbits-and-foxes ecology discussed in Problems 9.6 and 9.11 when we take the depletion of food into account. Then we have

$$\dot{x}_1 = kx_1 - ax_2 + ax_3,$$
$$\dot{x}_2 = -hx_2 + bx_1,$$
$$\dot{x}_3 = \beta x_3 - \gamma x_1,$$

where $x_3 = $ the amount of rabbit food per unit area. Again, assume that $k = 1$, $h = 3$, and $a = b = 2$. In this case $\beta = 1$ and $\alpha = 0.1$. Determine a suitable value for γ in order to eliminate the rabbits.

9.16. An improved model of the rabbits and foxes ecology of Problem 9.6 can be represented by the nonlinear equations

$$\dot{x}_1 = kx_1 - ax_1x_2,$$
$$\dot{x}_2 = -hx_2 + bx_1x_2.$$

The second terms of the right-hand side of each equation represents the probability of interaction, that is, a rabbit and a fox meeting. When $k = 1$, $h = 3$, and $a = b = 0.5$, determine the response of the system when $x_1(0) = x_2(0) = 10$.

9.17. The dynamics of a controlled submarine are significantly different from an aircraft, a missile, or a surface ship. This difference results primarily from the moment in the vertical plane due to the buoyancy effect. Therefore, it is interesting to consider the control of the depth of a submarine. The equations describing the dynamics of a submarine may be obtained by using Newton's laws and the angles defined in Fig. P9.17. In order to simplify the equa-

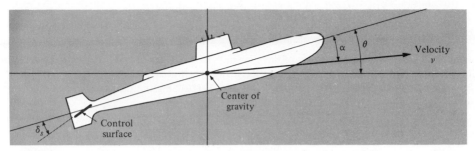

Figure P9.17

tions, we will assume that θ is a small angle and the velocity v is constant and equal to 25 ft/sec. The state variables of the submarine, considering only vertical control, are $x_1 = \theta$, $x_2 = d\theta/dt$ and $x_3 = \alpha$, where α is the angle of attack. Thus, the state vector differential equation for this system, when the submarine has an Albacore type hull, is

$$\dot{\mathbf{x}} = \begin{bmatrix} 0 & 1 & 0 \\ -0.0071 & -0.111 & 0.12 \\ 0 & 0.07 & -0.3 \end{bmatrix} \mathbf{x} + \begin{bmatrix} 0 \\ -0.095 \\ +0.072 \end{bmatrix} u(t),$$

where $u(t) = \delta_s(t)$ the deflection of the stern plane. (a) Determine whether the system is stable. (b) Using the discrete-time approximation, determine the response of the system to a stern plane step command of 0.285 degrees with the initial conditions equal to zero. Use a time increment of T equal to 2 sec. (c) (optional) Using a time increment of $T = 1$ sec and a digital computer, obtain the transient response for each state for 80 sec. Compare the response calculated for (b) and (c).

9.18. Consider the educated professional man whose economic status is such that he can easily purchase all the material necessities. His excess wages ω are divided into the following parts: one portion goes into increasing his possessions, and the second portion into increasing his social and economic status by joining clubs, entertaining socially, and similar activities [20,21,30]. The division of the excess wages is represented by the equations

$$\frac{dp}{dt} = K_p a \omega,$$

$$\frac{dL}{dt} = K_L (1 - a) \omega,$$

where K_p and K_L are the respective constants, p is the monetary value of the accrued possessions, and L is the value of the social and economic status on the social ladder. The variable J, the Jones variable, is a measure of the desire to maintain the status and is equal to

$$J = K_1 L - p.$$

The attempt to remain in competition with one's neighbors causes the man to attempt to increase his wages as

$$\frac{d\omega}{dt} = K_2 J.$$

(a) Draw the state variable flow graph and identify the state variables. (b) Write the state vector differential equation in matrix form and identify the coefficients of the matrix. (c) When $a = 0.5$, $K_p = K_L = 1$, and $K_1 = 1$, determine whether the system is stable by determining the characteristic roots of the system. The units are normalized so that time is in months and wages are in thousands of dollars. (d) The initial conditions of the system are $p(0) = 0$, $L(0) = 0$, and $\omega(0) = 1$ (thousand dollars). Utilizing the discrete approximation of the system, determine the value of $L(t)$ and $p(t)$ after 3 months.

9.19. As measured against the success of modeling physical systems, there has been only slight progress in the quantitative analysis of socioeconomic systems. A discrete-time model of a simple socioeconomic system has been proposed [9, 20, 21]. The system considered is a nonprofit organization receiving financial support from the constituents of the surrounding community and a national parent organization. A state variable of the organization is necessary to represent the capability of the organization to provide its service that is measured in terms of propensity. Another variable represents the economic flow analogous to force or current in mechanical or electrical systems, respectively. The discrete-time state equation of such a socioeconomic system is

$$\mathbf{x}(k+1) = \begin{bmatrix} 0.995 & a_{12} & 0 & 0 \\ 0 & a_{22} & 1 & -1 \\ 0 & 0.005 & 0 & 0 \\ -0.02 & 0 & 0 & 0 \end{bmatrix} \mathbf{x}(k) + \begin{bmatrix} 0 \\ 1 \\ 0 \\ 0 \end{bmatrix} u(k),$$

where $x_1(k)$ = physical facilities (propensity), $x_2(k)$ = liquid assets, $x_3(k)$ = income on investment of assets, $x_4(k)$ = operating costs, $u(k)$ = total income in the kth month. The coefficients of the $\psi(T)$ matrix are fixed on the basis of the financial policy of the organization. Therefore, the coefficient -0.02 indicates that the operating costs are directly proportional to the dollar value of the fixed assets, $x_1(k)$. The coefficients a_{12} and a_{22} are fixed by the governing board such that $a_{12} + a_{22} = 1$. (a) Calculate the response of this system to constantly increasing support, $u(k) = k$, when $\mathbf{x}(0) = 0$ and $a_{12} = a_{22} = 0.5$. (b) Calculate the response of this system when support is cut to zero, so that $u(k) = 0$, and $x_1(0) = 10,000$, $x_2(0) = 10,000$, $x_3(0) = 0$, $x_4(0) = 1000$. Again assume that $a_{12} = a_{22} = 0.5$. (c) Repeat part (b) when $a_{12} = 0.2$ and $a_{22} = 0.8$, so that the board pursues a conservative policy. Compare the responses of (b) and (c) for the loss of support.

9.20. Preliminary results show that lobsters can be raised—and their growth controlled—in heated water facilities [28]. It is desired to control the temperature in order to grow the lobsters more rapidly for sale. The cost of heating the water and maintaining the lobsters must be included in determining the economics of this operation. A linear state variable model is

$$\frac{dW}{dt} = aW + bu,$$

$$\frac{dC}{dt} = dW + ku,$$

where W = weight of the lobsters, C = cost of process, and u = water temperature. Determine the response of this system and select values for b, d and k such that the weight of the lobsters will increase more rapidly than the cost of the system.

9.21. The derivative of a state variable may be approximated by the equation

$$\dot{x}(t) \simeq \frac{1}{2T}[3x(k + 1) - 4x(k) + x(k - 1)].$$

This approximation of the derivative utilizes two past values to estimate the derivative, while Eq. (9.94) uses one past value of the state variable. Using this approximation for the derivative, repeat the calculations for Example 9.8. Compare the resulting approximation for $x_1(t)$, $T = 0.2$, with the results given in Table 9.1. Is this approximation more accurate?

9.22. Determine a state variable model of *one* of the processes listed below. Do not attempt to include the variables of secondary importance. Obtain the requirements on the coefficients of the system in order to maintain a stable system.

(a) Traffic control at an intersection based on the measurement of the arrival of vehicles at the intersection and using a four-way traffic signal light.

(b) The response of two interconnected power utility systems and the dependence upon load schedules and energy sources.

(c) A new business with limited capital has continuing expenses and invests continually in inventory and product development before there are sales of the product. Thus, cash on hand diminishes to a minimum at which the sales rate equals the operating expenses. If the minimum cash goes below a certain level the company is bankrupt. Obtain a model of the system and add a control for maintaining at least a minimum cash amount [24].

(d) Almost all agricultural crops suffer damage due to insects that eat or otherwise destroy the crop. Nature's way of keeping these pests in check is by exposing them to other insects who act as predators. Establish a model for one prey and one predator [16].

(e) Establish a model for one agricultural business (such as wheat) which includes the effect of government price supports and land bank controls [30].

9.23. There are several forms which are equivalent signal flow graph state models. Two equivalent state flow graph models for a fourth-order equation (Eq. 9.36) are shown in Fig. 9.7 and Fig. 9.9. Another alternative structure for a state flow graph model is shown in Fig. P9.23. In this case the system is second-order and the input-output transfer function is

$$G(s) = \frac{C(s)}{U(s)} = \frac{b_1 s + b_0}{s^2 + a_1 s + a_0}.$$

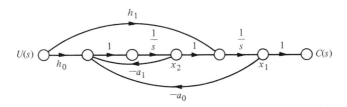

Figure P9.23

(a) Verify that the flow graph of Fig. P9.23 is in fact a model of $G(s)$. (b) Show that the vector differential equation representing the flow graph model of Fig. P9.23 is

$$\dot{x} = \begin{bmatrix} 0 & 1 \\ -a_0 & -a_1 \end{bmatrix} x + \begin{bmatrix} h_1 \\ h_0 \end{bmatrix} u(t),$$

where $h_1 = b_1$ and $h_0 = b_0 - b_1 a_1$.

9.24. A nuclear reactor which has been operating in equilibrium at a high thermal-neutron flux level is suddenly shut down. At shutdown, the density of xenon 135(X) and iodine 135(I) are 3×10^{15} and 7×10^{16} atoms per unit volume, respectively. The half lives of I 135 and Xe 135 nucleides are 6.7 and 9.2 hours, respectively. The decay equations are

$$\dot{I} = -\frac{0.693}{6.7} I,$$

$$\dot{X} = -\frac{0.693}{9.2} X - I.$$

Determine the concentrations of I 135 and Xe 135 as functions of time following shutdown by determining (a) the transition matrix and the system response, and (b) a discrete-time evaluation of the time response. Verify that the response of the system is that shown in Fig. P9.24.

9.25. Consider the following mathematical model of the social interaction of humans which is often called group dynamics. The four system variables of interest are (1) the intensity of interaction (or communication) among the members of the group, $x_1(t)$; (2) the amount of friendliness or group identification among group members, $x_2(t)$; (3) the total amount of activity carried on by a member of the group, $x_3(t)$; (4) the amount of activity imposed on it by its external environment, $u(t)$. A simple model might be represented by the set of equations [9]

$$\dot{x}_1 = a_1 x_2 + a_2 x_3,$$
$$\dot{x}_2 = b(x_1 - \beta x_2),$$
$$\dot{x}_3 = c_1(x_2 - \gamma x_3) + c_2(u - x_3).$$

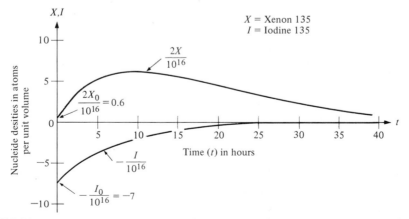

Figure P9.24

The first equation indicates that interaction results from friendliness or activity. The second equation indicates that friendliness will increase as the amount of interaction grows larger than friendliness. The third equation relates the effect of all variables on a change of activity. (a) Determine the requirements on the coefficients for a stable system. When the system is stable, the changes in the variables evidently decay to zero, and equilibrium is attained when $u(t) = 0$. Determine the values of the variables at equilibrium when $u(t) = 0$. Is this representative of the social disintegration of the group? Is an external force $u(t)$ required to maintain the group activity? (b) The problem of group morale has also been studied with the aid of this model. A group is said to have positive morale when the activity $x_3(t)$ exceeds that required by an external social force $u(t) = U$, where U is a constant. Determine the necessary relationship for the coefficients a_1, a_2, β, and γ for positive morale. (c) Determine the transient response of a group, such as a college social fraternity, which is highly active and is subjected to a high level of external social forces, $u(t) = U$. Assume that initially $x_1(0) = x_2(0) = x_3(0) = 0$, that $a_1 = a_2 = b = \beta = c_1 = c_2 = 1$, and $\gamma = -2$.

9.26. A very interesting game called Prisoner's Dilemma is illustrative of the utility of time-domain models of sociological behavior. Two prisoners, held incommunicado, are charged with the same crime. They can be convicted only if either confesses. The payoff, or performance index, associated with conviction on the basis of confessions by both is equal to minus one. If neither confesses, it is plus one. If only one confesses, he is set free for having turned state's evidence and given a reward of plus two in addition. The prisoner who holds out is convicted on the strength of the testimony and is given a more severe sentence than if he had confessed. His payoff is minus two. It is in the interest of each to confess irrespective of what the other does, but it is in their collective interest to hold out. There is no satisfactory solution to this paradoxical game. A simple model of cooperation of the prisoners has been developed where $c_1(t)$ = propensity or inclination for cooperation by the first prisoner and $c_2(t)$ = propensity for cooperation by the second prisoner as follows [25]:

$$\frac{dc_1}{dt} = \alpha_1 c_1 c_2 - \beta_1 c_1 (1 - c_2) - \gamma_1 c_2 (1 - c_1) + \delta_1 (1 - c_1)(1 - c_2),$$

$$\frac{dc_2}{dt} = \alpha_2 c_1 c_2 - \beta_2 c_2 (1 - c_1) - \gamma_2 c_1 (1 - c_2) + \delta_2 (1 - c_2)(1 - c_1).$$

The term with α represents the contribution due to cooperation and β represents the contribution due to unreciprocated cooperation. The terms due to γ represent a negative contribution due to a successful defection. We are unsure of the sign of δ_i representing the potential for possible defection. (a) Initially consider the system when

$$\alpha_i = \gamma_i = 1,$$
$$\beta_i = 2,$$
$$\delta_i = 1.$$

Determine the equilibrium points, that is, the points at which the state variables have ceased changing. (b) When $\alpha_i = 4$, $\beta_i = 3$, $\gamma_i = 2$, $\delta_i = -1$, determine the equilibrium points. (c) When $\alpha_1 = \alpha_2$, $\beta_1 = \beta_2$, $\gamma_1 = \gamma_2$, and $\delta_1 = \delta_2$, determine the requirement on the coefficients for the system to exhibit cooperation. (d) Consider the system of part (a) and determine the response of the system when

$$c_1(0) = c_2(0) = 0.5.$$

9.27. Consider the automatic ship steering system discussed in Problems 7.13 and 8.16. The state variable form of the system differential equation is

$$\dot{\mathbf{x}}(t) = \begin{bmatrix} -0.05 & -6 & 0 & 0 \\ -10^{-3} & -0.15 & 0 & 0 \\ 1 & 0 & 0 & 13 \\ 0 & 1 & 0 & 0 \end{bmatrix} \mathbf{x}(t) + \begin{bmatrix} -0.2 \\ 0.03 \\ 0 \\ 0 \end{bmatrix} \delta(t),$$

where $\mathbf{x}^T(t) = [\dot{v}, \omega_s, y, \theta]$. The state variables are $x_1 = \dot{v}$ = the transverse velocity; $x_2 = \omega_s$ = angular rate of ship's coordinate frame relative to response frame; $x_3 = y$ = deviation distance on an axis perpendicular to the track; $x_4 = \theta$ = deviation angle. (a) Determine whether the system is stable. (b) Feedback may be added so that

$$\delta(t) = -k_1 x_1 - k_3 x_3.$$

Determine whether this system is stable for suitable values of k_1 and k_3.

9.28. It is desirable to use well-designed controllers to maintain building temperature with solar collector space heating systems. One solar heating system may be described by [29]

$$\frac{dx_1}{dt} = 3x_1 + u_1 + u_2,$$

$$\frac{dx_2}{dt} = 2x_2 + u_2 + d,$$

where x_1 = temperature deviation from desired equilibrium and x_2 = temperature of the storage material (e.g., a water tank). Also u_1 and u_2 are the respective flow rates of conventional and solar heat, where the transport medium is forced air. A solar disturbance on the storage temperature is represented by d (e.g., overcast skies). Write the matrix equations and solve for the system response from equilibrium when $u_1 = 0$, $u_2 = 1$ and $d = 1$.

9.29. An important problem in modern urban society is the treatment of wastes. Biological treatment of urban wastes in sewerage treatment plants is useful for eliminating pollutants from waste water prior to discharging the water into the local rivers or lakes. A strain of bacteria is introduced into the waste flow that will grow at the expense of the pollutant and use the pollutant as the nutrient for its growth. The bacteria strain grows in the continuous flow of waste [22, 26]. The normalized material balance equations can be approximated by

$$\frac{dx}{dt} = ksx - Dx,$$

$$\frac{ds}{dt} = -ksx - Dx,$$

where x is the bacteria concentration used for the pollution removal, s is the pollutant serving as the nutrient substance for the bacteria, and D = rate of flow/volume, often called the dilution rate. (a) Find the equilibrium condition for this nonlinear system. (b) Find the response of the system for several units of time when $k = 10$ and $D = 1$. Since the above equations are normalized, time may be considered as units of minutes.

9.30. The dynamics of an armament competition between two nations can be approximated by two equations as follows [9]:

$$\dot{x} = ky - ax + u_1,$$
$$\dot{y} = qx - by + u_2,$$

where x and y represent the defensive weapons of each nation, and u_1 and u_2 represent the defense attitude factors. The coefficients are k, a, q and b. Determine the equilibrium points of the system, if it is stable. What are the conditions on the nations if it is to be a stabilized system?

9.31. For the fourth-order system of Problem 4.12, determine the state vector equations. Then for the approximate second-order model determine the state vector equations and compare with the fourth-order equations.

9.32. Consider a model of the interaction of the OPEC nations and the United States. The OPEC nations want to increase the price of their oil and maintain control over their destiny. The United States wishes to decrease the price of the imported oil and decrease the OPEC nation's control. The two state variables are price, p, and control, c. One model is then

$$p(k + 1) = p(k) - u_1(k) + u_2(k),$$
$$c(k + 1) = c(k) - u_1(k) + u_2(k),$$

where $u_1(k)$ = action by the United States, and $u_2(k)$ = action by OPEC. The United States selects a control action so that $u_1(k) = .5p(k)$, and OPEC selects $u_2(k) = .4c(k - 1)$. Examine the response of this system for several time periods. What will be the ultimate outcome if this model is a true representation? Assume that $p(0) = c(0) = 10$ and $c(-1) = 10$. Reexamine the situation if $u_2(k) = .6c(k)$.

REFERENCES

1. R. C. Dorf, *Time-Domain Analysis and Design of Control Systems,* Addison-Wesley, Reading, Mass., 1965.

2. C. M. Clare and D. K. Frederick, *Modeling and Analysis of Dynamic Systems,* Houghton Mifflin, Boston, 1978.

3. A. P. Sage, *Linear Systems Control,* Matrix Publishers, Champaign, Ill., 1978.

4. F. H. Raven, *Automatic Control Engineering,* 3rd ed., McGraw-Hill, New York, 1978; Ch. 9.

5. R. C. Dorf, *Matrix Algebra—A Programmed Introduction,* Wiley, New York, 1969.

6. C. R. Wylie, *Advance Engineering Mathematics,* 4th ed., McGraw-Hill, New York, 1975.

7. W. J. Gajda and W. E. Biles, *Engineering: Modeling and Computation,* Houghton Mifflin, New York, 1978.

8. B. C. Kuo, *Automatic Control Systems,* 3rd ed., Prentice-Hall, Englewood Cliffs, N.J., 1975.

9. M. Olinick, *An Introduction to Mathematical Models in the Social and Life Sciences,* Addison-Wesley, Reading, Mass., 1978.

10. H. Hemami, F. C. Weimer, and S. H. Koozekanani, "Some Aspects of the Inverted Pendulum Problem for Modeling of Locomotion Systems," *IEEE Transactions on Automatic Control,* December 1973; pp. 658–661.

11. H. Hemami, et al., "Biped Stability Considerations With Vestibular Models," *Proceedings of the 1977 JACC,* 1977; pp. 796–803.

12. G. F. Franklin and J. D. Powell, *Digital Control of Dynamic Systems,* Addison-Wesley, Reading, Mass, 1980.

13. C. Kallstrom, "Computing EXP(A) and SEXP(AS)ds," *Report 7309,* Lund Institute of Technology, Sweden, March 1973.

14. R. C. Dorf, *Introduction to Computers and Computer Science,* 2nd ed., Boyd and Fraser Publishing Co., San Francisco, 1977.

15. J. L. Melsa and S. K. Jones, *Computer Programs for Computational Assistance in the Study of Linear Control Theory,* McGraw-Hill, New York, 1973.

16. H. J. Gold, *Mathematical Modeling of Biological Systems*, Wiley, New York, 1977.

17. "An Undersea Robot," *Mini-Micro Systems,* May 1978; pp. 54–55.

18. A. K. Bhattacharya, *Foreign Trade and Economic Development,* Lexington Books, Lexington, Mass, 1976.

19. *Fluidics Quarterly,* **10,** 1978, Delbridge Publishing Co., Stanford, Cal.

20. J. N. Warfield, *Societal Systems: Planning, Policy and Complexity,* Wiley, New York, 1976.

21. I. W. Sandberg, "On the Mathematical Theory of Interaction in Social Groups," *IEEE Transactions on Systems, Man and Cybernetics,* September 1974; pp. 432–445.

22. P. R. Ehrlich, A. H. Ehrlich and J. P. Holdren, *Ecoscience: Population, Resources, Environment,* W. H. Freeman and Co., San Francisco, 1977.

23. S. E. Shladover, "Steering Controller Design for Automated Guideway Transit Vehicles," *J. of Dynamic Systems, Measurement, and Control,* March 1978, pp. 1–7.

24. N. J. Mass, "Economic Fluctuations: A Framework for Analysis and Policy Design," *IEEE Transactions on Systems, Man and Cybernetics,* June 1978, pp. 437–449.

25. A Rapoport and A. M. Chammah, *Prisoner's Dilemma,* University of Michigan Press, Ann Arbor, Mich., 1965.

26. S. Aborhey and D. Williamson, "State and Parameter Estimation of Microbial Growth Processes," *Automatica,* September 1978; pp. 493–498.

27. V. Jorgenson, "A Ball-Balancing System for Demonstration of Basic Concepts in the State Space Control Theory," *International J. Engineering Education,* **11,** 1974; pp. 367–376.

28. L. W. Botsford, H. E. Rauch, and R. A. Shleser, "Optimal Temperature Control of a Lobster Plant," *IEEE Transactions on Automatic Control,* October 1974; pp. 541–543.

29. G. R. Johnson, "A Microprocessor Based Solar Controller," *Proceedings of the IEEE Decision and Control Conference,* 1978; pp. 336–340.

30. J. D. Pitchford and S. J. Turnovsky, *Application of Control Theory to Economic Analysis,* Elsevier, New York, 1977.

31. D. G. Luenberger, *Introduction to Dynamic Systems,* Wiley, New York, 1979.

32. P. F. Blackman, *Introduction to State Variable Analysis,* Macmillan, London, 1977.

33. P. K. Sinha, "Control, Measurement and Design Aspects of Low-Speed Magnetically Suspended Vehicles," *Measurement and Control,* November 1978; pp. 427–435.

10 / The Design and Compensation of Feedback Control Systems

10.1 INTRODUCTION

The performance of a feedback control system is of primary importance. This subject was discussed at length in Chapter 4 and quantitative measures of performance were developed. We have found that a suitable control system is stable and that it results in an acceptable response to input commands, is less sensitive to system parameter changes, results in a minimum steady-state error for input commands, and, finally, is able to eliminate the effect of undesirable disturbances. A feedback control system that provides an optimum performance without any necessary adjustments is rare indeed. Usually one finds it necessary to compromise among the many conflicting and demanding specifications and to adjust the system parameters to provide a suitable and acceptable performance when it is not possible to obtain all the desired optimum specifications.

We have considered at several points in the preceding chapters the question of design and adjustment of the system parameters in order to provide a desirable response and performance. In Chapter 4, we defined and established several suitable measures of performance. Then, in Chapter 5, we determined a method of investigating the stability of a control system, since we recognized that a system is unacceptable unless it is stable. In Chapter 6, we utilized the root locus method to effect a design of a self-balancing scale (Section 6.4) and then illustrated a method of parameter design by using the root locus method (Section 6.5). Furthermore, in Chapters 7 and 8, we developed suitable measures of performance in terms of the frequency variable ω and utilized them to design several suitable control systems. Finally, using time-domain methods in Chapter 9, we investigated the selection of feedback parameters in order to stabilize a system. Thus, we have been considering the problems of the design of feedback control systems as an integral part of the subjects of the preceding chapters. It is now our purpose to study the question somewhat further and to point out several significant design and compensation methods.

We have found in the preceding chapters that it is often possible to adjust the system parameters in order to provide the desired system response. However, we often find that we are not able to simply adjust a system parameter and thus obtain the desired performance. Rather we are forced to reconsider the structure of the system and redesign the system in order to obtain a suitable one. That is, we must examine the scheme or plan of the system and obtain a new design or plan which results in a suitable system. Thus, *the design of a control system is concerned with the arrangement, or the plan, of the system structure and the selection of suitable components and parameters.* For example, if one desires a set of performance measures to be less than some specified values, often one encounters a conflicting set of requirements. Thus, if we wish a system to have a percent overshoot less than 20% and $\omega_n T_p = 3.3$, we obtain a conflicting requirement on the system damping ratio, ζ, as can be seen by examining Fig. 4.8. Now, if we are unable to relax these two performance requirements, we must alter the system in some way. Often the alteration or adjustment of a control system, in order to provide a suitable performance, is called *compensation;* that is, compensation is the adjustment of a system in order to make up for deficiencies or inadequacies. It is the purpose of this chapter to consider briefly the issue of the design and compensation of control systems.

In redesigning a control system in order to alter the system response, an additional component is inserted within the structure of the feedback system. It is this additional component or device that equalizes or compensates for the performance deficiency. The compensating device may be an electric, mechanical, hydraulic, pneumatic, or other type of device or network, and is often called a *compensator.* Commonly, an electric circuit serves as a compensator in many control systems. The transfer function of the compensator is designated as $G_c(s) = E_{out}(s)/E_{in}(s)$ and the compensator may be placed in a suitable location within the structure of the system. Several types of compensation are shown in Fig. 10.1 for a simple single-loop feedback control system. The compensator placed in the feedforward path is

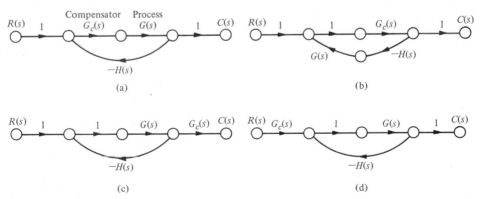

Fig. 10.1. Types of compensation. (a) Cascade compensation. (b) Feedback compensation. (c) Output or load compensation. (d) Input compensation.

called a cascade or series compensator. Similarly, the other compensation schemes are called feedback, output or load, and input compensation, as shown in Fig. 10.1(b), (c), and (d), respectively. The selection of the compensation scheme depends upon a consideration of the specifications, the power levels at various signal nodes in the system, and the networks available for use. It will not be possible for us to consider all the possibilities in this chapter, and the reader is referred to further work following the introductory material of this chapter [1, 2].

10.2 APPROACHES TO COMPENSATION

The performance of a control system may be described in terms of the time-domain performance measures or the frequency-domain performance measures. The performance of a system may be specified by requiring a certain peak time, T_p, maximum overshoot, and settling-time for a step input. Furthermore, it is usually necessary to specify the maximum allowable steady-state error for several test signal inputs and disturbance inputs. These performance specifications may be defined in terms of the desirable location of the poles and zeros of the closed-loop system transfer function, $T(s)$. Thus the location of the s-plane poles and zeros of $T(s)$ may be specified. As we found in Chapter 6, the locus of the roots of the closed-loop system may be readily obtained for the variation of one system parameter. However, when the locus of roots does not result in a suitable root configuration, one must add a compensating network (Fig. 10.1) in order to alter the locus of the roots as the parameter is varied. Therefore, one may utilize the root locus method and determine a suitable compensator network transfer function so that the resultant root locus results in the desired closed-loop root configuration.

Alternatively, one may describe the performance of a feedback control system in terms of frequency performance measures. Then a system may be described in terms of the peak of the closed-loop frequency response, M_{p_ω}, the resonant frequency, ω_r, the bandwidth, and the phase margin of the system. One may add a suitable compensation network, if necessary, in order to satisfy the system specifications. The design of the network $G_c(s)$, is developed in terms of the frequency response as portrayed on the polar plane, the Bode diagram, or the Nichols chart. Since a cascade transfer function is readily accounted for on a Bode plot by adding the frequency response of the network, we usually prefer to approach the frequency response methods by utilizing the Bode diagram.

Thus, the compensation of a system is concerned with the alteration of the frequency response or the root locus of the system in order to obtain a suitable system performance. For frequency response methods, we are concerned with altering the system so that the frequency response of the compensated system will satisfy the system specifications. Thus, in the case of the frequency response approach, one utilizes compensation networks to alter and reshape the frequency characteristics represented on the Bode diagram and Nichols chart.

Alternatively, the compensation of a control system may be accomplished in the

s-plane by root locus methods. For the case of the s-plane, the designer wishes to alter and reshape the root locus so that the roots of the system will lie in the desired position in the s-plane.

The time-domain method, expressed in terms of state variables, may also be utilized to design a suitable compensation scheme for a control system. Typically, one is interested in controlling the system with a control signal, $u(t)$, which is a function of several measurable state variables. Then one develops a state-variable controller which operates on the information available in measured form. This type of system compensation is quite useful for system optimization and will be considered briefly in this chapter.

We have illustrated several of the aforementioned approaches in the preceding chapters. In Example 6.5, we utilized the root locus method in considering the design of a feedback network in order to obtain a satisfactory performance. In Chapter 8, we considered the selection of the gain in order to obtain a suitable phase margin and, therefore, a satisfactory relative stability. Also, in Example 9.6, we compensated for the unstable response of the pendulum by controllng the pendulum with a function of several of the state variables of the system.

Quite often, in practice, the best and simplest way to improve the performance of a control system is to alter, if possible, the process itself. That is, if the system designer is able to specify and alter the design of the process which is represented by the transfer function $G(s)$, then the performance of the system may be readily improved. For example, in order to improve the transient behavior of a servomechanism position controller, one can often choose a better motor for the system. In the case of an airplane control system, one might be able to alter the aerodynamic design of the airplane and thus improve the flight transient characteristics. Thus, a control system designer should recognize that an alteration of the process may result in an improved system. However, often the process is fixed and unalterable or has been altered as much as is possible and is still found to result in an unsatisfactory performance. Then the addition of compensation networks becomes useful for improving the performance of the system. In the following sections we will assume that the process has been improved as much as possible and the $G(s)$ representing the process is unalterable.

It is the purpose of this chapter to further describe the addition of several compensation networks to a feedback control system. First we shall consider the addition of a so-called phase-lead compensation network and describe the design of the network by root locus and frequency response techniques. Then, using both the root locus and frequency response techniques, we shall describe the design of the integration compensation networks in order to obtain a suitable system performance. Finally, we shall determine an optimum controller for a system described in terms of state variables. While these three approaches to compensation are not intended to be discussed in a complete manner, the discussion that follows should serve as a worthwhile introduction to the design and compensation of feedback control systems.

10.3 CASCADE COMPENSATION NETWORKS

In this section, we shall consider the design of a cascade or feedback network as shown in Fig. 10.1(a) and Fig. 10.1(b), respectively. The compensation network, $G_c(s)$, is cascaded with the unalterable process $G(s)$ in order to provide a suitable loop transfer function $G_c(s)G(s)H(s)$. Clearly, the compensator $G_c(s)$ may be chosen to alter the shape of the root locus or the frequency response. In either case, the network may be chosen to have a transfer function

$$G_c(s) = \frac{K\prod_{i=1}^{M}(s + z_i)}{\prod_{j=1}^{N}(s + p_j)}. \tag{10.1}$$

Then the problem reduces to the judicious selection of the poles and zeros of the compensator. In order to illustrate the properties of the compensation network, we shall consider a first-order compensator. The compensation approach developed on the basis of a first-order compensator may then be extended to higher-order compensators.

Consider the first-order compensator with the transfer function

$$G_c(s) = \frac{K(s + z)}{(s + p)}. \tag{10.2}$$

The design problem becomes, then, the selection of z, p, and K in order to provide a suitable performance. When $|z| < |p|$, the network is called a *phase-lead network* and has a pole-zero s-plane configuration as shown in Fig. 10.2. If the pole was negligible, that is, $|p| \gg |z|$, and the zero occurred at the origin of the s-plane, we would have a differentiator so that

$$G_c(s) \simeq \left(\frac{K}{p}\right) s. \tag{10.3}$$

Thus a compensation network of the form of Eq. (10.2) is a differentiator type network. The differentiator network of Eq. (10.3) has a frequency characteristic as

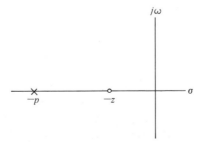

Fig. 10.2. The pole-zero diagram of the phase-lead network.

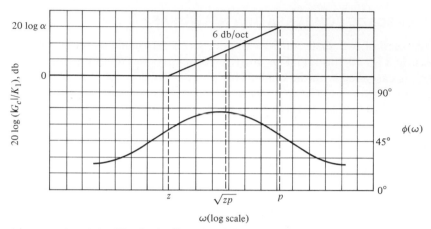

Fig. 10.3. The Bode diagram of the phase-lead network.

$$G_c(j\omega) = j\left(\frac{K}{p}\right)\omega = \left(\frac{K}{p}\omega\right)e^{+j90°} \tag{10.4}$$

and a phase angle of $+90°$, often called a phase-lead angle. Similarly, the frequency response of the differentiating network of Eq. (10.2) is

$$G_c(j\omega) = \frac{K(j\omega + z)}{(j\omega + p)} = \frac{(Kz/p)(j(\omega/z) + 1)}{(j(\omega/p) + 1)}$$

$$= \frac{K_1(1 + j\omega\alpha\tau)}{(1 + j\omega\tau)}, \tag{10.5}$$

where $\tau = 1/p$, $p = \alpha z$, and $K_1 = K/\alpha$. The frequency response of this phase-lead network is shown in Fig. 10.3. The angle of the frequency characteristic is

$$\phi(\omega) = \tan^{-1}\alpha\omega\tau - \tan^{-1}\omega\tau. \tag{10.6}$$

Since the zero occurs first on the frequency axis, we obtain a phase-lead characteristic as shown in Fig. 10.3. The slope of the asymptotic magnitude curve is $+6$ db/octave.

The phase-lead compensation transfer function can be obtained with the network shown in Fig. 10.4. The transfer function of this network is

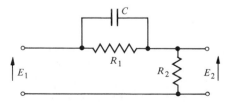

Fig. 10.4. A phase-lead network.

$$G_c(s) = \frac{E_2(s)}{E_1(s)} = \frac{R_2}{R_2 + \{R_1(1/Cs)/[R_1 + (1/Cs)]\}}$$

$$= \left(\frac{R_2}{R_1 + R_2}\right) \frac{(R_1Cs + 1)}{\{[R_1R_2/(R_1 + R_2)]Cs + 1\}}. \tag{10.7}$$

Therefore, we let

$$\tau = \frac{R_1R_2}{R_1 + R_2} C \quad \text{and} \quad \alpha = \frac{R_1 + R_2}{R_2}$$

and obtain the transfer function

$$G_c(s) = \frac{(1 + \alpha\tau s)}{\alpha(1 + \tau s)}, \tag{10.8}$$

which is equal to Eq. (10.5) when an additional cascade gain K is inserted.

The maximum value of the phase lead occurs at a frequency ω_m, where ω_m is the geometric mean of $p = 1/\tau$ and $z = 1/\alpha\tau$; that is, the maximum phase lead occurs halfway between the pole and zero frequencies on the logarithmic frequency scale. Therefore,

$$\omega_m = \sqrt{zp} = \frac{1}{\tau\sqrt{\alpha}}.$$

In order to obtain an equation for the maximum phase-lead angle, we rewrite the phase angle of Eq. (10.5) as

$$\phi = \tan^{-1} \frac{\alpha\omega\tau - \omega\tau}{1 + (\omega\tau)^2\alpha}. \tag{10.9}$$

Then, substituting the frequency for the maximum phase angle, $\omega_m = 1/\tau\sqrt{\alpha}$ we have

$$\tan \phi_m = \frac{(\alpha/\sqrt{\alpha}) - (1/\sqrt{\alpha})}{1 + 1}$$

$$= \frac{\alpha - 1}{2\sqrt{\alpha}}. \tag{10.10}$$

Since the $\tan \phi_m$ equals $(\alpha - 1)/2\sqrt{\alpha}$, we utilize the triangular relationship and note that

$$\sin \phi_m = \frac{\alpha - 1}{\alpha + 1}. \tag{10.11}$$

Equation (10.11) is very useful for calculating a necessary α ratio between the pole and zero of a compensator in order to provide a required maximum phase lead. A plot of ϕ_m versus α is shown in Fig. 10.5. Clearly, the phase angle readily obtainable from this network is not much greater than 70°. Also, since $\alpha = (R_1 + R_2)/R_2$, there are practical limitations on the maximum value of α that one should attempt to

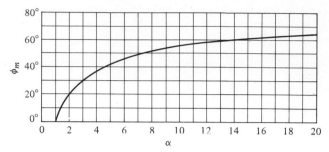

Fig. 10.5. The maximum phase angle ϕ_m versus α for a lead network.

obtain. Therefore, if one required a maximum angle of greater than 70°, two cascade compensation networks would be utilized. Then the equivalent compensation transfer function is $G_{c_1}(s)G_{c_2}(s)$ when the loading effect of $G_{c_2}(s)$ on $G_{c_1}(s)$ is negligible.

It is often useful to add a cascade compensation network which provides a phase-lag characteristic. The *phase-lag network* is shown in Fig. 10.6. The transfer function of the phase-lag network is

$$G_c(s) = \frac{E_2(s)}{E_1(s)} = \frac{R_2 + (1/Cs)}{R_1 + R_2 + (1/Cs)}$$
$$= \frac{R_2Cs + 1}{(R_1 + R_2)Cs + 1}. \tag{10.12}$$

When $\tau = R_2C$ and $\alpha = (R_1 + R_2)/R_2$, we have

$$G_c(s) = \frac{1 + \tau s}{1 + \alpha \tau s}$$
$$= \frac{1}{\alpha}\frac{(s + z)}{(s + p)}, \tag{10.13}$$

where $z = 1/\tau$ and $p = 1/\alpha\tau$. In this case, since $\alpha > 1$, the pole lies closest to the origin of the s-plane as shown in Fig. 10.7. This type of compensation network is often called an integrating network. The Bode diagram of the phase-lag network is obtained from the transfer function

$$G_c(j\omega) = \frac{1 + j\omega\tau}{1 + j\omega\alpha\tau} \tag{10.14}$$

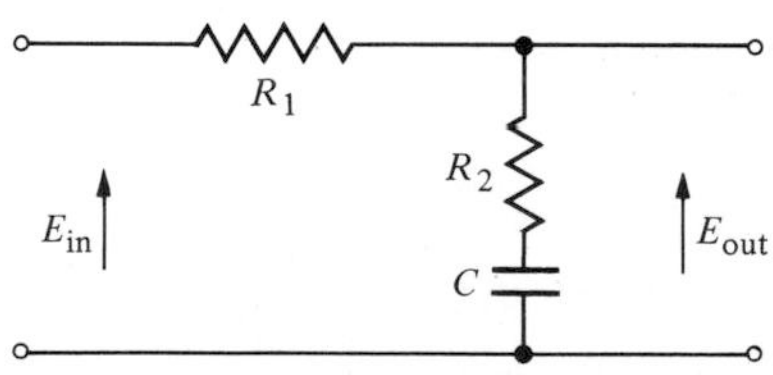

Fig. 10.6. A phase-lag network.

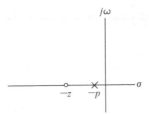

Fig. 10.7. The pole-zero diagram of the phase-lag network.

and is shown in Fig. 10.8. The form of the Bode diagram of the lag network is similar to that of the phase-lead network; the difference is the resulting attenuation and phase-lag angle instead of amplification and phase-lead angle. However, one notes that the shape of the diagrams of Figs. 10.3 and 10.8 are similar. Therefore, it can be shown that the maximum phase lag occurs at $\omega_m = \sqrt{zp}$.

In the succeeding sections, we wish to utilize these compensation networks in order to obtain a desired system frequency locus or s-plane root location. The lead network is utilized to provide a phase-lead angle and thus a satisfactory phase margin for a system. Alternatively, the use of the phase-lead network may be visualized on the s-plane as enabling one to reshape the root locus and thus provide the desired

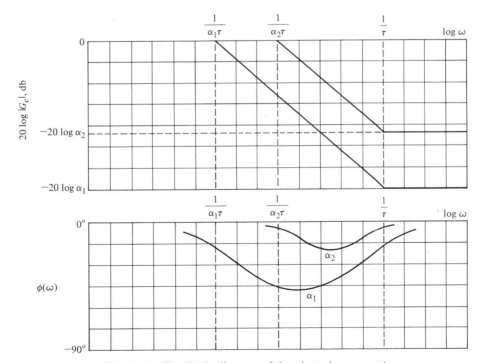

Fig. 10.8. The Bode diagram of the phase-lag network.

root locations. The phase-lag network is utilized not to provide a phase-lag angle, which is normally a destabilizing influence, but rather to provide an attenuation and increase the steady-state error constant [3]. These approaches to compensation utilizing the phase-lead and phase-lag networks will be the subject of the following four sections.

10.4 SYSTEM COMPENSATION ON THE BODE DIAGRAM USING THE PHASE-LEAD NETWORK

The Bode diagram is used in order to design a suitable phase-lead network in preference to other frequency response plots. The frequency response of the cascade compensation network is added to the frequency response of the uncompensated system. That is, since the total loop transfer function of Fig. 10.1(a) is $G_c(j\omega)G(j\omega)H(j\omega)$, we will first plot the Bode diagram for $G(j\omega)H(j\omega)$. Then one may examine the plot for $G(j\omega)H(j\omega)$ and determine a suitable location for p and z of $G_c(j\omega)$ in order to satisfactorily reshape the frequency response. The uncompensated $G(j\omega)$ is plotted with the desired gain to allow an acceptable steady-state error. Then the phase margin and the expected M_p are examined to find whether they satisfy the specifications. If the phase margin is not sufficient, phase lead may be added to the phase angle curve of the system by placing the $G_c(j\omega)$ in a suitable location. Clearly, in order to obtain maximum additional phase lead, we desire to place the network such that the frequency ω_m is located at the frequency where the magnitude of the compensated magnitude curves crosses the 0-db axis. (Recall the definition of phase margin.) The value of the added phase lead required allows us to determine the necessary value for α from Eq. (10.11) or Fig. 10.5. The zero $\omega = 1/\alpha\tau$ is located by noting that the maximum phase lead should occur at $\omega_m = \sqrt{zp}$, halfway between the pole and zero. Since the total magnitude gain for the network is $20 \log \alpha$, we expect a gain of $10 \log \alpha$ at ω_m. Thus we determine the compensation network by completing the following steps:

1. Evaluate the uncompensated system phase margin when the error constants are satisfied.

2. Allowing for a small amount of safety, determine the necessary additional phase lead, ϕ_m.

3. Evaluate α from Eq. (10.11).

4. Evaluate $10 \log \alpha$ and determine the frequency where the uncompensated magnitude curve is equal to $-10 \log \alpha$ db. This frequency is the new 0-db crossover frequency and ω_m simultaneously, since the compensation network provides a gain of $10 \log \alpha$ at ω_m.

5. Draw the compensated frequency response, check the resulting phase margin, and repeat the steps, if necessary. Finally, for an acceptable design, raise the gain of the amplifier in order to account for the attenuation $(1/\alpha)$.

Example 10.1.   Let us consider a single-loop feedback control system as shown in Fig. 10.1(a), where

$$G(s) = \frac{K_1}{s^2} \qquad (10.15)$$

and $H(s) = 1$. The uncompensated system is a type 2 system and at first appears to possess a satisfactory steady-state error for both step and ramp input signals. However, uncompensated, the response of the system is an undamped oscillation, since

$$T(s) = \frac{C(s)}{R(s)} = \frac{K_1}{s^2 + K_1}. \qquad (10.16)$$

Therefore, the compensation network is added so that the loop transfer function is $G_c(s)G(s)H(s)$. The specifications for the system are

Settling time, $T_s \leqslant 4$ sec,
Percent overshoot for a step input $\leqslant 20\%$.

Using Fig. 4.8, we estimate that the damping ratio should be $\zeta \geqslant 0.45$. The settling time requirement is

$$T_s = \frac{4}{\zeta \omega_n} = 4, \qquad (10.17)$$

and therefore

$$\omega_n = \frac{1}{\zeta} = \frac{1}{0.45} = 2.22.$$

Perhaps the simplest way to check the value of ω_n for the frequency response is to relate ω_n to the bandwidth and evaluate the bandwidth of the closed-loop system. For a closed-loop system with $\zeta = 0.45$, we estimate from Fig. 7.9 that $\omega_B = 1.36\omega_n$. Therefore, we require a closed-loop bandwidth $\omega_B = 1.36(2.22) = 3.02$. The bandwidth may be checked following compensation by utilizing the Nichols chart. For the uncompensated system, the bandwidth of the system is $\omega_B = 1.36\omega_n$ and $\omega_n = \sqrt{K}$. Therefore, a loop gain equal to $K = \omega_n^2 \approx 5$ would be sufficient. In order to provide a suitable margin for the settling time, we will select $K = 10$ in order to draw the Bode diagram of

$$GH(j\omega) = \frac{K}{(j\omega)^2}.$$

The Bode diagram of the uncompensated system is shown as solid lines in Fig. 10.9.

By using Eq. (8.58), the phase margin of the system is required to be approximately

$$\phi_{pm} = \frac{\zeta}{0.01} = \frac{0.45}{0.01} = 45°. \qquad (10.18)$$

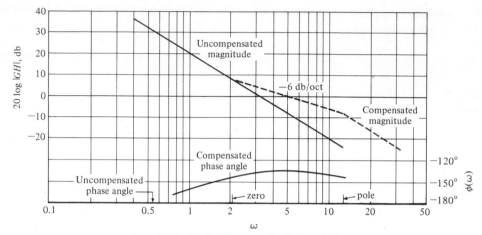

Fig. 10.9. Bode diagram for Example 10.1.

The phase margin of the uncompensated system is zero degrees since the double integration results in a constant 180° phase lag. Therefore we must add a 45° phase-lead angle at the crossover (0-db) frequency of the compensated magnitude curve. Evaluating the value of α, we have

$$\frac{\alpha - 1}{\alpha + 1} = \sin \phi_m$$
$$= \sin 45°, \tag{10.19}$$

and therefore $\alpha = 5.8$. In order to provide a margin of safety, we will use $\alpha = 6$. The value of $10 \log \alpha$ is then equal to 7.78 db. Then the lead network will add an additional gain of 7.78 db at the frequency ω_m, and it is desired to have ω_m equal to the compensated system crossover frequency. This is accomplished by drawing the compensated slope near the 0-db axis (the dotted line) so that the new crossover is ω_m and the dotted magnitude curve is 7.78 db above the uncompensated curve at the crossover frequency. Thus the compensated crossover frequency is located by evaluating the frequency where the uncompensated magnitude curve is equal to -7.78 db, which, in this case, is $\omega = 4.9$. Then the maximum phase-lead angle is added at $\omega = \omega_m = 4.9$ as shown in Fig. 10.9. The bandwidth of the compensated system may be obtained from the Nichols chart. For estimating the bandwidth, one may simply examine Fig. 8.22 and note that the -3-db line for the closed-loop system occurs when the magnitude of $GH(j\omega)$ is -6-db and the phase shift of $GH(j\omega)$ is approximately $-140°$. Therefore, in order to estimate the bandwidth from the open-loop diagram we will approximate the bandwidth as the frequency for which $20 \log |GH|$ is equal to -6 db. Therefore the bandwidth of the uncompensated system is approximately equal to $\omega_B = 4.4$, while the bandwidth of the compensated system is equal to $\omega_B = 8.4$. The lead compensation doubles the bandwidth in this case and the specification that $\omega_B > 3.02$ is satisfied. Therefore the compensation of the system is completed and the system specifications are satisfied. The total compensated loop transfer function is

$$G_c(j\omega)G(j\omega)H(j\omega) = \frac{10[(j\omega/2.1) + 1]}{(j\omega)^2[(j\omega/12.6) + 1]}. \tag{10.20}$$

The transfer function of the compensator is

$$\begin{aligned}
G_c(s) &= \frac{(1 + \alpha\tau s)}{\alpha(1 + \tau s)} \\
&= \frac{1}{6}\frac{[1 + (s/2.1)]}{[1 + (s/12.6)]}
\end{aligned} \tag{10.21}$$

in the form of Eq. (10.8). Since an attenuation of ⅙ results from the passive *RC* network, the gain of the amplifier in the loop must be raised by a factor of six so that the total dc loop gain is still equal to 10 as required in Eq. (10.20). When we add the compensation network Bode diagram to the uncompensated Bode diagram as in Fig. 10.9, we are assuming that we can raise the amplifier gain in order to account for this $1/\alpha$ attenuation. The pole and zero values are simply read from Fig. 10.9, noting that $p = \alpha z$.

Example 10.2. A feedback control system has a loop transfer function

$$GH(s) = \frac{K}{s(s + 2)}. \tag{10.22}$$

It is desired to have a steady-state error for a ramp input less than 5% of the magnitude of the ramp. Therefore, we require that

$$K_v = \frac{A}{e_{ss}} = \frac{A}{0.05A} = 20. \tag{10.23}$$

Furthermore, we desire that the phase margin of the system be at least 45°. The first step is to plot the Bode diagram of the uncompensated transfer function

$$\begin{aligned}
GH(j\omega) &= \frac{K_v}{j\omega(0.5j\omega + 1)} \\
&= \frac{20}{j\omega(0.5j\omega + 1)}
\end{aligned} \tag{10.24}$$

as shown in Fig. 10.10(a). The frequency at which the magnitude curve crosses the 0-db line is 6.2 rad/sec and the phase margin at this frequency is determined readily from the equation of the phase angle of $GH(j\omega)$ which is

$$\underline{/GH(j\omega)} = \phi(\omega) = -90° - \tan^{-1}(0.5\omega). \tag{10.25}$$

At the crossover frequency, $\omega = \omega_c = 6.2$ rad/sec, we have

$$\phi(\omega) = -162°, \tag{10.26}$$

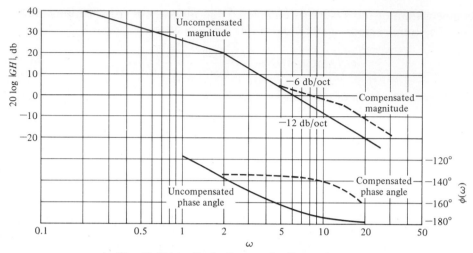

Fig. 10.10(a). Bode diagram for Example 10.2.

and therefore the phase margin is 18°. Using Eq. (10.25) to evaluate the phase margin is often easier than drawing the complete phase angle curve which is shown in Fig. 10.10(a). Thus we need to add a phase-lead network so that the phase margin is raised to 45° at the new crossover (0-db) frequency. Since the compensation crossover frequency is greater than the uncompensated crossover frequency, the phase lag of the uncompensated system is greater also. We shall account for this additional phase lag by attempting to obtain a maximum phase lead of 45° − 18° = 27° plus a small increment (10%) of phase lead to account for the added lag. Thus we will design a compensation network with a maximum phase lead equal to 27° + 3° = 30°. Then, calculating α, we obtain

$$\frac{\alpha - 1}{\alpha + 1} = \sin 30°$$
$$= 0.5, \tag{10.27}$$

and therefore $\alpha = 3$.

The maximum phase lead occurs at ω_m, and this frequency will be selected so that the new crossover frequency and ω_m coincide. The magnitude of the lead network at ω_m is $10 \log \alpha = 10 \log 3 = 4.8$ db. The compensated crossover frequency is then evaluated where the magnitude of $GH(j\omega)$ is -4.8 db and thus $\omega_m = \omega_c = 8.4$. Drawing the compensated magnitude line so that it intersects the 0-db axis at $\omega = \omega_c = 8.4$, we find that $z = 4.8$ and $p = \alpha z = 14.4$. Therefore the compensation network is

$$G_c(s) = \frac{1}{3} \frac{(1 + s/4.8)}{(1 + s/14.4)}. \tag{10.28}$$

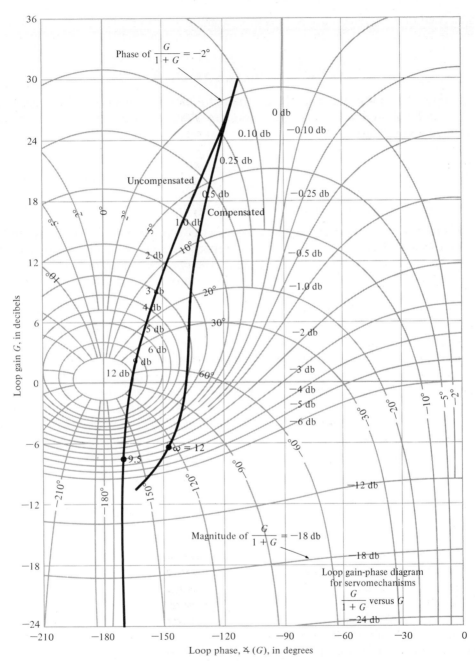

Fig. 10.10(b).   Nichols diagram for Example 10.2.

The total dc loop gain must be raised by a factor of 3 in order to account for the factor $1/\alpha = \frac{1}{3}$. Then the compensated loop transfer function is

$$G_c(s)GH(s) = \frac{20[(s/4.8) + 1]}{s(0.5s + 1)[(s/14.4) + 1]}. \tag{10.29}$$

In order to verify the final phase margin, we may evaluate the phase of $G_c(j\omega)GH(j\omega)$ at $\omega = \omega_c = 8.4$ and therefore obtain the phase margin. The phase angle is then

$$\phi(\omega_c) = -90^\circ - \tan^{-1} 0.5\omega_c - \tan^{-1} \frac{\omega_c}{14.4} + \tan^{-1} \frac{\omega_c}{4.8}$$
$$= -90^\circ - 76.5^\circ - 30.0^\circ + 60.2^\circ$$
$$= -136.3^\circ. \tag{10.30}$$

Therefore the phase margin for the compensated system is 43.7°. If we desire to have exactly 45° phase margin, we would repeat the steps with an increased value of α; for example, with $\alpha = 3.5$. In this case, the phase lag increased by 7° between $\omega = 6.2$ and $\omega = 8.4$, and therefore the allowance of 3° in the calculation of α was not sufficient.

The Nichols diagram for the compensated and uncompensated system is shown on Fig. 10.10(b). The reshaping of the frequency response locus is clear on this diagram. One notes the increased phase margin for the compensated system as well as the reduced magnitude of M_{p_ω}, the maximum magnitude of the closed-loop frequency response. In this case, M_{p_ω} has been reduced from an uncompensated value of +12 db to a compensated value of approximately +3.2 db. Also, we note that the closed-loop 3-db bandwidth of the compensated system is equal to 12 rad/sec compared with 9.5 rad/sec for the uncompensated system.

Examining both Examples 10.1 and 10.2 we note that the system design is satisfactory when the asymptotic curve for the magnitude $20 \log |GG_c|$ crosses the 0 db line with a slope of -6 db/octave.

10.5 COMPENSATION ON THE s-PLANE USING THE PHASE-LEAD NETWORK

The design of a phase-lead compensation network may also be readily accomplished on the s-plane. The phase-lead network has a transfer function

$$G_c(s) = \frac{[s + (1/\alpha\tau)]}{[s + (1/\tau)]} = \frac{(s + z)}{(s + p)}, \tag{10.31}$$

where α and τ are defined for the RC network in Eq. (10.7). The locations of the zero and pole are selected in order to result in a satisfactory root locus for the compensated system. The specifications of the system are used to specify the desired location of the dominant roots of the system. The s-plane root locus method is as follows:

1. List the system specifications and translate these specifications into a desired root location for the dominant roots.

2. Sketch the uncompensated root locus and determine whether the desired root locations can be realized with an uncompensated system.

3. If the compensator is necessary, place the zero of the phase-lead network directly below the desired root location.

4. Determine the pole location so that the total angle at the desired root location is 180° and therefore is on the compensated root locus.

5. Evaluate the total system gain at the desired root location and then calculate the error constant.

6. Repeat the steps if the error constant is not satisfactory.

Therefore, we first locate our desired dominant root locations so that the dominant roots satisfy the specifications in terms of ζ and ω_n as shown in Fig. 10.11(a). The root locus of the uncompensated system is sketched as illustrated in Fig. 10.11(b). Then the zero is added to provide a phase lead of $+90°$ by placing it directly below the desired root location. Actually, some caution must be maintained

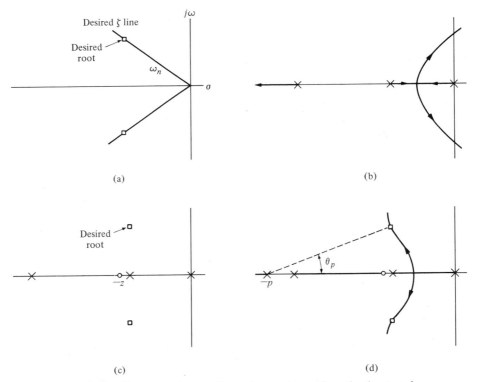

Fig. 10.11. Compensation on the s-plane using a phase-lead network.

since the zero must not alter the dominance of the desired roots; that is, the zero should not be placed nearer the origin than the second pole on the real axis or a real root near the origin will result and will dominate the system response. Thus, in Fig. 10.11(c), we note that the desired root is directly above the second pole, and we place the zero z somewhat to the left of the pole.

Then the real root will be near the real zero, the coefficient of this term of the partial fraction expansion will be relatively small, and thus the response due to this real root will have very little effect on the overall system response. Nevertheless, the designer must be continually aware that the compensated system response will be influenced by the roots and zeros of the system and the dominant roots will not by themselves dictate the response. It is usually wise to allow for some margin of error in the design and to test the compensated system using a digital simulation (e.g., CSMP simulation).

Since the desired root is a point on the root locus when the final compensation is accomplished, we expect the algebraic sum of the vector angles to be 180° at that point. Thus we calculate the angle from the pole of compensator, θ_p, in order to result in a total angle of 180°. Then, locating a line at an angle θ_p intersecting the desired root, we are able to evaluate the compensator pole, p, as shown in Fig. 10.11(d).

The advantage of the s-plane method is the ability of the designer to specify the location of the dominant roots and, therefore, the dominant transient response. The disadvantage of the method is that one cannot directly specify an error constant (for example, K_v) as in the Bode diagram approach. After the design is completed, one evaluates the gain of the system at the root location, which depends upon p and z, and then calculates the error constant for the compensated system. If the error constant is not satisfactory, one must repeat the design steps and alter the location of the desired root as well as the location of the compensator pole and zero. We shall reconsider the two examples we completed in the preceding section and design a compensation network using the root locus (s-plane) approach.

Example 10.3.   Let us reconsider the system of Example 10.1 where the open-loop uncompensated transfer function is

$$GH(s) = \frac{K_1}{s^2}.$$
(10.32)

The characteristic equation of the uncompensated system is

$$1 + GH(s) = 1 + \frac{K_1}{s^2} = 0,$$
(10.33)

and the root locus is the $j\omega$-axis. Therefore we desire to compensate this system with a network, $G_c(s)$, where

$$G_c(s) = \frac{s + z}{s + p}$$
(10.34)

and $|z| < |p|$. The specifications for the system are

Settling time, $T_s \leq 4$ sec,
Percent overshoot for a step input $\leq 30\%$.

Therefore the damping ratio should be $\zeta \geq 0.35$. The settling time requirement is

$$T_s = \frac{4}{\zeta\omega_n} = 4,$$

and therefore $\zeta\omega_n = 1$. Thus we will choose a desired dominant root location as

$$r_1, \hat{r}_1 = -1 \pm j2 \tag{10.35}$$

as shown in Fig. 10.12 (thus $\zeta = 0.45$).

Now, we place the zero of the compensator directly below the desired location at $s = -z = -1$ as shown in Fig. 10.12. Then, measuring the angle at the desired root, we have

$$\phi = -2(116°) + 90° = -142°.$$

Therefore, in order to have a total of $180°$ at the desired root, we evaluate the angle from the undetermined pole, θ_p, as

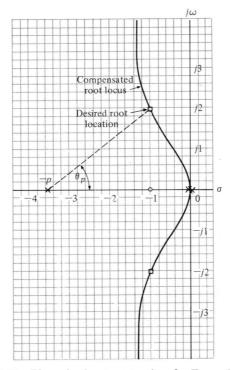

Fig. 10.12. Phase-lead compensation for Example 10.3.

$$-180° = -142° - \theta_p \qquad (10.36)$$

or $\theta_p = 38°$. Then a line is drawn at an angle $\theta_p = 38°$ intersecting the desired root location and the real axis as shown in Fig. 10.12. The point of intersection with the real axis is then $s = -p = -3.6$. Therefore, the compensator is

$$G_c(s) = \frac{s + 1}{s + 3.6}, \qquad (10.37)$$

and the compensated transfer function for the system is

$$GH(s)G_c(s) = \frac{K_1(s + 1)}{s^2(s + 3.6)}. \qquad (10.38)$$

The gain K_1 is evaluated by measuring the vector lengths from the poles and zeros to the root location. Hence

$$K_1 = \frac{(2.23)^2(3.25)}{2} = 8.1. \qquad (10.39)$$

Finally, the error constants of this system are evaluated. We find that this system with two open-loop integrations will result in a zero steady-state error for a step and ramp input signal. The acceleration constant is

$$K_a = \frac{8.1}{3.6} = 2.25. \qquad (10.40)$$

The steady-state performance of this system is quite satisfactory and, therefore, the compensation is complete. When we compare the compensation network evaluated by the s-plane method with the network obtained by using the Bode diagram approach, we find that the magnitudes of the poles and zeros are different. However, the resulting system will have the same performance and we need not be concerned with the difference. In fact, the difference arises from the arbitrary design step (Number 3), which places the zero directly below the desired root location. If we placed the zero at $s = -2.1$, we would find that the pole evaluated by the s-plane method is approximately equal to the pole evaluated by the Bode diagram approach.

The specifications for the transient response of this system were originally expressed in terms of the overshoot and the settling time of the system. These specifications were translated, on the basis of an approximation of the system by a second-order system, to an equivalent ζ and ω_n and therefore a desired root location. However, the original specifications will be satisfied only if the roots selected are dominant. The zero of the compensator and the root resulting from the addition of the compensator pole result in a third-order system with a zero. The validity of approximating this system with a second-order system without a zero is dependent upon the validity of the dominance assumption. Often, the designer will simulate the final design by using an analog computer or a digital computer and obtain the actual transient response of the system. In this case, an analog computer simulation of the system resulted in an overshoot of 40% and a settling time of 3.8 sec for a step input. These values compare moderately well with the specified values of 30% and 4 sec and justify the utilization of the dominant root specifications. The difference in the overshoot from the specified value is due to the third root which is not negligible. Thus, again we find that the specification of dominant roots is a useful approach, but must be utilized with caution and understanding. A second attempt to obtain a compensated system with an overshoot of 30% would utilize a compensator with a zero at -2 and then calculate the necessary pole location to yield the desired root locations for the dominant roots. This approach would move the third root farther to the left in the s-plane, reduce the effect of the third root on the transient response, and reduce the overshoot.

Example 10.4.   Now let us reconsider the system of Example 10.2 and design a compensator based on the s-plane approach. The open-loop system transfer function is

$$GH(s) = \frac{K}{s(s + 2)}.$$ (10.41)

It is desired that the damping ratio of the dominant roots of the system be $\zeta = 0.45$ and that the velocity error constant be equal to 20. In order to satisfy the error constant requirement, the gain of the uncompensated system must be $K = 40$. When $K = 40$, the roots of the uncompensated system are

$$s^2 + 2s + 40 = (s + 1 + j6.25)(s + 1 - j6.25).$$ (10.42)

The damping ratio of the uncompensated roots is approximately 0.16, and therefore a compensation network must be added. In order to achieve a rapid settling time, we will select the real part of the desired roots as $\zeta\omega_n = 4$ and therefore $T_s = 1$ sec. Also, the natural frequency of these roots is fairly large, $\omega_n = 9$; hence the velocity constant should be reasonably large. The location of the desired roots is shown on Fig. 10.13 for $\zeta\omega_n = 4$, $\zeta = 0.45$, and $\omega_n = 9$.

The zero of the compensator is placed at $s = -z = -4$, directly below the desired root location. Then the angle at the desired root location is

$$\phi = -116° - 104° + 90° = -130°.$$ (10.43)

Therefore the angle from the undetermined pole is determined from

$$-180° = -130° - \theta_p,$$

and thus $\theta_p = 50°$. This angle is drawn to intersect the desired root location, and p is evaluated as $s = -p = -10.6$ as shown in Fig. 10.13. The gain of the compensated system is then

$$K = \frac{9(8.25)(10.4)}{8} = 96.5$$ (10.44)

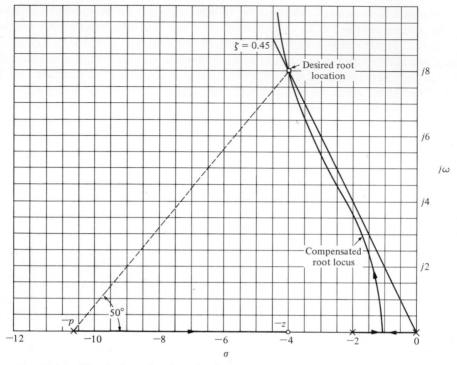

Fig. 10.13. The design of a phase-lead network on the s-plane for Example 10.4.

The compensated system is then

$$G_c(s)GH(s) = \frac{96.5(s + 4)}{s(s + 2)(s + 10.6)}. \tag{10.45}$$

Therefore the velocity constant of the compensated system is

$$K_v = \lim_{s \to 0} s\{G(s)H(s)G_c(s)\} = \frac{96.5(4)}{2(10.6)} = 18.2. \tag{10.46}$$

The velocity constant of the compensated system is less than the desired value of 20. Therefore, one must repeat the design procedure for a second choice of a desired root. If we choose $\omega_n = 10$, the process may be repeated and the resulting gain K will be increased. The compensator pole and zero location will also be altered. Then the velocity constant may be again evaluated. We will leave it as an exercise for the reader to show that for $\omega_n = 10$, the velocity constant is $K_v = 22.7$ when $z = 4.5$ and $p = 11.6$.

Finally, for the compensation network of Eq. (10.45), we have

$$G_c(s) = \frac{s + 4}{s + 10.6} = \frac{(s + 1/\alpha\tau)}{(s + 1/\tau)}. \tag{10.47}$$

The design of an RC-lead network as shown in Fig. 10.4 follows directly from Eqs. (10.47) and (10.7), and is

$$G_c(s) = \left(\frac{R_2}{R_1 + R_2} \right) \frac{(R_1 Cs + 1)}{(R_1 R_2/(R_1 + R_2))Cs + 1}. \qquad (10.48)$$

Thus, in this case, we have

$$\frac{1}{R_1 C} = 4$$

and

$$\alpha = \frac{R_1 + R_2}{R_2} = \frac{10.6}{4}.$$

Then, choosing $C = 1$ μf, we obtain $R_1 = 250,000$ ohms and $R_2 = 152,000$ ohms.

The phase-lead compensation network is a useful compensator for altering the performance of a control system. The phase-lead network adds a phase-lead angle in order to provide adequate phase margin for feedback systems. Using an s-plane design approach, the phase-lead network may be chosen in order to alter the system root locus and place the roots of the system in a desired position in the s-plane. When the design specifications include an error constant requirement, the Bode diagram method is more suitable, since the error constant of a system designed on the s-plane must be ascertained following the choice of a compensator pole and zero. Therefore, the root locus method often results in an iterative design procedure when the error constant is specified. On the other hand, the root locus is a very satisfactory approach when the specifications are given in terms of overshoot and settling time, thus specifying the ζ and ω_n of the desired dominant roots in the s-plane. The use of a lead network compensator always extends the bandwidth of a feedback system which may be objectionable for systems subjected to large amounts of noise. Also, lead networks are not suitable for providing high steady-state accuracy systems requiring very high error constants. In order to provide large error constants, typically K_p and K_v, one must consider the use of integration-type compensation networks, and, therefore, this will be the subject of concern in the following section.

10.6 SYSTEM COMPENSATION USING INTEGRATION NETWORKS

For a large percentage of control systems, the primary objective is to obtain a high steady-state accuracy. Furthermore, it is desired to maintain the transient performance of these systems within reasonable limits. As we found in Chapters 3 and 4, the steady-state accuracy of many feedback systems may be increased by increasing the amplifier gain in the forward channel. However, the resulting transient response may be totally unacceptable, if not even unstable. Therefore it is often necessary to introduce a compensation network in the forward path of a feedback control system in order to provide a sufficient steady-state accuracy.

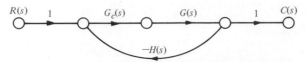

Fig. 10.14. A single-loop feedback control system.

Consider the single-loop control system shown in Fig. 10.14. The compensation network is to be chosen in order to provide a large error constant. The steady-state error of this system is

$$\lim_{t \to \infty} e(t) = \lim_{s \to 0} s \left[\frac{R(s)}{1 + G_c(s)G(s)H(s)} \right]. \tag{10.49}$$

We found in Section 3.5 that the steady-state error of a system depends upon the number of poles at the origin for $G_c(s)G(s)H(s)$. A pole at the origin may be considered an integration and therefore the steady-state accuracy of a system ultimately depends upon the number of integrations in the transfer function $G_c(s)G(s)H(s)$. If the steady-state accuracy is not sufficient, we will introduce an integration-type network $G_c(s)$ in order to compensate for the lack of integration in the original transfer function $G(s)H(s)$.

One form of controller available and widely used in industrial process control is called a *three-mode controller* or *process controller*. This controller has a transfer function

$$\frac{U(s)}{E(s)} = G_c(s) = K_p + \frac{K_1}{s} + K_D s. \tag{10.50}$$

The controller provides a proportional term, an integration term, and a derivative term. The equation for the output in the time domain is

$$u(t) = K_p e(t) + K_1 \int e(t)\, dt + K_D \frac{de(t)}{dt}. \tag{10.51}$$

The three mode controller is also called a PID controller since it contains a proportional, an integration, and a derivative term. The transfer function of the derivative term is actually

$$G_d(s) = \frac{K_D s}{\tau_d s + 1}, \tag{10.52}$$

but usually τ_d is much smaller than the time constants of the process itself and may be neglected.

For an example, let us consider a temperature control system where the transfer function of the heat process is $G(s) = K_1 / (\tau_1 s + 1)$ and the measurement transfer function is

$$H(s) = \frac{1}{\tau_2 s + 1}.$$

The steady-state error of the uncompensated system is then

$$\lim_{t \to \infty} e(t) = \lim_{s \to 0} s \left\{ \frac{A/s}{1 + G(s)H(s)} \right\}$$

$$= \frac{A}{1 + K_1}, \qquad (10.53)$$

where $R(s) = A/s$, a step input signal. Clearly, in order to obtain a small steady-state error (less than 0.05 A, for example), the magnitude of the gain K_1 must be quite large. However, when K_1 is quite large, the transient performance of the system will very likely be unacceptable. Therefore, we must consider the addition of a compensation transfer function $G_c(s)$ as shown in Fig. 10.14. In order to eliminate the steady-state error of this system, we might choose the compensation as

$$G_c(s) = K_2 + \frac{K_3}{s} = \frac{K_2 s + K_3}{s}. \qquad (10.54)$$

This compensation may be readily constructed by using an integrator and an amplifier and adding their output signals. Now, the steady-state error for a step input of the system is always zero, since

$$\lim_{t \to \infty} e(t) = \lim_{s \to 0} s \, \frac{A/s}{1 + G_c(s)GH(s)}$$

$$= \lim_{s \to 0} \frac{A}{1 + [(K_2 s + K_3)/s]\{K_1/[(\tau_1 s + 1)(\tau_2 s + 1)]\}}$$

$$= 0. \qquad (10.55)$$

The transient performance can be adjusted to satisfy the system specifications by adjusting the constants K_1, K_2, and K_3. The adjustment of the transient response is perhaps best accomplished by using the root locus methods of Chapter 6 and drawing a root locus for the gain $K_2 K_1$ after locating the zero $s = -K_3/K_2$ on the s-plane by the method outlined for the s-plane in the preceding section.

The addition of an integration as $G_c(s) = K_2 + (K_3/s)$ may also be used to reduce the steady-state error for a ramp input, $r(t) = t$, $t \geq 0$. For example, if the uncompensated system $GH(s)$ possessed one integration, the additional integration due to $G_c(s)$ would result in a zero steady-state error for a ramp input. In order to illustrate the design of this type of integration compensation, we will consider a temperature control system in some detail.

Example 10.5. The uncompensated loop transfer function of a temperature control system is

$$GH(s) = \frac{K_1}{(2s + 1)(0.5s + 1)}, \qquad (10.56)$$

where K_1 may be adjusted. In order to maintain zero steady-state error for a step input, we will add the compensation network

$$G_c(s) = K_2 + \frac{K_3}{s}$$
$$= K_2 \left(\frac{s + K_3/K_2}{s} \right). \qquad (10.57)$$

Furthermore, the transient response of the system is required to have an overshoot less than or equal to 10%. Therefore, the dominant complex roots must be on (or below) the $\zeta = 0.6$ line as shown in Fig. 10.15. We will adjust the compensator zero so that the real part of the complex roots is $\zeta \omega_n = 0.75$ and thus the settling time is $T_s = 4/\zeta \omega_n = \frac{16}{3}$ sec. Now, as in the preceding section, we will determine the location of the zero, $z = -K_3/K_2$, by assuring that the angle at the desired root is $-180°$. Therefore, the sum of the angles at the desired root is

$$-180° = -127° - 104° - 38° + \theta_z,$$

where θ_z is the angle from the undetermined zero. Therefore, we find that $\theta_z = +89°$ and the location of the zero is $z = -0.75$. Finally, in order to determine the gain at the desired root, we evaluate the vector lengths from the poles and zeros and obtain

$$K = K_1 K_2 = \frac{1.25(1.06)1.6}{.95} = 2.23.$$

The compensated root locus and the location of the zero are shown in Fig. 10.15. It should be noted that the zero, $z = -K_3/K_2$, should be placed to the left of the pole at $s = -0.5$ in order to ensure that the complex roots dominate the transient response. In fact, the third root of the compensated system of Fig. 10.15 may be determined as $s = -1.0$, and therefore this real root is only $\frac{4}{3}$ times the real part of the complex roots. Thus, while complex roots dominate the response of the system,

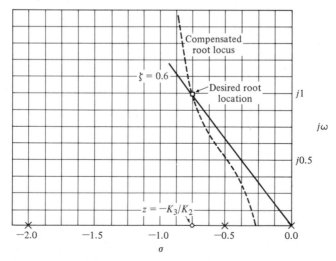

Fig. 10.15. The s-plane design of an integration compensator.

the equivalent damping of the system is somewhat less than $\zeta = 0.60$ due to the real root and zero.

10.7 COMPENSATION ON THE s-PLANE USING A PHASE-LAG NETWORK

The phase-lag RC network of Fig. 10.6 is an integration-type network and may be used to increase the error constant of a feedback control system. We found in Section 10.3 that the transfer function of the RC phase-lag network is of the form

$$G_c(s) = \frac{1}{\alpha} \frac{(s + z)}{(s + p)}, \tag{10.58}$$

as given in Eq. (10.13), where

$$z = \frac{1}{\tau} = \frac{1}{R_2 C}, \qquad \alpha = \frac{R_1 + R_2}{R_2}, \qquad p = \frac{1}{\alpha \tau}.$$

The steady-state error of an uncompensated system is

$$\lim_{t \to 0} e(t) = \lim_{s \to 0} s \left\{ \frac{R(s)}{1 + GH(s)} \right\}. \tag{10.59}$$

Then, for example, the velocity constant of a type-one system is

$$K_v = \lim_{s \to 0} s\{GH(s)\} \tag{10.60}$$

as shown in Section 4.4. Therefore, if $GH(s)$ is written as

$$GH(s) = \frac{K \prod_{i=1}^{M} (s + z_i)}{s \prod_{j=1}^{Q} (s + p_j)}, \tag{10.61}$$

we obtain the velocity constant

$$K_v = \frac{K \prod_{i=1}^{M} z_i}{\prod_{j=1}^{Q} p_j}. \tag{10.62}$$

We will now add the integration type phase-lag network as a compensator and determine the compensated velocity constant. If the velocity constant of the uncompensated system (Eq. 10.62) is designated as $K_{v_{\text{uncomp}}}$, we have

$$K_{v_{\text{comp}}} = \lim_{s \to 0} s\{G_c(s)GH(s)\} = \lim_{s \to 0} (G_c(s)) K_{v_{\text{uncomp}}}$$

$$= \left(\frac{z}{p}\right) \left(\frac{1}{\alpha}\right) K_{v_{\text{uncomp}}} = \left(\frac{z}{p}\right) \left(\frac{K}{\alpha}\right) \left(\frac{\prod z_i}{\prod p_j}\right). \tag{10.63}$$

The gain on the compensated root locus at the desired root location will be (K/α). Now, if the pole and zero of the compensator are chosen so that $|z| = \alpha |p| < 1$, the resultant K_v will be increased at the desired root location by the ratio $z/p = \alpha$. Then, for example, if $z = 0.1$ and $p = 0.01$, the velocity constant of the desired root location will be increased by a factor of 10. However, if the compensator pole and zero appear relatively close together on the s-plane, their effect on the location of the desired root will be negligible. Therefore the compensator pole-zero combination near the origin may be used to increase the error constant of a feedback system by the factor α while altering the root location very slightly. The factor α does have an upper limit, typically about 100, since the required resistors and capacitors of the network become excessively large for a higher α. For example, when $z = 0.1$ and $\alpha = 100$, we find from Eq. (10.58) that

$$z = 0.1 = \frac{1}{R_2 C}$$

and

$$\alpha = 100 = \frac{R_1 + R_2}{R_2}.$$

If we let $C = 10$ μf, then $R_2 = 1$ megohm and $R_1 = 99$ megohms. As we increase α, we increase the magnitude of R_1 required. However, we should note that an attenuation, α, of 1000 or more may be obtained by utilizing pneumatic process controllers which approximate a phase-lag characteristic (Fig. 10.8).

The steps necessary for the design of a phase-lag network on the s-plane are as follows:

1. Obtain the root locus of the uncompensated system.

2. Determine the transient performance specifications for the system and locate suitable dominant root locations on the uncompensated root locus that will satisfy the specifications.

3. Calculate the loop gain at the desired root location and, thus, the system error constant.

4. Compare the uncompensated error constant with the desired error constant and calculate the necessary increase that must result from the pole-zero ratio of the compensator, α.

5. With the known ratio of the pole-zero combination of the compensator, determine a suitable location of the pole and zero of the compensator so that the compensated root locus will still pass through the desired root location.

The fifth requirement can be satisfied if the magnitude of the pole and zero is less than one and they appear to merge as measured from the desired root location. The pole and zero will appear to merge at the root location if the angles from the compensator pole and zero are essentially equal as measured to the root location. One method of locating the zero and pole of the compensator is based on the requirement that the difference between the angle of the pole and the angle of the zero as measured at the desired root is less than 2°. An example will illustrate this approach to the design of a phase-lag compensator.

Example 10.6. Consider the uncompensated system of Example 10.2, where the uncompensated open-loop transfer function is

$$GH(s) = \frac{K}{s(s + 2)}. \tag{10.64}$$

It is required that the damping ratio of the dominant complex roots is 0.45, while a system velocity constant equal to 20 is attained. The uncompensated root locus is a vertical line at $s = -1$ and results in a root on the $\zeta = 0.45$ line at $s = -1 \pm j2$ as shown in Fig. 10.16. Measuring the gain at this root, we have $K = (2.24)^2 = 5$. Therefore the velocity constant of the uncompensated system is

$$K_v = \frac{K}{2} = \frac{5}{2} = 2.5.$$

Thus the ratio of the zero to the pole of the compensator is

$$\left|\frac{z}{p}\right| = \alpha = \frac{K_{v_{comp}}}{K_{v_{uncomp}}} = \frac{20}{2.5} = 8. \tag{10.65}$$

Examining Fig. 10.17, we find that we might set $z = -0.1$ and then $p = -0.1/8$. The difference of the angles from p and z at the desired root is approximately one degree, and therefore, $s = -1 \pm j2$ is still the location of the dominant roots. A

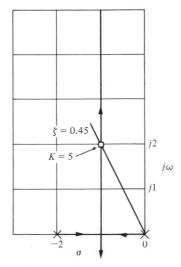

Fig. 10.16. Root locus of the uncompensated system of Example 10.6.

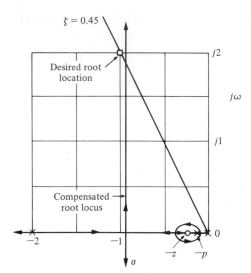

Fig. 10.17. Root locus of the compensated system of Example 10.6. Note that the actual root will differ from the desired root by a slight amount. The vertical portion of the locus leaves the σ axis at $\sigma = -0.95$.

sketch of the compensated root locus is shown as a heavy line in Fig. 10.17. Therefore the compensated system transfer function is

$$G_c(s)GH(s) = \frac{5(s + 0.1)}{s(s + 2)(s + 0.0125)},\qquad(10.66)$$

where $(K/\alpha) = 5$ or $K = 40$ in order to account for the attenuation of the lag network.

Example 10.7. Let us now consider a system which is difficult to compensate by a phase-lead network. The open-loop transfer function of the uncompensated system is

$$GH(s) = \frac{K}{s(s + 10)^2}.\qquad(10.67)$$

It is specified that the velocity constant of this system be equal to 20, while the damping ratio of the dominant roots be equal to 0.707. The gain necessary for a K_v of 20 is

$$K_v = 20 = \frac{K}{(10)^2}$$

or $K = 2000$. However, using Routh's criterion, we find that the roots of the characteristic equation lie on the $j\omega$-axis at $\pm j10$ when $K = 2000$. Clearly, the roots of

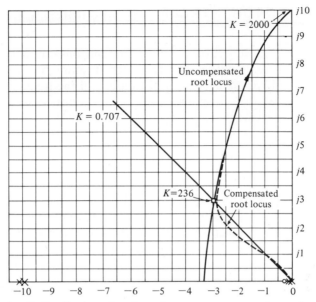

Fig. 10.18. Design of a phase-lag compensator on the s-plane.

the system when the K_v-requirement is satisfied are a long way from satisfying the damping ratio specification, and it would be dificult to bring the dominant roots from the $j\omega$-axis to the $\zeta = 0.707$ line by using a phase-lead compensator. Therefore, we will attempt to satisfy the K_v- and ζ-requirements by using a phase-lag network. The uncompensated root locus of this system is shown in Fig. 10.18 and the roots are shown when $\zeta = 0.707$ and $s = -2.9 \pm j2.9$. Measuring the gain at these roots, we find that $K = 236$. Therefore, the necessary ratio of zero to pole of the compensator is

$$\alpha = \left|\frac{z}{p}\right| = \frac{2000}{236} = 8.5.$$

Therefore we will choose $z = 0.1$ and $p = 0.1/9$ in order to allow a small margin of safety. Examining Fig. 10.18, we find that the difference between the angle from the pole and zero of $G_c(s)$ is negligible. Therefore the compensated system is

$$G_c(s)GH(s) = \frac{236(s + 0.1)}{s(s + 10)^2(s + 0.0111)}, \tag{10.68}$$

where $(K/\alpha) = 236$ and $\alpha = 9$.

The design of an integration compensator in order to increase the error constant of an uncompensated control system is particularly illustrative using s-plane and root locus methods. We shall now turn to similarly useful methods of designing integration compensation using Bode diagrams.

10.8 COMPENSATION ON THE BODE DIAGRAM USING A PHASE-LAG NETWORK

The design of a phase-lag RC network suitable for compensating a feedback control system may be readily accomplished on the Bode diagram. The advantage of the Bode diagram is again apparent for we will simply add the frequency response of the compensator to the Bode diagram of the uncompensated system in order to obtain a satisfactory system frequency response. The transfer function of the phase-lag network written in Bode diagram form is

$$G_c(j\omega) = \frac{1 + j\omega\tau}{1 + j\omega\alpha\tau} \qquad (10.69)$$

as we found in Eq. (10.14). The Bode diagram of the phase-lag network is shown in Fig. 10.8 for two values of α. On the Bode diagram, the pole and zero of the compensator have a magnitude much smaller than the smallest pole of the uncompensated system. Thus the phase lag is not the useful effect of the compensator, but rather it is the attenuation $-20 \log \alpha$ which is the useful effect for compensation. The phase-lag network is used to provide an attenuation and, therefore, to lower the 0-db (crossover) frequency of the system. However, at lower crossover frequencies, we usually find that the phase margin of the system is increased and our specifications may be satisfied. The design procedure for a phase-lag network on the Bode diagram is as follows:

1. Draw the Bode diagram of the uncompensated system with the gain adjusted for the desired error constant.

2. Determine the phase margin of the uncompensated system and, if it is insufficient, proceed with the following steps.

3. Determine the frequency where the phase margin requirement would be satisfied if the magnitude curve crossed the 0-db line at this frequency, ω_c'. (Allow for 5° phase lag from the phase-lag network when determining the new crossover frequency.)

4. Place the zero of the compensator one decade below the new crossover frequency ω_c' and thus ensure only 5° of lag at ω_c' (see Fig. 10.8).

5. Measure the necessary attenuation at ω_c' in order to ensure that the magnitude curve crosses at this frequency.

6. Calculate α by noting that the attenuation is $-20 \log \alpha$.

7. Calculate the pole as $\omega_p = 1/\alpha\tau = \omega_z/\alpha$ and the design is completed.

An example of this design procedure will illustrate that the method is simple to carry out in practice.

Example 10.8. Let us reconsider the system of Example 10.6 and design a phase-lag network so that the desired phase margin is obtained. The uncompensated transfer function is

$$GH(j\omega) = \frac{K}{j\omega(j\omega + 2)} - \frac{K_v}{j\omega(0.5j\omega + 1)}, \tag{10.70}$$

where $K_v = K/2$. It is desired that $K_v = 20$ while a phase margin of 45° is attained. The uncompensated Bode diagram is shown as a solid line in Fig. 10.19. The uncompensated system has a phase margin of 20°, and the phase margin must be increased. Allowing 5° for the phase-lag compensator, we locate the frequency ω where $\phi(\omega) = -130°$, which is to be our new crossover frequency ω'_c. In this case, we find that $\omega'_c = 1.5$, which allows for a small margin of safety. The attenuation necessary to cause ω'_c to be the new crossover frequency is equal to 20 db, accounting for a 2-db difference between the actual and asymptotic curves. Then we find that 20 db $= 20 \log \alpha$, or $\alpha = 10$. Therefore the zero is one decade below the crossover, or $\omega_z = \omega'_c/10 = 0.15$, and the pole is at $\omega_p = \omega_z/10 = 0.015$. The compensated system is then

$$G_c(j\omega)GH(j\omega) = \frac{20(6.66j\omega + 1)}{j\omega(0.5j\omega + 1)(66.6j\omega + 1)}. \tag{10.71}$$

The frequency response of the compensated system is shown in Fig. 10.19 with dotted lines. It is evident that the phase lag introduces an attenuation which lowers the crossover frequency and, therefore, increases the phase margin. Note that the phase angle of the lag network has almost totally disappeared at the crossover frequency ω'_c. As a final check, we numerically evaluate the phase margin at $\omega'_c = 1.5$ and find that $\phi_{pm} = 45°$, which is the desired result. Using the Nichols chart, we

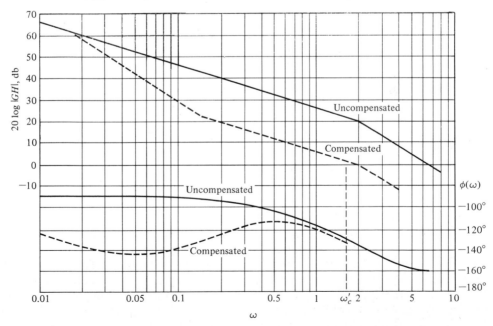

Fig. 10.19. Design of a phase-lag network on the Bode diagram for Example 10.8.

find that the closed-loop bandwidth of the system has been reduced from $\omega = 10$ rad/sec for the uncompensated system to $\omega = 2.5$ rad/sec for the compensated system.

Example 10.9. Let us reconsider the system of Example 10.7 which is

$$GH(j\omega) = \frac{K}{j\omega(j\omega + 10)^2}$$

$$= \frac{K_v}{j\omega(0.1j\omega + 1)^2}$$

$$(10.72)$$

where $K_v = K/100$. A velocity constant of K_v equal to 20 is specified. Furthermore, a damping ratio of 0.707 for the dominant roots is required. From Fig. 8.18, we estimate that a phase margin of 65° is required. The frequency response of the uncompensated system is shown in Fig. 10.20. The phase margin of the uncompensated system is zero degrees. Allowing 5° for the lag network, we locate the frequency where the phase is $-110°$. This frequency is equal to 1.74, and therefore we shall attempt to locate the new crossover frequency at $\omega_c' = 1.5$. Measuring the necessary attenuation at $\omega = \omega_c'$, we find that 23 db is required; $23 = 20 \log \alpha$, or $\alpha = 14.2$. The zero of the compensator is located one decade below the crossover frequency and thus

$$\omega_z = \frac{\omega_c'}{10} = 0.15.$$

The pole is then

$$\omega_p = \frac{\omega_z}{\alpha} = \frac{0.15}{14.2}.$$

Therefore the compensated system is

$$G_c(j\omega)GH(j\omega) = \frac{20(6.66j\omega + 1)}{j\omega(0.1j\omega + 1)^2 (94.6j\omega + 1)}.$$

$$(10.73)$$

The compensated frequency response is shown in Fig. 10.20. As a final check, we numerically evaluate the phase margin at $\omega_c' = 1.5$ and find that $\phi_{pm} = 67°$, which is within the specifications.

Therefore, a phase-lag compensation network may be used to alter the frequency response of a feedback control system in order to attain satisfactory system performance. Examining both Examples 10.8 and 10.9, we note again that the system design is satisfactory when the asymptotic curve for the magnitude of the compensated system crosses the 0 db line with a slope of -6 db/octave. The attenuation of the phase-lag network reduces the magnitude of the crossover (0-db) frequency to a point where the phase margin of the system is satisfactory. Thus, in contrast to the phase-lead network, the phase-lag network reduces the closed-loop bandwidth of the system as it maintains a suitable error constant.

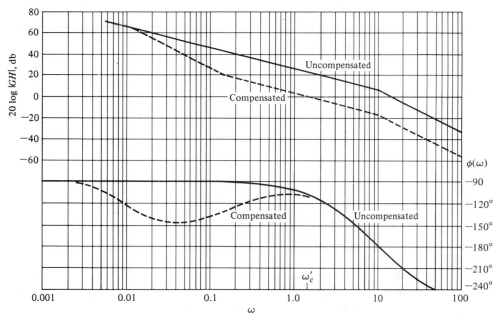

Fig. 10.20. Design of a phase-lag network on the Bode diagram for Example 10.9.

One might ask, why do we not place the compensator zero more than one decade below the new crossover ω'_c (see item 4 of the design procedure) and thus ensure less than 5° of lag at ω_c due to the compensator? This question may be answered by considering the requirements placed on the resistors and capacitors of the lag network by the values of the poles and zeros (see Eq. 10.12). As the magnitudes of the pole and zero of the lag network are decreased, the magnitudes of the resistors and the capacitor required increase proportionately. The zero of the lag compensator in terms of the circuit components is $z = 1/R_2C$, and the α of the network is $\alpha = (R_1 + R_2)/R_2$. Thus, considering the preceding example, 10.9, we require a zero at $z = 0.15$ which can be obtained with $C = 1$ μf and $R_2 = 6.66$ megohms. However, for $\alpha = 14$, we require a resistance R_1 of $R_1 = R_2(\alpha - 1) = 88$ megohms. Clearly, a designer does not wish to place the zero z further than one decade below ω'_c and thus require larger values of R_1, R_2, and C.

The phase-lead compensation network alters the frequency response of a network by adding a positive (leading) phase angle and, therefore, increases the phase margin at the crossover (0-db) frequency. It becomes evident that a designer might wish to consider using a compensation network which provided the attenuation of a phase-lag network and the lead-phase angle of a phase-lead network. Such a network exists and is called a lead-lag network and is shown in Fig. 10.21. The transfer function of this network is

$$\frac{E_2(s)}{E_1(s)} = \frac{(R_1C_1s + 1)(R_2C_2s + 1)}{R_1R_2C_1C_2s^2 + (R_1C_1 + R_1C_2 + R_2C_2)s + 1}. \tag{10.74}$$

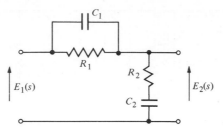

Fig. 10.21. An *RC* lead-lag network.

When $\alpha\tau_1 = R_1C_1$, $\beta\tau_2 = R_2C_2$, $\tau_1\tau_2 = R_1R_2C_1C_2$, we note that $\alpha\beta = 1$ and then Eq. (10.74) is

$$\frac{E_2(s)}{E_1(s)} = \frac{(1 + \alpha\tau_1 s)(1 + \beta\tau_2 s)}{(1 + \tau_1 s)(1 + \tau_2 s)},$$ (10.75)

where $\alpha > 1$, $\beta < 1$. The first terms in the numerator and denominator, which are a function of τ_1, provide the phase-lead portion of the network. The second terms which are a function of τ_2, provide the phase-lag portion of the compensation network. The parameter β is adjusted to provide suitable attenuation of the low frequency portion of the frequency response, and the parameter α is adjusted to provide an additional phase lead at the new crossover (0-db) frequency. Alternatively, the compensation may be designed on the s-plane by placing the lead pole and zero compensation in order to locate the dominant roots in a desired location. Then the phase-lag compensation is used to raise the error constant at the dominant root location by a suitable ratio, $1/\beta$. The design of a phase lead-lag compensator follows the procedures already discussed, and the reader is referred to further literature illustrating the utility of lead-lag compensation [2, 3].

10.9. COMPENSATION ON THE BODE DIAGRAM USING ANALYTICAL AND COMPUTER METHODS

It is desirable to use computers, when appropriate, to assist the designer in the selection of the parameters of a compensator. The development of algorithms for computer-added design is an important alternative approach to the trial-and-error methods considered in earlier sections. By the use of compensators, computer programs have been developed for the selection of suitable parameter values based on satisfaction of frequency response criteria such as phase margin [16, 17].

An analytical technique of selecting the parameters of a lead or lag network has been developed for Bode diagrams [18, 19]. For a single-stage compensator

$$G_c(s) = \frac{1 + \alpha\tau s}{1 + \tau s},$$ (10.76)

where $\alpha < 1$ yields a lag compensator and $\alpha > 1$ yields a lead compensator. The phase contribution of the compensator at the desired crossover frequency ω_c (see Eq. 10.9) is

$$p = \tan\phi = \frac{\alpha\omega_c\tau - \omega_c\tau}{1 + (\omega_c\tau)^2\alpha}. \tag{10.77}$$

The magnitude M (in db) of the compensator at ω_c is

$$c = 10^{M/10} = \frac{1 + (\omega_c\alpha\tau)^2}{1 + (\omega_c\tau)^2}. \tag{10.78}$$

Eliminating $\omega_c\tau$ from Eqs. 10.77 and 10.78, we obtain the nontrivial solution equation for α as

$$(p^2 - c + 1)\alpha^2 + 2p^2c\alpha + p^2c^2 + c^2 - c = 0. \tag{10.79}$$

For a single-stage compensator, it is necessary that $c > p^2 + 1$. If we solve for α from Eq. 10.79, we can obtain τ from

$$\tau = \frac{\alpha}{\omega_c}\sqrt{\frac{1 - c}{c - \alpha^2}}. \tag{10.80}$$

The design steps for a lead compensator are:

1. Select the desired ω_c.
2. Determine the phase margin desired and therefore the required phase ϕ for Eq. 10.77.
3. Verify that the phase lead is applicable, $\phi > 0$ and $M > 0$.
4. Determine whether a single stage will be sufficient when $c > p^2 + 1$.
5. Determine α from Eq. 10.79.
6. Determine τ from Eq. 10.80.

If one needs to design a single lag compensator, then $\phi < 0$ and $M < 0$ (step 3). Also, step 4 will require $c < [1/(1 + p^2)]$. Otherwise the method is the same.

Example 10.10.   Let us reconsider the system of Example 10.1 and design a lead network by the analytical technique. Examine the uncompensated curves in Fig. 10.9. We select $\omega_c = 5$. Then, as before, we desire a phase margin of 45°. The compensator must yield this phase, so

$$p = \tan 45° = 1. \tag{10.81}$$

The required magnitude contribution is 8 db or $M = 8$, so that

$$c = 10^{8/10} = 6.31. \tag{10.82}$$

Using c and p we obtain

$$-4.31\alpha^2 + 12.62\alpha + 73.32 = 0. \tag{10.83}$$

Solving for α we obtain $\alpha = 5.84$. Solving Eq. (10.80), we obtain $\tau = 0.510$. Therefore the compensator is

$$G_c(s) = \frac{1 + 2.98s}{1 + 0.51s}. \tag{10.84}$$

The pole is equal to 1.96 and the zero is 0.336. This design is similar to that obtained by the iteration technique of Section 10.4.

10.10 THE DESIGN OF CONTROL SYSTEMS IN THE TIME-DOMAIN

The design of automatic control systems is an important function of control engineering. The purpose of design is to realize a system with practical components which will provide the desired operating performance. The desired performance can be readily stated in terms of time-domain performance indices. For example, the maximum overshoot and rise time for a step input are valuable time-domain indices. In the case of steady-state and transient performance, the performance indices are normally specified in the time domain and, therefore, it is natural that we wish to develop design procedures in the time domain.

The performance of a control system may be represented by integral performance measures as we found in Section 4.5. Therefore, the design of a system must be based on minimizing a performance index such as the integral of the squared error (ISE) as in Section 4.5. Systems which are adjusted to provide a minimum performance index are often called *optimum control systems*. We shall consider, in this section, the design of an optimum control system where the system is described by a state variable formulation.

However, before proceeding to the specifics, we should note that we did design a system in the time domain in Example 9.6. In this example, we considered the unstable portion of an inverted pendulum system and developed a suitable feedback control so that the system was stable. This design was based on measuring the state variables of the system and using them to form a suitable control signal $u(t)$ so that the system was stable. In this section, we shall again consider the measurement of the state variables and their use in developing a control signal $u(t)$ so that the performance of the system is optimized.

The performance of a control system, written in terms of the state variables of a system, may be expressed in general as

$$J = \int_0^{t_f} g(\mathbf{x}, \mathbf{u}, t)\, dt, \tag{10.85}$$

where $\mathbf{x}$ equals the state vector and $\mathbf{u}$ equals the control vector.*

We are interested in minimizing the error of the system and, therefore, when the desired state vector is represented as $\mathbf{x}_d = \mathbf{0}$, we are able to consider the error as identically equal to the value of the state vector. That is, we desire the system to be

---

* Note that J is used to denote the performance index, instead of I, which was used in Chapter 4. This will enable the reader to readily distinguish the performance index from the identity matrix which is represented by the bold-faced capital $\mathbf{I}$.

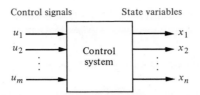

Fig. 10.22. A control system in terms of x and u.

at equilibrium, $\mathbf{x} = \mathbf{x}_d = \mathbf{0}$, and any deviation from equilibrium is considered an error. Therefore we will consider, in this section, the design of optimum control systems using *state-variable feedback* and error-squared performance indices [1, 2, 4, 5].

The control system which we will consider is shown in Fig. 10.22 and may be represented by the vector differential equation

$$\dot{\mathbf{x}} = \mathbf{Ax} + \mathbf{Bu}. \tag{10.86}$$

We will select a feedback controller so that $\mathbf{u}$ is some function of the measured state variables $\mathbf{x}$ and therefore

$$\mathbf{u} = \mathbf{h}(\mathbf{x}). \tag{10.87}$$

For example, one might use

$$u_1 = k_1 x_1,$$
$$u_2 = k_2 x_2,$$
$$\vdots$$
$$u_m = k_m x_m. \tag{10.88}$$

Alternatively, one might choose the control vector as

$$u_1 = k_1(x_1 + x_2),$$
$$u_2 = k_2(x_2 + x_3), \tag{10.89}$$
$$\vdots$$

The choice of the control signals is somewhat arbitrary and depends partially upon the actual desired performance and the complexity of the feedback structure allowable. Often we are limited in the number of state variables available for feedback, since we are only able to utilize measurable state variables.

Now, in our case, we limit the feedback function to a linear function so that $\mathbf{u} = \mathbf{Hx}$ where $\mathbf{H}$ is an $m \times n$ matrix. Therefore, in expanded form, we have

$$
\begin{bmatrix} u_1 \\ u_2 \\ \vdots \\ u_m \end{bmatrix} = \begin{bmatrix} h_{11} & \cdots & h_{1n} \\ \vdots & & \vdots \\ h_{m1} & \cdots & h_{mn} \end{bmatrix} \begin{bmatrix} x_1 \\ x_2 \\ \vdots \\ x_n \end{bmatrix} \tag{10.90}
$$

Then, substituting Eq. (10.90) into Eq. (10.86), we obtain

$$\dot{x} = Ax + BHx = Dx,$$ (10.91)

where D is the $n \times n$ matrix resulting from the addition of the elements of A and BH.

Now, returning to the error-squared performance index, we recall from Section 4.5 that the index for a single state variable, x_1, is written as

$$J = \int_0^{t_f} (x_1(t))^2 \, dt.$$ (10.92)

A performance index written in terms of two state variables would then be

$$J = \int_0^{t_f} (x_1^2 + x_2^2) dt.$$ (10.93)

Therefore, since we wish to define the performance index in terms of an integral of the sum of the state variables squared, we will utilize the matrix operation

$$x^T x = [x_1, x_2, x_3, \ldots, x_n] \begin{bmatrix} x_1 \\ x_2 \\ \vdots \\ x_n \end{bmatrix} = (x_1^2 + x_2^2 + x_3^2 + \ldots + x_n^2), \quad (10.94)$$

where x^T indicates the transpose of the x matrix.* Then the general form of the performance index, in terms of the state vector, is

$$J = \int_0^{t_f} (x^T x) \, dt.$$ (10.95)

Again considering Eq. (10.95), we will let the final time of interest be $t_f = \infty$. In order to obtain the minimum value of J, we postulate the existence of an exact differential so that

$$\frac{d}{dt} (x^T P x) = -x^T x,$$ (10.96)

where P is to be determined. A symmetric P matrix will be used in order to simplify the algebra without any loss of generality. Then, for a symmetric P matrix, $p_{ij} = p_{ji}$. Completing the differentiation indicated on the left-hand side of Eq. (10.96), we have

$$\frac{d}{dt} (x^T P x) = \dot{x}^T P x + x^T P \dot{x}.$$

---

*The matrix operation $x^T x$ is discussed in Appendix C, Section C.4.

Then, substituting Eq. (10.91), we obtain

$$\frac{d}{dt}(\mathbf{x}^T\mathbf{P}\mathbf{x}) = (\mathbf{D}\mathbf{x})^T\mathbf{P}\mathbf{x} + \mathbf{x}^T\mathbf{P}(\mathbf{D}\mathbf{x})$$

$$= \mathbf{x}^T\mathbf{D}^T\mathbf{P}\mathbf{x} + \mathbf{x}^T\mathbf{P}\mathbf{D}\mathbf{x} \qquad (10.97)$$

$$= \mathbf{x}^T(\mathbf{D}^T\mathbf{P} + \mathbf{P}\mathbf{D})\mathbf{x},$$

where $(\mathbf{D}\mathbf{x})^T = \mathbf{x}^T\mathbf{D}^T$ by the definition of the transpose of a product. If we let $(\mathbf{D}^T\mathbf{P} + \mathbf{P}\mathbf{D}) = -\mathbf{I}$, then Eq. (10.97) becomes

$$\frac{d}{dt}(\mathbf{x}^T\mathbf{P}\mathbf{x}) = -\mathbf{x}^T\mathbf{x}, \qquad (10.98)$$

which is the exact differential we are seeking. Substituting Eq. (10.98) into Eq. (10.95), we obtain

$$J = \int_0^\infty -\frac{d}{dt}(\mathbf{x}^T\mathbf{P}\mathbf{x})\,dt$$

$$= -\mathbf{x}^T\mathbf{P}\mathbf{x}\Big|_0^\infty$$

$$= \mathbf{x}^T(0)\mathbf{P}\mathbf{x}(0). \qquad (10.99)$$

In the evaluation of the limit at $t = \infty$, we have assumed that the system is stable and hence $\mathbf{x}(\infty) = 0$ as desired. Therefore, in order to minimize the performance index J, we consider the two equations

$$J = \int_0^\infty \mathbf{x}^T\mathbf{x}\,dt = \mathbf{x}^T(0)\mathbf{P}\mathbf{x}(0) \qquad (10.100)$$

and

$$\mathbf{D}^T\mathbf{P} + \mathbf{P}\mathbf{D} = -\mathbf{I}. \qquad (10.101)$$

The design steps are then as follows:

1. Determine the matrix $\mathbf{P}$ which satisfies Eq. (10.101), where $\mathbf{D}$ is known.

2. Minimize J by determining the minimum of Eq. (10.100).

Example 10.11. Consider the control system shown in Fig. 10.23 in signal-flow graph form. The state variables are identified as x_1 and x_2. The performance of this

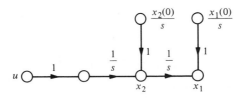

Fig. 10.23. The signal-flow graph of the control system of Example 10.10.

system is quite unsatisfactory since an undamped response results for a step input or disturbance signal. The vector differential equation of this system is

$$\frac{d}{dt}\begin{bmatrix} x_1 \\ x_2 \end{bmatrix} = \begin{bmatrix} 0 & 1 \\ 0 & 0 \end{bmatrix}\begin{bmatrix} x_1 \\ x_2 \end{bmatrix} + \begin{bmatrix} 0 \\ 1 \end{bmatrix}u(t), \tag{10.102}$$

where

$$\mathbf{A} = \begin{bmatrix} 0 & 1 \\ 0 & 0 \end{bmatrix}$$

We will choose a feedback control system so that

$$u(t) = -k_1 x_1 - k_2 x_2, \tag{10.103}$$

and therefore the control signal is a linear function of the two state variables. The sign of the feedback is negative in order to provide negative feedback. Then Eq. (10.102) becomes

$$\dot{x}_1 = x_2, \\ \dot{x}_2 = -k_1 x_1 - k_2 x_2, \tag{10.104}$$

or, in matrix form, we have

$$\dot{\mathbf{x}} = \mathbf{Dx} \\ = \begin{bmatrix} 0 & 1 \\ -k_1 & -k_2 \end{bmatrix}\mathbf{x}. \tag{10.105}$$

We note that x_1 would represent the position of a position-control system and the transfer function of the system would be $G(s) = 1/Ms^2$, where $M = 1$ and the friction is negligible. In any case, in order to avoid needless algebraic manipulation, we will let $k_1 = 1$ and determine a suitable value for k_2 so that the performance index is minimized. Then, writing Eq. (10.101), we have

$$\mathbf{D}^T\mathbf{P} + \mathbf{PD} = -\mathbf{I},$$

$$\begin{bmatrix} 0 & -1 \\ 1 & -k_2 \end{bmatrix}\begin{bmatrix} p_{11} & p_{12} \\ p_{12} & p_{22} \end{bmatrix} + \begin{bmatrix} p_{11} & p_{12} \\ p_{12} & p_{22} \end{bmatrix}\begin{bmatrix} 0 & 1 \\ -1 & -k_2 \end{bmatrix} = \begin{bmatrix} -1 & 0 \\ 0 & -1 \end{bmatrix}. \tag{10.106}$$

Completing the matrix multiplication and addition, we have

$$-p_{12} - p_{12} = -1, \\ p_{11} - k_2 p_{12} - p_{22} = 0, \\ p_{12} - k_2 p_{22} + p_{12} - k_2 p_{22} = -1. \tag{10.107}$$

Then, solving these simultaneous equations, we obtain

$$p_{12} = \frac{1}{2}, \quad p_{22} = \frac{1}{k_2}, \quad p_{11} = \frac{k_2^2 + 2}{2k_2}.$$

The integral performance index is then

$$J = \mathbf{x}^T(0)\mathbf{P}\mathbf{x}(0), \tag{10.108}$$

and we shall consider the case where each state is initially displaced one unit from equilibrium so that $\mathbf{x}^T(0) = [1, 1]$. Therefore Eq. (10.108) becomes

$$J = [1, 1] \begin{bmatrix} p_{11} & p_{12} \\ p_{12} & p_{22} \end{bmatrix} \begin{bmatrix} 1 \\ 1 \end{bmatrix}$$

$$= [1, 1] \begin{bmatrix} (p_{11} + p_{12}) \\ (p_{12} + p_{22}) \end{bmatrix}$$

$$= (p_{11} + p_{12}) + (p_{12} + p_{22}) = p_{11} + 2p_{12} + p_{22}. \tag{10.109}$$

Substituting the values of the elements of $\mathbf{P}$, we have

$$J = \frac{k_2^2 + 2}{2k_2} + 1 + \frac{1}{k_2}$$

$$= \frac{k_2^2 + 2k_2 + 4}{2k_2}. \tag{10.110}$$

In order to minimize as a function of k_2, we take the derivative with respect to k_2 and set it equal to zero as follows:

$$\frac{\partial J}{\partial k_2} = \frac{2k_2(2k_2 + 2) - 2(k_2^2 + 2k_2 + 4)}{(2k_2)^2} = 0, \tag{10.111}$$

and therefore $k_2^2 = 4$ and $k_2 = 2$ when J is a minimum. The minimum value of J is obtained by substituting $k_2 = 2$ into Eq. (10.110) and thus we obtain

$$J_{\min} = 3.$$

The system matrix $\mathbf{D}$, obtained for the compensated system, is then

$$\mathbf{D} = \begin{bmatrix} 0 & 1 \\ -1 & -2 \end{bmatrix} \tag{10.112}$$

The characteristic equation of the compensated system is therefore

$$\det [\lambda\mathbf{I} - \mathbf{D}] = \det \begin{bmatrix} \lambda & -1 \\ 1 & \lambda + 2 \end{bmatrix}$$

$$= \lambda^2 + 2\lambda + 1. \tag{10.113}$$

Since this is a second-order system, we note that the characteristic equation is of the form $(s^2 + 2\zeta\omega_n s + \omega_n^2)$, and therefore the damping ratio of the compensated system is $\zeta = 1.0$. This compensated system is considered to be an optimum system in that the compensated system results in a minimum value for the performance index. Of course, we recognize that this system is only optimum for the specific set of initial conditions that were assumed. The compensated system is shown in Fig.

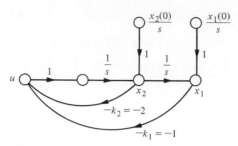

Fig. 10.24. The compensated control system of Example 10.10.

10.24. A curve of the performance index as a function of k_2 is shown in Fig. 10.25. It is clear that this system is not very sensitive to changes in k_2 and will maintain a near minimum performance index if the k_2 is altered some percentage. We define the sensitivity of an optimum system as

$$S_k^{\text{opt}} = \frac{\Delta J/J}{\Delta k/k},\tag{10.114}$$

where k is the design parameter. Then, for this example, we have $k = k_2$ and therefore

$$S_{k_2}^{\text{opt}} \simeq \frac{0.08/3}{0.5/2} = 0.107.\tag{10.115}$$

Example 10.12. Now let us reconsider the system of the previous example where both the feedback gains, k_1 and k_2, are unspecified. In order to simplify the algebra, without any loss in insight into the problem, let us set $k_1 = k_2 = k$. The reader may prove that if k_1 and k_2 are unspecified then $k_1 = k_2$ when the minimum of the

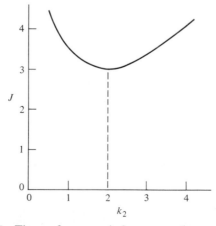

Fig. 10.25. The performance index versus the parameter k_2.

performance index (Eq. 10.100) is obtained. Then for the system of the previous example, Eq. (10.105) becomes

$$\dot{x} = Dx$$
$$= \begin{bmatrix} 0 & 1 \\ -k & -k \end{bmatrix} x. \tag{10.116}$$

In order to determine the P matrix, we utilize Eq. (10.101), which is

$$D^T P + PD = -I. \tag{10.117}$$

Solving the set of simultaneous equations resulting from Eq. (10.117), we find that

$$P_{12} = \frac{1}{2k}, \quad P_{22} = \frac{(k+1)}{2k^2}, \quad P_{11} = \frac{(1+2k)}{2k}.$$

Let us consider the case where the system is initially displaced one unit from equilibrium so that $x^T(0) = [1, 0]$. Then the performance index (Eq. 10.100) becomes

$$J = \int_0^\infty x^T x \, dt = x^T(0)Px(0) = P_{11}. \tag{10.118}$$

Thus the performance index to be minimized is

$$J = P_{11} = \frac{(1+2k)}{2k} = 1 + \frac{1}{2k}. \tag{10.119}$$

Clearly, the minimum value of J is obtained when k approaches infinity; the result is $J_{min} = 1$. A plot of J versus k is shown in Fig. 10.26. This plot illustrates that the

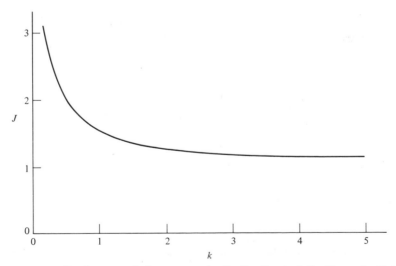

Fig. 10.26. Performance index versus the feedback gain k for Example 10.11.

performance index approaches a minimum asymptotically as k approaches an infinite value. Now we recognize that in providing a very large gain k, we cause the feedback signal

$$u(t) = -k(x_1(t) + x_2(t))$$

to be very large. However, we are restricted to realizable magnitudes of the control signal $u(t)$. Therefore we must introduce a *constraint* on $u(t)$ so that the gain k is not made too large. Then, for example, if we establish a constraint on $u(t)$ so that

$$|u(t)| \leqslant 50, \qquad (10.120)$$

we require that the maximum acceptable value of k in this case

$$k_{max} = \frac{|u|_{max}}{x_1(0)} = 50. \qquad (10.121)$$

Then the minimum value of J is

$$\begin{aligned} J_{min} &= 1 + \frac{1}{2k_{max}} \\ &= 1.01, \end{aligned} \qquad (10.122)$$

which is sufficiently close to the absolute minimum of J in order to satisfy our requirements.

Upon examination of the performance index (Eq. 10.95), we recognize that the reason the magnitude of the control signal is not accounted for in the original calculations is that $u(t)$ is not included within the expression for the performance index. However, there are many cases where we are concerned with the expenditure of the control signal energy. For example, in a space vehicle attitude control system, $[u(t)]^2$ represents the expenditure of jet fuel energy and must be restricted in order to conserve the fuel energy for long periods of flight. In order to account for the expenditure of the energy of the control signal, we will utilize the performance index

$$J = \int_0^\infty (\mathbf{x}^T \mathbf{I} \mathbf{x} + \lambda \mathbf{u}^T \mathbf{u}) dt, \qquad (10.123)$$

where λ is a scalar weighting factor and $\mathbf{I} =$ identity matrix. The weighting factor λ will be chosen so that the relative importance of the state variable performance is contrasted with the importance of the expenditure of the system energy resource which is represented by $\mathbf{u}^T \mathbf{u}$. As in the previous paragraphs we will represent the state variable feedback by the matrix equation

$$\mathbf{u} = \mathbf{H} \mathbf{x} \qquad (10.124)$$

and the system with this state variable feedback as

$$\begin{aligned} \dot{\mathbf{x}} &= \mathbf{A} \mathbf{x} + \mathbf{B} \mathbf{u} \\ &= \mathbf{D} \mathbf{x}. \end{aligned} \qquad (10.125)$$

Now, substituting Eq. (10.124) into Eq. (10.123), we have

$$J = \int_0^\infty (\mathbf{x}^T \mathbf{Ix} + \lambda(\mathbf{Hx})^T(\mathbf{Hx}))\, dt$$

$$= \int_0^\infty [\mathbf{x}^T(\mathbf{I} + \lambda \mathbf{H}^T \mathbf{H})\mathbf{x}]\, dt$$

$$= \int_0^\infty \mathbf{x}^T \mathbf{Qx}\, dt, \tag{10.126}$$

where $\mathbf{Q} = (\mathbf{I} + \lambda \mathbf{H}^T \mathbf{H})$ is an $n \times n$ matrix. Following the development of Eqs. (10.95) through (10.99), we postulate the existence of an exact differential so that

$$\frac{d}{dt}(\mathbf{x}^T \mathbf{Px}) = -\mathbf{x}^T \mathbf{Qx}. \tag{10.127}$$

Then in this case we require that

$$\mathbf{D}^T \mathbf{P} + \mathbf{PD} = -\mathbf{Q}, \tag{10.128}$$

and thus we have as before (Eq. 10.99)

$$J = \mathbf{x}^T(0)\mathbf{Px}(0). \tag{10.129}$$

Now the design steps are exactly as for Eqs. (10.100) and (10.101) with the exception that the left side of Eq. (10.128) equals $-\mathbf{Q}$ instead of $-\mathbf{I}$. Of course, if $\lambda = 0$, Eq. (10.128) reduces to Eq. (10.101). Now, let us reconsider the previous example when λ is other than zero and account for the expenditure of control signal energy.

Example 10.13. Let us reconsider the system of the previous example which is shown in Fig. 10.23. For this system we use a state variable feedback so that

$$\mathbf{u} = \mathbf{Hx}$$
$$= \begin{bmatrix} k & 0 \\ 0 & k \end{bmatrix} \begin{bmatrix} x_1 \\ x_2 \end{bmatrix} = k\mathbf{Ix}. \tag{10.130}$$

Therefore, the matrix $\mathbf{Q}$ is then

$$\mathbf{Q} = (\mathbf{I} + \lambda \mathbf{H}^T \mathbf{H})$$
$$= (\mathbf{I} + \lambda k^2 \mathbf{I}) \tag{10.131}$$
$$= (1 + \lambda k^2)\mathbf{I}.$$

As in the previous example we will let $\mathbf{x}^T(0) = [1, 0]$ so that $J = p_{11}$. We evaluate p_{11} from Eq. (10.128) as

$$\mathbf{D}^T \mathbf{P} + \mathbf{PD} = -\mathbf{Q}$$
$$= -(1 + \lambda k^2)\mathbf{I}. \tag{10.132}$$

Thus we find that

$$J = p_{11} = (1 + \lambda k^2)\left(1 + \frac{1}{2k}\right), \tag{10.133}$$

and we note that the right-hand side of Eq. (10.133) reduces to Eq. (10.119) when $\lambda = 0$. Now the minimum of J is found by taking the derivative of J, which is

$$\frac{dJ}{dk} = 2\lambda k + \frac{\lambda}{2} - \frac{1}{2k^2} = \frac{4\lambda k^3 + \lambda k^2 - 1}{2k^2} = 0. \qquad (10.134)$$

Therefore, the minimum of the performance index occurs when $k = k_{min}$, where k_{min} is the solution of Eq. (10.134).

A simple method of solution for Eq. (10.134) is the Newton-Raphson method illustrated in Section 5.4. Let us complete this example for the case where the control energy and the state variables squared are equally important so that $\lambda = 1$. Then Eq. (10.134) becomes $4k^3 + k^2 - 1 = 0$, and using the Newton-Raphson method, we find that $k_{min} = 0.555$. The value of the performance index J obtained with k_{min} is considerably greater than that of the previous example, since the expenditure of energy is equally weighted as a cost. The plot of J versus k for this case is shown in Fig. 10.27. Also the plot of J versus k for Example 10.11 is shown for comparison on Fig. 10.27. It has become clear from this and the previous examples that the actual minimum obtained depends upon the initial conditions, the definition of the performance index, and the value of the scalar factor λ.

The design of several parameters may be accomplished in a similar manner to that illustrated in the examples. Also, the design procedure can be carried out for higher-order systems. However, one must then consider the use of a digital computer to determine the solution of Eq.(10.101) in order to obtain the **P** matrix. Also, the computer would provide a suitable approach for evaluating the minimum value of J for the several parameters. The newly emerging field of adaptive and optimal

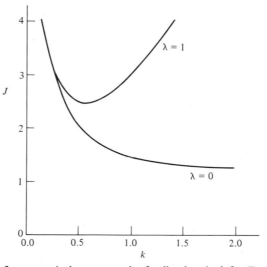

Fig. 10.27. Performance index versus the feedback gain k for Example 10.12.

control systems is based on the formulation of the time-domain equations and the determination of an optimum feedback control signal $\mathbf{u}(t)$ [5, 6, 10]. The design of control systems using time-domain methods will continue to develop in the future and will provide the control engineer with many interesting challenges and opportunities.

10.11 STATE-VARIABLE FEEDBACK

In the previous section we considered the use of state-variable feedback in achieving optimization of a performance index. In this section we will use *state-variable feedback* in order to achieve the desired pole location of the closed-loop transfer function $T(s)$. The approach is based on the feedback of all the state variables, and therefore

$$\mathbf{u} = \mathbf{Hx}. \tag{10.135}$$

When using this state-variable feedback, the roots of the characteristic equation are placed where the transient performance meets the desired response.

As an example of state-variable feedback, consider the feedback system shown in Fig. 10.28. This position control uses a field controlled motor, and the transfer function was obtained in Section 2.5 as

$$G(s) = \frac{K}{s(s + f/J)(s + R_f/L_f)}, \tag{10.136}$$

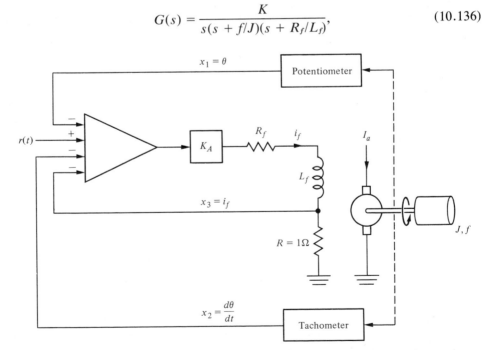

Fig. 10.28. A position control system with state-variable feedback.

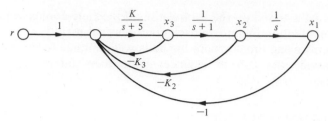

Fig. 10.29. The signal flow graph of the state-variable feedback system.

where $K = K_a K_m / JL_f$. For our purposes we will assume that $f/J = 1$ and $R_f/L_f = 5$. As shown in Fig. 10.28, the system has feedback of the three state variables: position, velocity, and field current. We will assume that the feedback constant for the position is equal to 1, as shown in Fig. 10.29, which provides a signal-flow graph representation of the system. Without state-variable feedback of x_2 and x_3, we set $K_3 = K_2 = 0$ and we have

$$G(s) = \frac{K}{s(s + 1)(s + 5)}. \tag{10.137}$$

This system will become unstable when $K \geqslant 30$. However, with variable feedback of all the state variables we can assure that the system is stable and set the transient performance of the system to a desired performance.

In general, the state-variable feedback signal-flow graph can be converted to the block diagram form shown in Fig. 10.30. The transfer function $G(s)$ remains unaffected (as in Eq. 10.137) and the $H(s)$ accounts for the state variable feedback. Therefore,

$$H(s) = K_3 \left[s^2 + \left(\frac{K_3 + K_2}{K_3} \right) s + \frac{1}{K_3} \right] \tag{10.138}$$

and

$$G(s)H(s) = \frac{M[s^2 + Qs + (1/K_3)]}{s(s + 1)(s + 5)}, \tag{10.139}$$

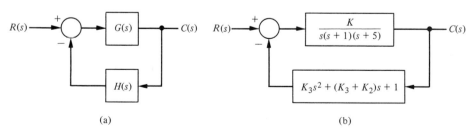

(a)   (b)

Fig. 10.30. An equivalent block diagram representation of the state-variable feedback system.

where $M = KK_3$ and $Q = (K_3 + K_2)/K_3$. Since K_3 and K_2 may be set independently, the designer may select the location of the zeros of $G(s)H(s)$.

As an illustration, let us choose the zeros of $GH(s)$ so that they cancel the real poles of $G(s)$. We set the numerator polynomial

$$H(s) = K_3\left(s^2 + Qs + \frac{1}{K_3}\right)$$
$$= K_3(s + 1)(s + 5). \tag{10.140}$$

This requires $K_3 = \frac{1}{5}$ and $Q = 6$, which sets $K_2 = 1$. Then

$$GH(s) = \frac{M(s + 1)(s + 5)}{s(s + 1)(s + 5)}, \tag{10.141}$$

where $M = KK_3$. The closed-loop transfer function is then

$$\frac{C(s)}{R(s)} = T(s) = \frac{G(s)}{1 + G(s)H(s)} = \frac{K}{(s + 1)(s + 5)(s + M)}. \tag{10.142}$$

Therefore, while we could choose $M = 10$ which would ensure the stability of the system, the closed-loop response of the system will be dictated by the poles at $s = -1$ and $s = -5$. Therefore we will usually choose the zeros of $GH(s)$ in order to achieve closed-loop roots in a desirable location in the left-hand plane and assure system stability.

Example 10.14. Let us again consider the system of Fig. 10.30(b) and set the zeros of $GH(s)$ at $s = -4 + j2$ and $s = -4 - j2$. Then the numerator of $GH(s)$ will be

$$H(s) = K_3\left(s^2 + Qs + \frac{1}{K_3}\right)$$
$$= K_3(s + 4 + j2)(s + 4 - j2)$$
$$= K_3(s^2 + 8s + 20). \tag{10.143}$$

Therefore $K_3 = \frac{1}{20}$ and $Q = 8$ resulting in $K_2 = \frac{7}{20}$. The resulting root locus for

$$G(s)H(s) = \frac{M(s^2 + 8s + 20)}{s(s + 1)(s + 5)} \tag{10.144}$$

is shown in Fig. 10.31. The system is stable for all values of gain $M = KK_3$. For $M = 10$ the complex roots have $\zeta = 0.73$, so that we might expect an overshoot for a step input of approximately 5%. The settling time will be approximately 1 second. The closed-loop transfer function is

$$\frac{C(s)}{R(s)} = T(s) = \frac{G(s)}{1 + G(s)H(s)}$$
$$= \frac{200}{(s + 3.45 + j3.2)(s + 3.45 - j3.2)(s + 9.1)}. \tag{10.145}$$

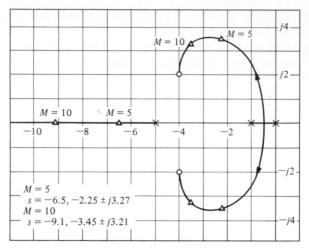

Fig. 10.31. The compensated system root locus.

An alternative approach is to set the closed-loop roots of $1 + G(s)H(S) = 0$ at desired locations and then solve for the gain values of K, K_3 and K_2 that are required. For example, if we desire closed-loop roots at $s = -10$, $s = -5 + j$ and $s = -5-j$ we have the characteristic equation

$$q(s) = (s + 10)(s^2 + 10s + 26)$$
$$= s^3 + 20s^2 + 126s + 260 = 0. \qquad (10.146)$$

Since

$$1 + G(s)H(s) = s(s + 1)(s + 5) + M\left(s^2 + Qs + \frac{1}{K_3}\right) = 0, \qquad (10.147)$$

we equate Eq. (10.146) and Eq. (10.147), obtaining $M = 14$, $Q = 121$, $K_3 = 14/260$ and $K_2 = 6.462$.

In many cases the state variables are available and we can use state variable feedback to obtain a stable, well-compensated system.

10.12 SUMMARY

In this chapter we have considered several alternative approaches to the design and compensation of feedback control systems. In the first two sections, we discussed the concepts of design and compensation and noted the several design cases which we completed in the preceding chapters. Then, the possibility of introducing cascade compensation networks within the feedback loops of control systems was examined. The cascade compensation networks are useful for altering the shape of the root locus or frequency response of a system. The phase-lead network and the phase-lag network were considered in detail as candidates for system compensators. Then, system compensation was studied by using a phase-lead s-plane network on the Bode diagram and the root locus s-plane diagram successively. System compensation using integration networks and phase-lag networks was also considered on the Bode diagram and the s-plane. We noted that the phase-lead compensator increases the phase margin of the system and thus provides additional stability. When the design specifications include an error constant, the design of a phase lead network is more readily accomplished on the Bode diagram. Alternatively, when an error constant is not specified, but the settling time and overshoot for a step input are specified, the design of a phase-lead network is more readily carried

Table 10.1 A Summary of the Characteristics of Phase-Lead and Phase-Lag Compensation Networks

	Compensation	
	Phase-lead	Phase-lag
Approach	Addition of phase-lead angle near the crossover frequency or to yield the desired dominant roots in the s-plane	Addition of phase-lag to yield an increased error constant while maintaining the desired dominant roots in the s-plane or phase margin on the Bode diagram
Results	1. Increases system bandwidth 2. Increases gain at higher frequencies	1. Decreases system bandwidth
Advantages	1. Yields desired response 2. Faster dynamic response	1. Suppresses high frequency noise 2. Reduces the steady-state error
Disadvantages	1. Requires additional amplifier gain 2. Increases bandwidth and thus susceptibility to noise 3. May require large values of components for the RC network	1. Slows down transient response 2. May require large values of components for the RC network
Applications	1. When fast transient response is desired	1. When error constants are specified
Not applicable	1. When phase decreases rapidly near the crossover frequency	1. When no low frequency range exists where the phase is equal to the desired phase margin

out on the s-plane. When large error constants are specified for a feedback system, it is usually easier to compensate the system by using integration (phase-lag) networks. We also noted that the phase-lead compensation increases the system bandwidth, while the phase-lag compensation decreases the system bandwidth. The bandwidth often may be an important factor when noise is present at the input and generated within the system. Also we noted that a satisfactory system is obtained when the asymptotic course for magnitude of the compensated system crosses the 0 db line with a slope of -6 db/octave. The characteristics of the phase-lead and phase-lag compensation networks are summarized in Table 10.1. Also, the design of control systems in the time domain was briefly examined. Specifically, the optimum design of a system using state-variable feedback and an integral performance index was considered. Finally, the s-plane design of systems utilizing state-variable feedback was examined.

PROBLEMS

10.1. The design of the Lunar Excursion Module (LEM) is an interesting control problem [6]. The Apollo 11 lunar landing vehicle is shown in Fig. P10.1(a). The attitude control system for the lunar vehicle is shown in Fig. P10.1(b). The vehicle damping is negligible and the attitude is controlled by gas jets. The torque, as a first approximation, will be considered to be proportional to the signal $V(s)$ so that $T(s) = K_2 V(s)$. The loop gain may be selected by the designer in order to provide a suitable damping. A damping ratio of $\zeta = 0.5$ with a settling time less than 2 sec is required. Using a lead-network compensation, select the necessary compensator $G_c(s)$ by using (a) frequency response techniques, and (b) root locus methods.

10.2. A magnetic tape-recorder transport for modern computers requires a high-accuracy, rapid-response control system. The requirements for a specific transport are as follows: (1) the tape must stop or start in 3 msec; (2) it must be possible to read 45,000 characters per second. This system was discussed in Problem 6.11. It is desired to set $J = 5 \times 10^{-3}$, and K_a is set on the basis of the maximum error allowable for a velocity input. In this case, it is desired to maintain a steady-state speed error of less than 2%. However, it is not possible to use a tachometer in this case and thus $K_2 = 0$. In order to provide a suitable performance, a compensator $G_c(s)$ is inserted in cascade between the photocell transducer and the amplifier. Select a compensator $G_c(s)$ so that the overshoot of the system for a step input is less than 30%.

10.3. A simplified version of the attitude rate control for the F-94 or X-15 type aircraft is shown in Fig. P10.3. When the vehicle is flying at four times the speed of sound (Mach 4) at an altitude of 100,000 feet, the parameters are

$$\frac{1}{\tau_a} = 0.04, \qquad K_1 = 0.02, \qquad \zeta\omega_a = 0.04, \qquad \omega_a = 2.$$

Design a compensation network so that the complex poles have approximately a $\zeta = 0.707$ and $\omega_n = 3$.

10.4. Magnetic particle clutches are useful actuator devices for high power requirements, since they can typically provide a 200-watt mechanical power output. The particle clutches

(a)

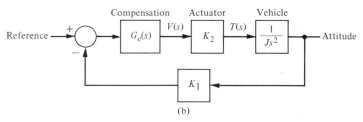

(b)

Fig. P10.1. Apollo 11 lunar excursion module viewed from the command ship. Inside the LEM were astronauts Neil Armstrong and Edwin Aldrin, Jr. The LEM landed on the moon on July 20, 1969. (Photo courtesy of NASA Manned Spacecraft Center.)

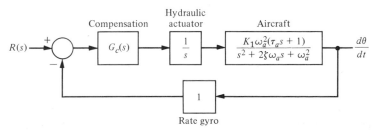

Figure P10.3

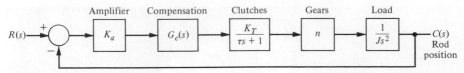

Figure P10.4

provide a high torque-to-inertia ratio and fast time constant response. A particle clutch positioning system for nuclear reactor rods is shown in Fig. P10.4. The motor drives two counter-
rotating clutch housings. Clutch housing are geared through parallel gear trains and the direction of the servo output is dependent upon the clutch which is energized. The time constant
of a 200-watt clutch is $\tau = \frac{1}{40}$ sec. The constants are such that $K_T n/J = 1$. It is desired that
the maximum overshoot for a step input is in the range of 10% to 20%. Design a compensating
network so that the system is adequately stabilized. The settling time of the system should
be less than or equal to 1 sec. This system requires two compensation networks in cascade.

10.5. A stabilized precision rate table uses a precision tachometer and a dc direct-drive
torque motor as shown in Fig. P10.5. It is desired to maintain a high steady-state accuracy
for the speed control. In order to obtain a zero steady-state error for a step command design,
select a proportional plus integral compensator as discussed in Section 10.6. Select the appropriate gain constants so that the system has an overshoot of approximately 10% and a settling
time in the range of 0.5 to 1 sec.

10.6. Repeat Problem 10.5 by using a lead network compensator and compare the results.

10.7. The primary control loop of a nuclear power plant includes a time delay due to the
time necessary to transport the fluid from the reactor to the measurement point [7]. (See Fig.
P10.7.) The transfer function of the controller is

$$G_1(s) = \left(K_1 + \frac{K_2}{s} \right).$$

The transfer function of the reactor and time delay is

$$G(s) = \frac{e^{-sT}}{\tau s + 1},$$

where $T = 0.5$ sec and $\tau = 0.2$ sec. Using frequency response methods, design the controller
so that the overshoot of the system is less than 25%. Estimate the settling time of the system
designed.

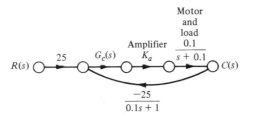

Figure P10.5

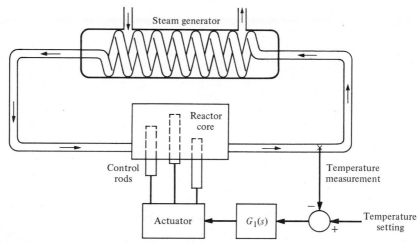

Figure P10.7

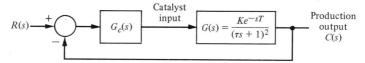

Figure P10.8

10.8. A chemical reactor process whose production rate is a function of catalyst addition is shown in block diagram form in Fig. P10.8 [8, 9]. The time delay is $T = 10$ min and the time constant, τ, is approximately 10 min. The gain of the process is $K = 1$. Design a compensation by using Bode diagram methods in order to provide a suitable system response. It is desired to have a steady-state error for a step input, $R(s) = A/s$, less than 0.05 A. For the system with the compensation added, estimate the settling time of the system.

10.9. A numerical path-controlled machine turret lathe is an interesting problem in attaining sufficient accuracy. A block diagram of a turret lathe control system is shown in Fig. P10.9. The gear ratio is $n = 0.1$, $J = 10^{-3}$, and $f = 10^{-2}$. It is necessary to attain an accuracy of 5 $\times 10^{-4}$ in., and therefore a steady-state position accuracy of 1% is specified for a ramp input. Design a cascade compensator to be inserted before the silicon-controlled rectifiers in order to provide a response to a step-command with an overshoot less than 2%. A suitable damping ratio for this system is 0.8. The gain of the silicon-controlled rectifiers is $K_R = 5$. Design a suitable compensator by using (a) the Bode diagram method, and (b) the s-plane method.

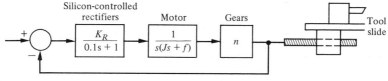

Figure P10.9

(a)

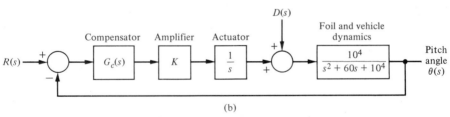

(b)

Fig. P10.10. The HS Denison hydrofoil. (Photo Courtesy of Grumman Aircraft Engineering Corp.)

10.10. The HS Denison, shown in Fig. P10.10(a), is a large hydrofoil seacraft built by Grumman Corp. for the U. S. Maritime Administration. The Denison is an 80-ton hydrofoil capable of operating in seas ranging to 9 ft in amplitude at a speed of 60 knots as a result of the utilization of an automatic stabilization control system. Stabilization is achieved by means of flaps on the main foils and the incidence of the aft foil. The stabilization control system maintains a level flight through rough seas. Thus, a system which minimizes deviations from a constant lift force or equivalently minimizes the pitch angle θ has been designed [11, 12]. A block diagram of the lift control system is shown in Fig. P10.10(b). The desired response of the system to wave disturbance is a constant-level travel of the craft. Establish a set of reasonable specifications and design a compensator $G_c(s)$ so that the performance of the system is suitable. Assume that the disturbance is due to waves with a frequency $\omega = 1$ rad/sec.

10.11. A first-order system is represented by the time-domain differential equation

$$\dot{x} = 3x + 2u.$$

A feedback controller is to be designed where

$$u(t) = -kx$$

and the desired equilibrium condition is $x(t) = 0$ as $t \to \infty$. The performance integral is defined as

$$J = \int_0^\infty x^2 \, dt,$$

and the initial value of the state variable is $x(0) = 1$. Obtain the value of k in order to make J a minimum. Is this k physically realizable? Select a practical value for the gain k and evaluate the performance index with that gain. Is the system stable without the feedback due to $u(t)$?

10.12. In order to account for the expenditure of energy and resources the control signal is often included in the performance integral. Then, the system may not utilize an unlimited control signal $u(t)$. One suitable performance index which includes the effect of the magnitude of the control signal is

$$J = \int_0^\infty (x^2(t) + \lambda u^2(t)) dt.$$

(a) Repeat Problem 10.11 for this performance index. (b) If $\lambda = 1$, obtain the value of k which minimizes the performance index. Calculate the resulting minimum value of J.

10.13. An unstable economic system is described by the vector differential equation [13]

$$\frac{d}{dt} \begin{bmatrix} x_1 \\ x_2 \end{bmatrix} = \begin{bmatrix} 1 & 0 \\ -1 & 2 \end{bmatrix} \begin{bmatrix} x_1 \\ x_2 \end{bmatrix} + \begin{bmatrix} 1 \\ 1 \end{bmatrix} u(t).$$

Both state variables are measurable, and so the control signal is set as $u(t) = -k(x_1 + x_2)$. Following the method of Section 10.10, design gain k so that the performance index is minimized. Evaluate the minimum value of the performance index. Determine the sensitivity of the performance to a change in k. Assume that the initial conditions are

$$\mathbf{x}(0) = \begin{bmatrix} 1 \\ 1 \end{bmatrix}.$$

Is the system stable without the feedback signals due to $u(t)$?

10.14. Determine the feedback gain k of Example 10.12 which minimizes the performance index

$$J = \int_0^\infty \mathbf{x}^T \mathbf{x} \, dt$$

when $\mathbf{x}^T(0) = [1, 1]$. Plot the performance index J versus the gain k.

10.15. Determine the feedback gain k of Example 10.13 which minimizes the performance index

$$J = \int_0^\infty (\mathbf{x}^T\mathbf{x} + \mathbf{u}^T\mathbf{u}) \, dt$$

when $\mathbf{x}^T(0) = [1, 1]$. Plot the performance index J versus the gain k.

10.16. For the solutions of Problems 10.13, 10.14, and 10.15 determine the roots of the closed-loop optimal control system. Note how the resulting closed-loop roots depend upon the performance index selected.

10.17. A system has the vector differential equation as given in Eq. (10.101). It is desired that both state variables are used in the feedback so that $u(t) = -k_1x_1 - k_2x_2$. Also, it is desired to have a natural frequency, ω_n, for this system equal to 2. Find a set of gains k_1 and k_2 in order to achieve an optimal system when J is given by Eq. (10.95). Assume $\mathbf{x}^T(0) = [1, 0]$.

10.18. A control system has a plant with a third-order transfer function:

$$G(s) = \frac{K}{s(s + 10)(s + 1000)}.$$

It is desired that the overshoot be approximately 7.5% for a step input and the settling time of the system be 400 milliseconds. Find a suitable phase-lead compensator by using root locus methods. Let the zero of the compensator be located at $s = -20$ and determine the compensator pole. Determine the resulting system K_v.

10.19. Now that expenditures for space exploration have leveled off and the last manned Apollo flight to the moon has been successfully completed, NASA is developing remote manipulators. These manipulators can be used to extend the hand and the power of man through space by means of radio. A concept of a remote manipulator is shown in Fig. P10.19(a). The closed loop control is shown schematically in Fig. P10.19(b). Assuming an average distance of 238,855 miles from earth to moon, the time delay in transmission of a communication signal τ is 1.28 seconds. The operator uses a control stick to remotely control the manipulator placed on the moon to assist in geological experiments, and the TV display to access the response of the manipulator [6]. (a) Set the gain K_1 so that the system has a phase margin of approximately 20°. Evaluate the percentage steady-state error for this system for a step input. (b) In order to reduce the steady-state error for a position command input to 9%, add a lag compensation network in cascade with K_1.

10.20. In Problem 2.26 we found that the output line pressure as a function of the demand flow and control flow was

$$P_4(s) = \frac{1}{Cs(1 + 2/\tau s)} \left[\frac{Q_1(s)}{\tau s} - Q_3(s) \left(1 + \frac{1}{\tau s} + R_2Cs + \frac{2R_2C}{\tau}\right) \right].$$

Let $C = 0.5\text{ft}^5/\text{lb}$, $R_2 = 200$ lb-sec/ft^5, and $\tau = 333$ sec. (a) Select a controller $Q_1(s)/Q_3(s)$ of the form

$$\frac{Q_1(s)}{Q_3(s)} = (1 + k_1 s)$$

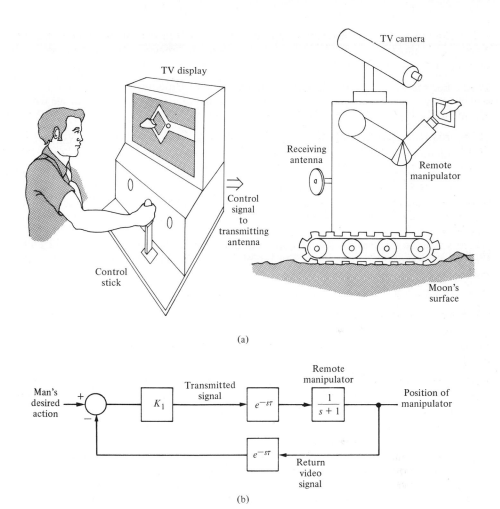

(a)

(b)

Fig. P10.19. (a) Conceptual diagram of a remote manipulator on the moon controlled by a man on the earth. (b) Feedback diagram of the remote manipulator control system with $\tau =$ transmission time delay of the video signal.

so as to maintain p_4 at a desired level. Determine a suitable value for k_1. (b) Determine the improved performance if the system includes a measurement of p_4 and therefore

$$Q_1(s) = (1 + k_2 s)(Q_3(s) - k_4 P_4(s)).$$

10.21. An uncompensated control system with unity feedback has a plant transfer function

$$G(s) = \frac{K}{s(s/2 + 1)(s/6 + 1)}.$$

It is desired to have a velocity error constant of $K_v = 20$. It is also desired to have a phase margin approximately to 45° and a closed-loop bandwidth greater than $\omega = 4$ rad/sec. Use two identical cascaded phase lead networks to compensate the system.

10.22. For the system of Problem 10.21, design a phase lag network to yield the desired specifications, with the exception that a bandwidth equal to or greater than 2 rad/sec will be acceptable.

10.23. For the system of Problem 10.21 we wish to achieve the same phase margin and K_v, but in addition we wish to limit the bandwidth to less than 10 rad/sec but greater than 2 rad/sec. Utilize a lead-lag compensation network to compensate the system. The lead-lag network could be of the form

$$G_c(s) = \frac{(1 + s/10a)(1 + s/b)}{(1 + s/a)(1 + s/10b)},$$

where a is to be selected for the lag portion of the compensator and b is to be selected for the lead portion of the compensator. The ratio α is chosen to be 10 for both the lead and lag portions.

10.24. For the system of Example 10.10, determine the optimum value for k_2 when $k_1 = 1$ and $x^T(0) = [1, 0]$.

10.25. In Chapter Four we considered the usefulness of lower-order models for higher-order systems. Consider the third-order transfer function of Example 4.6 and its second-order model. The transfer function is the plant of a closed-loop system with negative unity feedback and a cascade amplifier K. It is desired to obtain a position constant equal to 9. Evaluate and compare the resulting closed-loop responses of the second- and third-order models. Design a lead compensator using the second-order model so that the settling time is equal to or less than two seconds.

10.26. The possibility of overcoming wheel friction, wear, and vibration by contactless suspension for passenger-carrying mass-transit vehicles is being investigated throughout the world [33]. One design uses a magnetic suspension with an attraction force between the vehicle and the guideway with an accurately controlled airgap. A system is shown in Fig. P10.26 which incorporates feedback compensation. Using root-locus methods, select a suitable valve for K_1 and b so the system has a damping ratio for the underdamped roots of $\zeta =$

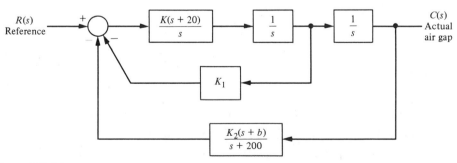

Figure P10.26.

0.50. Assume, if appropriate, that the pole of the airgap feedback loop ($s = -200$) may be neglected.

10.27. A computer uses a printer as a fast output device. It is desirable to maintain accurate position control while moving the paper rapidly through the printer [15]. Consider a system with unity feedback and a transfer function for the motor and amplifier as

$$G(s) = \frac{0.15}{s(s + 1)(5s + 1)}.$$

Design a lead network compensator so that the system bandwidth is 0.75 rad/sec and the phase margin is 30°. Use a lead network with $\alpha = 10$.

REFERENCES

1. A. P. Sage, Linear Systems Control, *Matrix Publishers,* Champaign, Ill., 1978.

2. J. J. D'Azzo and C. H. Houpis, *Linear Control System Analysis and Design,* McGraw-Hill, New York, 1975.

3. S. M. Shinners, "How to Approach the Stability Analysis and Compensation of Control Systems," *Control Engineering,* May 1978; pp. 62–67.

4. A. P. Sage and C. C. White, *Optimum Systems Control,* 2nd ed., Prentice-Hall, Englewood Cliffs, N.J., 1977.

5. R. C. Dorf, *Time Domain Analysis and Design of Control Systems,* Addison-Wesley, Reading, Mass., 1965.

6. E. Heer, "New Luster for Space Robots and Automation," *Astronautics and Aeronautics,* September 1978; pp. 48–60.

7. R. C. Dorf, *Energy, Resources and Policy,* Addison-Wesley, Reading, Mass., 1978; Ch. 12.

8. C. D. Johnson, *Process Control Instrumentation Technology,* Wiley, New York, 1977.

9. F. R. Groves, Jr., "New Ideas in Practical Control Schemes and Tuning Techniques," *ISA Transactions,* **17,** 2, 1978; pp. 9–19.

10. E. A. Farag, "Up-Grading of Control Systems Using Servo-Compensators," *Proceedings of 1978 JACC, Instrument Society of America,* 1978; pp. 229–234.

11. P. Lafrance, "Ship Hydrodynamics," *Physics Today,* June 1978; pp. 34–42.

12. R. M. Rose, "The Rough Water Performance of the HS Denison," *American Institute of Aeronautics and Astronautics, Paper No. 64–197,* May 1964.

13. N. J. Mass, "Economic Fluctuations: A Framework for Analysis and Policy Design," *IEEE Transactions on Systems, Man and Cybernetics,* June 1978; pp. 437–449.

14. D. V. Dalsen, "Need an Active Filter? Try These Design Aids," *Electronic Design News,* November 5, 1978; pp. 105–110.

15. T. J. Cameron and M. H. Dost, "Paper Servo Design for a High Speed Printer Using Simulation," *IBM J. Research and Development,* January 1978; pp. 19–25.

16. T. C. Coffey, "Automated Frequency Domain Synthesis of Multiloop Control Systems," *AIAA Journal,* **8,** 10, 1970; pp. 48–54.

17. J. R. Mitchell and W. L. McDaniel, Jr., "A Computerized Compensator Design Algorithm with Launch Vehicle Applications," *IEEE Transactions on Automatic Control,* June 1976; pp. 366–371.

18. W. R. Wakeland, "Bode Compensator Design," *IEEE Transactions on Automatic Control,* October 1976; pp. 771–773.

19. J. R. Mitchell, "Comments on Bode Compensator Design," *IEEE Transactions on Automatic Control,* October 1977; pp. 869–870.

11 / Digital Control Systems

11.1 INTRODUCTION

The use of a digital computer as a compensator device has grown during the past decade as the price and reliability of digital computers have improved dramatically [1,2]. A block diagram of a single-loop digital control system is shown in Fig. 11.1. The digital computer in this system configuration receives the error, $e(t)$, and performs calculations in order to provide an output $u^*(t)$. The computer may be programmed to provide an output, $u^*(t)$, so that the performance of the process is near or equal to the desired performance. Many computers are able to receive and manipulate several inputs, so a digital computer control system can often be a multivariable system.

11.2 DIGITAL COMPUTER CONTROL SYSTEM APPLICATIONS

A more complete block diagram of a computer control system is shown in Fig. 11.2. This diagram recognizes that a digital computer receives and operates on signals in digital (numerical) form, as contrasted to continuous signals [3]. The measurement data are converted from analog form to digital form by means of the converter shown in Fig. 11.2. After the digital computer has processed the inputs, it provides an output in digital form. This output is then converted to analog form by the digital to analog converter shown in Fig. 11.2.

The total number of computer control systems installed in industry has grown over the past two decades, as shown in Fig. 11.3 [3]. There are currently approximately one million control systems using a computer, although the computer size and power may vary significantly. If we consider only computer control systems of a relatively complex nature, such as chemical process control or aircraft control, the number of computer control systems is approximately one-hundred thousand.

A digital computer consists of a control processing unit (CPU), input-output units, and a memory unit. The size and power of a computer will vary according to

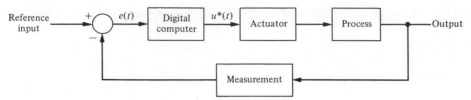

Fig. 11.1. A block diagram of a computer control system.

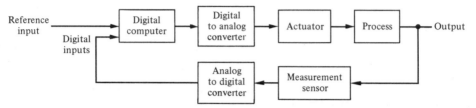

Fig. 11.2. A block diagram of a computer control system including the signal converters.

the size, speed, and power of the CPU, as well as the size, speed, and organization of the memory unit. Small computers called *minicomputers* have grown increasingly common since 1965. A typical minicomputer would be about 3 cm × 10 cm × 15 cm, with a word length of 16 binary digits and a memory of 8,000 words, and would cost about $10,000.

Since 1975, small inexpensive computers called *microcomputers,* which use an eight-bit word and cost about $1,000, have become readily available. These systems use a microprocessor as a CPU and are slower in performing operations than minicomputers. Therefore, the nature of the control task, the extent of the data required in memory, and the speed of calculation required will dictate the selection of the computer within the range of available computers. Figure 11.4 shows the price of

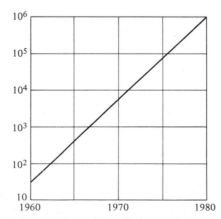

Fig. 11.3. The total number of installed computer control systems.

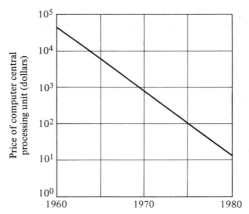

Fig. 11.4. The price of the central processing unit of a small computer. The curve from 1975 on indicates the price of a microprocessor.

the CPU for a small computer from 1960 to 1975; the curve from 1975 onward indicates the price of the microprocessor.

The size of computers and the cost for the active logic devices used to construct them have both declined exponentially since 1960 (see Fig. 11.5). The active components per cubic centimeter have increased so that the actual computer can be reduced in size. Also the speed of computers has increased exponentially, as is aptly illustrated by Fig. 11.6. It is estimated that the sale of minicomputers and microcomputers in 1978 accounted for over 1.2 billion dollars in the U.S. With the availability of fast, low-priced, and relatively small-sized computers, much of the

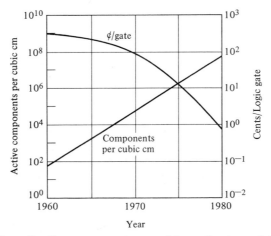

Fig. 11.5. The number of active components per cubic centimeter and the cost per logic gate for the period 1960 to 1980.

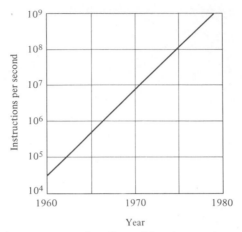

Fig. 11.6. The instructions per second performed by a computer available in a year during the period 1960 to 1980.

control of industrial and commercial processes is moving toward the use of computers within the control system. A small microcomputer is shown in Fig. 11.7.

Digital control systems are used in many applications: for machine tools, metal working processes, chemical processes, aircraft control, and automobile traffic control, among others [4,5,6,7]. An example of a computer control system used in the metals industry is shown in Fig. 11.8. Automatic computer controlled systems are

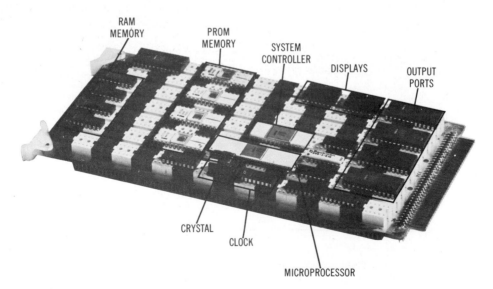

Fig. 11.7. A small microcomputer built on a breadboard. (Courtesy of Western Electric)

Fig. 11.8. The first continuous cold reduction steel mill with computer automation and drive systems in the United States was put into operation at National Steel's Weirton, West Virginia, plant in November 1975. The computer-controlled system changes the thickness and the width of the steel as it continuously passes through the rolls. The control room is shown with the steel rolls visible to the right through the window. The continuous operation of the mill permits a 50% increase in productivity compared to a traditional noncontinuous mill. (Courtesy of National Steel Corporation.)

Fig. 11.9. Low-cost digital closed-loop control positioning of a dc motor for industrial applications. The controller is a digital minicomputer. (Courtesy of Control Systems Research, Inc.)

used for purposes as diverse as measuring the objective refraction of the human eye and controlling the engine spark timing or air-fuel ratio of automobile engines [12]. The latter innovations are necessary to reduce automobile emissions and increase fuel economy. A small computer controller for a dc motor is shown in Fig. 11.9.

11.3 AUTOMATIC ASSEMBLY AND ROBOTS

Automatic handling equipment for home, school, and industry is particularly useful for hazardous, repetitive, dull, or simple tasks. Machines that automatically load and unload, cut, weld, or cast are used by industry in order to obtain [13,14,15] accuracy, safety, economy, and productivity. The use of computers integrated with machines that perform tasks as a human worker does has been foreseen by several authors. In his famous 1923 play, entitled "R.U.R." [16], Karel Capek called artificial workers *robots,* deriving the word from the Czech noun *robota,* meaning "work." Two modern robots that appeared in the 1977 film "Star Wars" are shown in Fig. 11.10.

Robots are programmable computers integrated with machines. They often substitute for human labor in specific repeated tasks. Some devices even have anthropomorphic mechanisms, including what we might recognize as mechanical arms, wrists, and hands [17,18,19]. Robots may be used extensively in space exploration and assembly [20]. They can be flexible, accurate aids on assembly lines, as shown in Fig. 11.11.

Approximately 4000 robots have been installed for industrial uses over the past decade, and about 1000 new robots are now being installed annually. An experimental robot assembly station is shown in Fig. 11.12.

Fig. 11.10. Two modern robots appeared in the film "Star Wars." R2-D2 (left) and C-3PO were the able assistants and supporters of Luke and Leia, the hero and heroine of the film. (Courtesy of Twentieth Century Fox-Film Corporation.)

11.4 SAMPLED-DATA SYSTEMS

Computers are interconnected to the actuator and process by means of signal converters. The output of the computer is processed by a digital-to-analog converter. We will assume that all the numbers that enter or leave the computer do so at the same fixed period T, called the *sampling period*. Thus, for example, the reference input shown in Fig. 11.13 is a sequence of sample values $r(kT)$. The variables $r(kT)$,

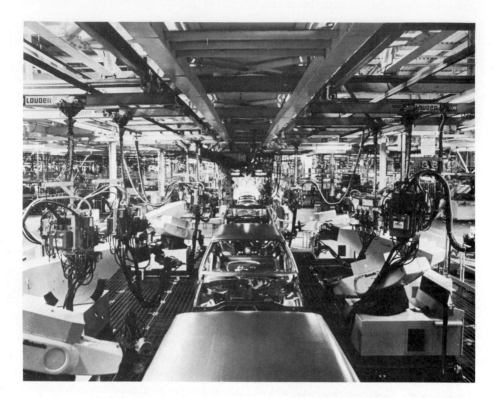

Fig. 11.11. At its Mirafiori auto assembly plant, Fiat of Italy uses 190 Unimate Robots for welding and loading and unloading tasks. The robots are shown here on both sides of the welding line. (Courtesy of Unimation, Inc.)

$m(kT)$ and $u(kT)$ are discrete signals in contrast to $m(t)$ and $c(t)$, which are continuous functions of time.

The sampler is basically a switch that closes every T seconds for one instant of time. For an ideal sampler the input is $r(t)$ and the output is $r^*(t)$, where nT is the current sample time and the current value of $r^*(t)$ is $r(nT)$. We then have $r^*(t) = r(nT)\,\delta\,(t - nT)$ where δ is the impulse function. The closed-loop computer system shown in Fig. 11.13 illustrates the use of continuous and discrete signals. If the sampling period T is chosen very small compared to the time constants of the process, then the system is essentially continuous, and methods used in previous chapters prevail. Normally, however, the sampling period is of the same magnitude as the time constants and the effects of sampling must be accounted for.

Let us assume that we sample a signal $r(t)$, as shown in Fig. 11.14, and obtain $r^*(t)$. Then we portray the series for $r^*(t)$ as shown in Fig. 11.15, where an ideal sampler implies that the output, $r^*(t)$, is a string of impulses starting at $t = 0$, spaced at T seconds, and of amplitude $r(kT)$. If we hold the output amplitude constant at $r(kT)$ during the following T seconds, we use the value until a new update

Fig. 11.12. A programmable robot assembly station can assemble the 17 parts of a commercial automobile alternator in two minutes, 42 seconds. At the far right is a control box through which the robot can be taught a sequence of moves that are recorded in the memory of a minicomputer. (Courtesy of *Scientific American*. Photo by Ben Rose.)

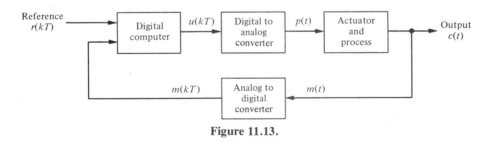

Figure 11.13.

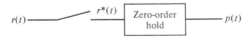

Fig. 11.14. A sampler and zero-order hold circuit.

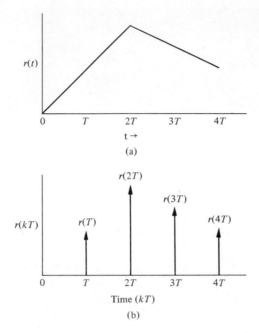

(a)

(b)

Fig. 11.15. (a) An input signal $r(t)$ and (b) the sampled signal $r^*(t) = \displaystyle\sum_{k=0}^{\infty} r(kT)\delta(t - kT)$.

sample occurs. The signal is then said to be held constant by a zero-order hold, as shown in Fig. 11.14. The response of a zero-order hold is shown in Fig. 11.16.

A sampler and zero-order hold can accurately follow the input signal if T is small compared to the transient changes in the signal. The response of a sampler and zero-order hold to a step input of magnitude A is shown in Fig. 11.17. The response of a sampler and zero-order hold for a ramp input is shown in Fig. 11.18.

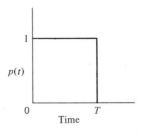

Time

Fig. 11.16. The response of a zero-order hold to an impulse input $r(kT) = 1$ when $k = 0$ and equals zero when $k \neq 0$, so that $r^*(t) = r(0)\,\delta(t)$.

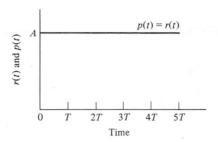

Fig. 11.17. The response of a sampler and zero-order hold for a step input of magnitude A.

Finally, the response of a sampler and zero-order hold for an exponentially decaying signal is shown in Fig. 11.19 for two values of sampling period. Clearly the output $p(t)$ will approach the input $r(t)$ as T approaches zero (as we sample very often).

The precision of the digital computer and the associated signal converters is limited (see Fig. 11.13). The *precision* is the degree of exactness or discrimination with which a quantity is stated. The precision of the computer is limited by a finite word length. For example, many minicomputers have a word length of 16 binary digits. The precision of the analog-to-digital converter is limited by an ability to store its output only in digital logic composed of a finite number of binary digits. The converted signal, $m(kT)$, is then said to include an *amplitude quantization error*. When the quantization error and the error due to computer finite word size are small relative to the amplitude of the signal, then the system is sufficiently precise and the precision limitations may be neglected.

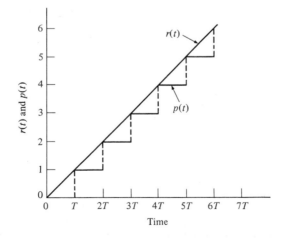

Fig. 11.18. The response of a sampler and zero-order hold for a ramp input $r(t) = t$.

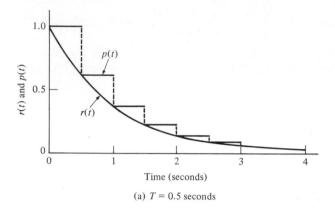

(a) $T = 0.5$ seconds

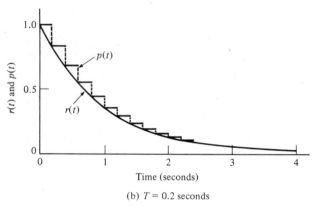

(b) $T = 0.2$ seconds

Fig. 11.19. The response of a sampler and zero-order hold to an input $r(t) = e^{-t}$ for two values of sampling period T.

11.5 COMPUTER PROGRAMS FOR CONTROL

If the computer is simply to yield a gain to the closed loop, then the program can be written in FORTRAN or BASIC as

$$E = R - M$$
$$U = K * E \tag{11.1}$$

using the definition of the variables as given in Fig. 11.13.

A BASIC program that implements both a gain and an integral of the error is [21]

```
10   E = R − M
20   I = I + E * T
30   U = K1*E + K2*I
```

$$\begin{array}{ll} 40 & \text{WAIT T} \\ 50 & \text{GO TO 10} \end{array} \qquad (11.2)$$

The approximate integration is achieved in line 20 of the computer, while $u(kT)$ is generated in line 30. An integration can be written during the period T as

$$\int_{t_1}^{t_1+T} e(t)dt \approx e(kT)*T, \qquad (11.3)$$

where $t_1 = kT$. The gain for the integration is K2 in line 30.

In Section 10.6 we introduced the concept of three mode controllers, often called PID controllers. These controllers include a proportional term, an integration term, and a derivative term. Using a computer equation, a derivative term may be implemented as follows:

$$\begin{array}{l} \text{U} = \text{KD}*(\text{E}-\text{E1})/\text{T} \\ \text{E1} = \text{E} \end{array} \qquad (11.4)$$

where KD is the gain of the derivative term, and T is the sampling period. The variable E1 is the value of the previous sample of the error. A complete three term controller yields:

$$u(t) = K_1 e(t) + K_2 \int e(t)dt + K_D \frac{de(t)}{dt}. \qquad (11.5)$$

A computer program for the three term PID controller is then

$$\begin{array}{ll} 10 & \text{E} = \text{R} - \text{M} \\ 20 & \text{I} = \text{I} + \text{E} * \text{T} \\ 30 & \text{D} = \text{KD} * (\text{E}-\text{E1})/\text{T} \\ 40 & \text{E1} = \text{E} \\ 50 & \text{U} = \text{K1}*\text{E} + \text{K2}*\text{I} + \text{D} \\ 60 & \text{DAC} = \text{U} \\ 70 & \text{WAIT T} \\ 80 & \text{GO TO 10} \end{array} \qquad (11.6)$$

Line 60 provides the calculated $u(kT)$ to the digital-to-analog converter. Of course the computer can also be used to approximate computer nonlinear equations among other functions.

11.6 THE z-TRANSFORM

Since the output of the ideal sampler, $r*(t)$, is a series of impulses with values $r(kT)$, we have

$$r*(t) = \sum_{k=0}^{\infty} r(kT)\delta(t - kT) \qquad (11.7)$$

for a signal for $t > 0$. Using the Laplace transform, we have

$$\mathcal{L}\{r^*(t)\} = \sum_{k=0}^{\infty} r(kT)e^{-kTs}. \tag{11.8}$$

We now have an infinite series which involves factors of e^{Ts} and its powers. We define

$$z = e^{sT}, \tag{11.9}$$

where this relationship involves a conform mapping from the s-plane to the z-plane. We then define a new transform, called the z-transform, so that

$$\mathcal{Z}\{r(t)\} = \mathcal{Z}\{r^*(t)\} = \sum_{k=0}^{\infty} r(kT)z^{-k}. \tag{11.10}$$

As as example, let's determine the z-transform of the unit step function $u(t)$ (not to be confused with the control signal $u(t)$). We obtain

$$\mathcal{Z}\{u(t)\} = \sum_{k=0}^{\infty} u(kT)z^{-k}$$

$$= \sum_{k=0}^{\infty} z^{-k}. \tag{11.11}$$

This series may be written in closed form as †

$$U(z) = \frac{1}{1 - z^{-1}}$$

$$= \frac{z}{z - 1}. \tag{11.12}$$

In general we will define the z-transform of a function $f(t)$ as

$$\mathcal{Z}\{f(t)\} = F(z) = \sum_{k=0}^{\infty} f(kT)z^{-k}. \tag{11.13}$$

Example 11.1. Let us determine the z-transform of $f(t) = e^{at}$ for $t \geq 0$. Then

$$\mathcal{Z}\{e^{-at}\} = F(z)$$

$$= \sum_{k=0}^{\infty} e^{-akT}z^{-k}$$

$$= \sum_{k=0}^{\infty} (ze^{+aT})^{-k}. \tag{11.14}$$

---

†Recall that the infinite geometric series may be written as follows:

$$(1 - bx)^{-1} = 1 + bx + (bx)^3 + (bx)^2 + \ldots, \text{ if } (bx)^2 < 1.$$

Again this series may be written in closed form as

$$F(z) = \frac{1}{1 - (ze^{aT})^{-1}}$$

$$= \frac{z}{z - e^{-aT}}. \tag{11.15}$$

In general we may show that

$$Z\{e^{-at}f(t)\} = F(e^{aT}z).$$

Example 11.2. Let us determine the z-transform of $f(t) = \sin \omega t$ for $t \geq 0$. We may write $\sin \omega t$ as

$$\sin \omega t = \frac{e^{jwT} - e^{-jwT}}{2j}.$$

Therefore

$$\{\sin \omega t\} = \left\{ \frac{e^{jwT}}{2j} - \frac{e^{-jwT}}{2j} \right\}. \tag{11.16}$$

Then

$$F(z) = \frac{1}{2j} \left(\frac{z}{z - e^{jwT}} - \frac{z}{z - e^{-jwT}} \right)$$

$$= \frac{1}{2j} \left(\frac{z(e^{jwT} - e^{-jwT})}{z^2 - z(e^{jwT} + e^{-jwT}) + 1} \right)$$

$$= \frac{z \sin \omega T}{z^2 - 2z\cos \omega T + 1}. \tag{11.17}$$

A table of z-transforms is given in Table 11.1. A table of properties of the z-transform is given in Table 11.2. As in the case of Laplace transforms, we are ultimately interested in the output of the system, $c(t)$. Therefore we must use an inverse transform to obtain $c(t)$ from $C(z)$. We may obtain the output by (1) expanding $C(z)$ in a power series, (2) expanding $C(z)$ into partial fractions and using Table 11.1 to obtain the inverse of each term, or (3) obtaining the inverse z-transform by an inversion integral. We will limit our methods to (1) and (2) in this limited discussion.

Example 11.3. Let us consider the system shown in Fig. 11.20 for $T = 1$. From the block diagram we obtain

$$G(s) = \frac{C(s)}{R*(s)} = \frac{1 - e^{-sT}}{s^2(s + 1)}. \tag{11.18}$$

Table 11.1 z-Transforms

$X(t)$	$X(s)$	$X(z)$
$\delta(t) = \begin{cases} 1, & t = 0 \\ 0, & t = kT,\ k \neq 0 \end{cases}$	1	1
$\delta(t - kT) = \begin{cases} 1, & t = kT \\ 0, & t \neq kT \end{cases}$	e^{-kTs}	z^{-k}
$u(t)$, unit step	$1/s$	$\dfrac{z}{z - 1}$
t	$1/s^2$	$\dfrac{Tz}{(z - 1)^2}$
e^{-at}	$\dfrac{1}{s + a}$	$\dfrac{z}{z - e^{-aT}}$
$1 - e^{-at}$	$\dfrac{1}{s(s + a)}$	$\dfrac{(1 - e^{-aT})z}{(z - 1)(z - e^{-aT})}$
$\sin \omega t$	$\dfrac{\omega}{s^2 + \omega^2}$	$\dfrac{z \sin \omega T}{z^2 - 2z \cos \omega T + 1}$
$\cos \omega t$	$\dfrac{s}{s^2 + \omega^2}$	$\dfrac{z(z - \cos \omega T)}{z^2 - 2z \cos \omega T + 1}$
$e^{-at} \sin \omega t$	$\dfrac{\omega}{(s + a)^2 + \omega^2}$	$\dfrac{ze^{-aT} \sin \omega T)}{z^2 - 2ze^{-aT} \cos \omega T + e^{-2aT}}$
$e^{-at} \cos \omega t$	$\dfrac{s + a}{(s + a)^2 + \omega^2}$	$\dfrac{z^2 - ze^{-aT} \cos \omega T}{z^2 - 2ze^{-aT} \cos \omega T + e^{-2aT}}$

Table 11.2 Properties of the z-Transform

	$x(t)$	$X(z)$
1.	$kx(t)$	$KX(z)$
2.	$x_1(t) + x_2(t)$	$X_1(z) + X_2(z)$
3.	$x(t + T)$	$zX(z) - zx(0)$
4.	$tx(t)$	$-Tz\dfrac{dX(z)}{dz}$
5.	$e^{-at}x(t)$	$X(ze^{aT})$
6.	$x(0)$, initial value	$\lim\limits_{z \to \infty} X(z)$ if the limit exists
7.	$x(\infty)$, final value	$\lim\limits_{z \to 1} (z - 1)X(z)$ if the limit exists and the system is stable; i.e., all poles of $(z - 1)X(z)$ are inside unit circle $|z| = 1$ on z-plane.

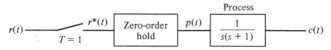

Fig. 11.20. An open-loop sampled data system.

Expanding into partial fractions, we have

$$G(s) = (1 - e^{-sT}) \left(\frac{1}{s^2} - \frac{1}{s} + \frac{1}{s+1} \right). \tag{11.19}$$

Therefore

$$\begin{aligned}
G(z) &= Z\{G(s)\} \\
&= (1 - z^{-1}) \left(\frac{1}{s^2} - \frac{1}{s} + \frac{1}{s+1} \right) \\
&= (1 - z^{-1}) \left(\frac{Tz}{(z-1)^2} - \frac{z}{z-1} + \frac{z}{z - e^{-T}} \right) \\
&= \left(\frac{(ze^{-T} - z + Tz) + (1 - e^{-T} - Te^{-T})}{(z-1)(z - e^{-T})} \right). \tag{11.20}
\end{aligned}$$

When $T = 1$, we have

$$\begin{aligned}
G(z) &= \frac{ze^{-1} + 1 - 2e^{-1}}{(z-1)(z - e^{-1})} \\
&= \frac{0.3678z + 0.2644}{(z-1)(z - 0.3678)} \\
&= \frac{0.3678z + 0.2644}{z^2 - 1.3678z + 0.3678}. \tag{11.21}
\end{aligned}$$

The response of this system to a unit impulse is obtained for $R(z) = 1$ so that $C(z) = G(z) \cdot 1$. We may obtain $C(z)$ by dividing the denominator into the numerator as:

$$
z^2 - 1.3678z + .3678 \div
\begin{array}{l}
0.3678z^{-1} + 0.7675z^{-2} + 0.9145z^{-3} + \cdots \\ \hline
0.3678z \quad\quad + 0.2644 \\
0.3678z \quad\quad - 0.5031 \quad\quad + \; .1353z^{-1} \\ \hline
\quad\quad\quad + 0.7675 \quad\quad - \; .1353z^{-1} \\
\quad\quad\quad + 0.7675 \quad\quad - 1.0497z^{-1} + 0.2823z^{-2} \\ \hline
\quad\quad\quad\quad\quad\quad\quad\quad\quad 0.9145z^{-1} - 0.2823z^{-2}
\end{array} \tag{11.22}
$$

This calculation yields the response at the sampling instants and may be carried as far as is needed for $C(z)$. In this case we have obtained $C(kT)$ as follows: $C(0) = 0$, $C(T) = 0.3678$, $C(2T) = 0.7675$, and $C(3T) = 0.9145$.

Example 11.4. Now let us consider the closed-loop system as shown in Fig. 11.21. We have already obtained $G(z)$. Following the block diagram algebra of earlier chapters, we find that

$$\frac{C(z)}{R(z)} = \frac{G(z)}{1 + G(z)}. \tag{11.23}$$

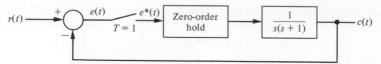

Fig. 11.21. A closed-loop sampled data system.

Since for T = 1, Eq. (11.21) yields $G(z)$, we substitute, obtaining

$$\frac{C(z)}{R(z)} = \frac{0.3678z + 0.2644}{z^2 - z + 0.6322}.$$ (11.24)

$$R(z) = \frac{z}{z - 1}.$$ (11.25)

Then

$$C(z) = \frac{z(0.3678z + 0.2644)}{(z - 1)(z^2 - z + 0.6322)}$$

$$= \frac{0.3678z^2 + 0.2644z}{z^3 - 2z^2 + 1.6322z - 0.6322}.$$

Completing the division, we have

$$C(z) = 0.3678z^{-1} + z^{-2} + 1.4z^{-3} + 1.4z^{-4} + 1.147z^{-5} \ldots.$$ (11.26)

The values of $C(kT)$ are shown in Fig. 11.22, using the symbol □. The complete response of the sampled-data closed-loop system is shown (obtained by advanced

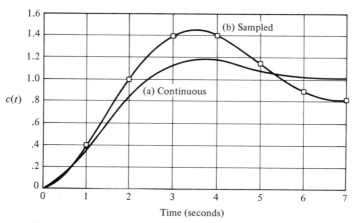

Fig. 11.22. The response of a second-order system: (a) continuous, (b) sampled.

methods) and contrasted to the response of a continuous system (when $T = 0$). The overshoot of the sampled system is 45%, in contrast to 17% for the continuous system. Furthermore, the settling time of the sampled system is twice as great as that of the continuous system.

11.7 STABILITY ANALYSIS IN THE z-PLANE

A linear continuous feedback control system is stable if all poles of the closed-loop transfer function, $T(s)$, lie in the left half of the s-plane. The z-plane is related to the s-plane by the transformation

$$z = e^{sT}$$
$$= e^{(\sigma + j\omega)T}. \tag{11.27}$$

We may also write this relationship as

$$|z| = e^{\sigma T}$$

and

$$z = \omega T. \tag{11.28}$$

In the left-hand s-plane, $\sigma < 0$ and therefore the related magnitude of z varies between 0 and 1. Therefore the imaginary axis of the s-plane corresponds to the unit circle in the z-plane and the inside of the unit circle corresponds to the left half of the s-plane.

Therefore we may state that a *sampled system is stable if all the poles of the closed-loop transfer function, T(z), lie within the unit circle.*

Example 11.5.   Let us consider the system shown in Fig. 11.23, when $T = 1$ and

$$G(s) = \frac{K}{s(s + 1)} \tag{11.29}$$

Then, recalling Eq. (11.21), we note that

$$G(z) = \frac{K(0.3678z + 0.2644)}{z^2 - 1.3678z + 0.3678} = \frac{K(az + b)}{z^2 - (1 + a)z + a}, \tag{11.30}$$

where $a = 0.3678$ and $b = 0.2644$.

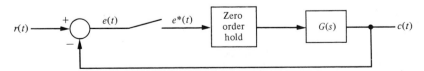

Fig. 11.23. A closed-loop sampled system.

The poles of the closed-loop transfer function $T(z)$ are the roots of the equation $1 + G(z) = 0$. We call $q(z) = 1 + G(z) = 0$ the characteristic equation. Therefore, we obtain

$$q(z) = 1 + G(z) = z^2 - (1 + a)z + a + Kaz + Kb = 0. \qquad (11.31)$$

When $K = 1$, we have

$$\begin{aligned} q(z) &= z^2 - z + 0.6322 \\ &= (z + 0.50 + j0.60)(z + 0.50 - j0.60) = 0. \end{aligned} \qquad (11.32)$$

Therefore the system is stable, since the roots lie within the unit circle. When $K = 10$, we have

$$\begin{aligned} q(z) &= z^2 + 4.952z + 0.368 \\ &= (z + 0.076)(z + 4.876) = 0, \end{aligned} \qquad (11.33)$$

and the system is unstable, since one root lies outside the unit circle. This system is stable for $0 < K < 2.32$.

We notice that a second-order sampled system can be unstable with increasing gain where a second-order continuous system is stable for all values of gain (assuming both the poles of the open-loop system lie in the left half s-plane).

11.8 PERFORMANCE OF A SAMPLED-DATA SECOND-ORDER SYSTEM

Let us consider the performance of a sampled second-order system with a zero-order hold, as shown in Fig. 11.23, when [22]

$$G(s) = \frac{K}{s(\tau s + 1)}. \qquad (11.34)$$

We then obtain $G(z)$ for the unspecified sampling period T as

$$G(z) = \frac{K[(z - E)[T - \tau(z - 1)] + \tau(z - 1)^2]}{(z - 1)(z - E)}, \qquad (11.35)$$

where $E = e^{-T/\tau}$. The stability of the system is analyzed by considering the characteristic equation

$$\begin{aligned} q(z) = z^2 + z[K[T - \tau(1 - E)] - (1 + E)] \\ + K[\tau(1 - E) - TE] + E = 0. \end{aligned} \qquad (11.36)$$

Since the polynomial $q(z)$ is a quadratic and has real coefficients, the necessary and sufficient conditions for $q(z)$ to have all its roots within the unit circle are

$$|q(0)| < 1, \qquad q(1) > 0, \qquad q(-1) > 0.$$

Table 11.3 Maximum Gain for a Second-Order Sampled System

T/τ	0	0.1	0.5	1	2
Maximum K_τ	∞	20.4	4.0	2.32	1.45

Using these conditions we have established the necessary conditions as

$$K_\tau < \frac{1 - E}{1 - E - (T/\tau)E}, \tag{11.37}$$

$$K_\tau < \frac{2(1 + E)}{(T/\tau)(1 + E) - 2(1 - E)}, \tag{11.38}$$

and $K > 0$, $T > 0$. For this system we may calculate the maximum gain permissible for a stable system. The maximum gain allowable is given in Table 11.3 for several values of T/τ. If the computer system has sufficient speed of computation and data handling, it is possible to set $T/\tau = 0.1$ and obtain system characteristics approaching those of a continuous system.

The maximum overshoot of the second-order system is shown in Fig. 11.24. Notice that the maximum overshoot can exceed 100% and the system is still stable.

The performance criteria, integral square error, can be written as

$$I = \frac{1}{\tau} \int_0^\infty e^2(t) \, dt \tag{11.39}$$

The loci of this criteria is given in Fig. 11.25. Note that there exists a line which represents the minimum of I as shown in the figure.

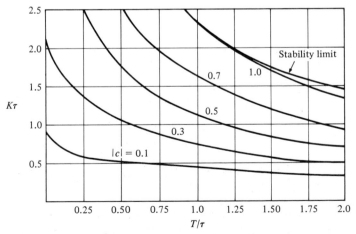

Fig. 11.24. The maximum overshoot, $|c|$, for a second-order sampled system for a unit step input.

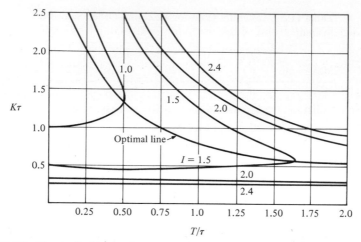

Fig. 11.25. The loci of integral square error for a second-order sampled system.

The steady-state error for a unit ramp input, $r(t) = t$, is shown in Fig. 11.26. For a given T/τ we can reduce the steady-state error, but then the system yields greater overshoot and settling time for a step input.

Example 11.6. Let us consider a closed-loop sampled system as shown in Fig. 11.23 when

$$G(s) = \frac{K}{s(0.1s + 1)(0.005s + 1)} \tag{11.40}$$

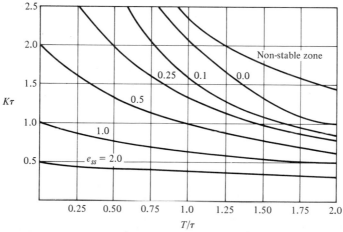

Fig. 11.26. The steady-state error of a second-order sampled system for a ramp input, $r(t) = t, t > 0$.

and we need to select T and K for suitable performance. As an approximation, we neglect the effects of the time constant $\tau_2 = 0.005$ seconds since it is only 5% of the primary time constant $\tau_1 = 0.1$. Then we can use Figs. 11.24, 11.25, and 11.26 to select K and T. Limiting the overshoot to 30% for a step input, we select $T/\tau = 0.25$, yielding $K\tau = 1.4$. For these values, the steady-state error to a unit ramp input is approximately 0.6.

Since $\tau = 0.1$, we then set $T = 0.025$ seconds and $K = 14$. The sampling rate is then required to be 40 samples per second, which will require a relatively fast computer.

11.9 CLOSED-LOOP SYSTEMS WITH DIGITAL COMPUTER COMPENSATION

A closed-loop sampled system with a digital computer used to improve the performance is shown in Fig. 11.27. The closed-loop transfer function is

$$\frac{C(z)}{R(z)} = T(z) = \frac{G(z)D(z)}{1 + G(z)D(z)} \tag{11.41}$$

The transfer function of the computer is represented by

$$\frac{E(z)}{U(z)} = D(z) \tag{11.42}$$

In our prior calculations the $D(z)$ could be represented simply by a gain K. As an illustration of the power of the computer as a compensator, we will reconsider the second-order system

$$G(z) = \frac{1}{s(s + 1)} \qquad \text{when T = 1.}$$

Then (see Eq. 11.18)

$$G(z) = \frac{0.3678(z + 0.7189)}{(z - 1)(z - 0.368)}. \tag{11.43}$$

If we select

$$D(z) = \frac{K(z - 0.3678)}{(z + r)}, \tag{11.44}$$

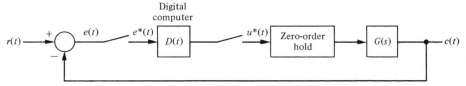

Fig. 11.27. A closed-loop system with a digital computer.

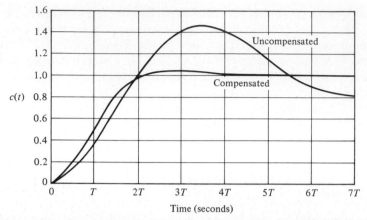

Fig. 11.28. The response of a second-order system to a unit step input.

we cancel the pole of $G(z)$ at $z = 0.3678$ and have two parameters, r and K, to set. If we select

$$D(z) = \frac{1.359(z - 0.3678)}{(z + 0.240)}, \qquad (11.45)$$

we have

$$G(z)D(z) = \frac{0.50(z + 0.7189)}{(z + 1)(z + 0.240)}. \qquad (11.46)$$

If we calculate the response of the system to a unit step, we find that the output is equal to the input at the fourth sampling instant and thereafter. The responses for $K = 1$ of both the uncompensated and the compensated system are shown in Fig. 11.28. The overshoot of the compensated system is 4%, while the overshoot of the uncompensated system is 45%. It is beyond the objective of this book to discuss methods for the analytical selection of the parameters of $D(z)$, and the reader is referred to advanced texts [1,4,23].

11.10 SUMMARY

The use of a digital computer as the compensation device for a closed-loop control system has grown during the past decade as the price and reliability of computers have improved dramatically. A computer can be used to complete many calculations during the sampling interval, T, and to provide an output signal which is used to drive an actuator of a process. Computer control is used for chemical processes, aircraft control, machine tools, and many common processes today. Computers may also be used in both anthropomorphic and zoomorphic (Fig.11.29) devices commonly called robots.

The z-transform may be used to analyze the stability and response of a sampled

Fig. 11-29. A floppy-eared mechanical rodent that can find its way through a maze is demonstrating that the days of the useful robot may not be far off. The self-contained mouse has a microcomputer "brain" which completely maps the maze in two passes. On the third run it goes from start to finish without bumping into a wall or making a wrong turn. The mouse was built for an electronics-maze contest by Battelle's Pacific Northwest Laboratory. Beams of infrared light are used to identify pathways through the maze. The mouse moves on two wheels actuated by stepping motors driven by the microcomputer. (Courtesy of Battelle's Pacific Northwest Laboratories.)

system and to design appropriate systems incorporating a computer. Computer control systems have become increasingly common as low cost computers have become readily available.

PROBLEMS

11.1. The input to a sampler is $r(t) = \sin \omega t$, where $\omega = 1/\pi$. Plot the input to the sampler and the output $r^*(t)$ for the first two seconds when $T = 0.25$ seconds.

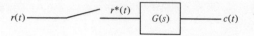

Figure P11.3

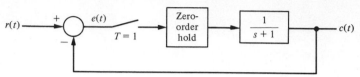

Figure P11.4

11.2. The input to a sampler is $r(t) = \sin \omega t$, where $\omega = 1/\pi$. The output of the sampler enters a zero-order hold, as shown in Fig. 11.14. Plot the output of the hold circuit, $p(t)$, for the first two seconds when $T = 0.25$ seconds.

11.3. A unit ramp, $r(t) = t, t > 0$, is used as an input to a process where $G(s) = 1/(s + 1)$ as shown in Fig. P11.3. Determine the output $c(kT)$ for the first four sampling instants.

11.4. A closed-loop system has a hold circuit and process as shown in Fig. P11.4. Determine $G(z)$.

11.5. For the system in Problem 11.4, let $r(t)$ be a unit step input and calculate the response of the system by synthetic division.

11.6. For the output of the system in Problem 11.4, find the initial and final values of the output directly from $C(z)$.

11.7. A closed-loop system is shown in Fig. P11.7. The transfer function is $G(s) = K/[s(.5s + 1)]$. Select a gain K and the sampling period T so that the overshoot is limited to 0.3 for a unit step input and the steady-state error for a unit ramp input is less than 1.0.

11.8. Consider the computer compensated system shown in Fig. 11.27 when $T = 1$ and

$$G(s) = \frac{1}{s(s + 1)}.$$

Select the parameters K and r of $D(z)$ when

$$D(z) = \frac{K(z - 0.3678)}{(z + r)}.$$

Select within the range: $1 < K < 2$ and $0 < r < 1$.

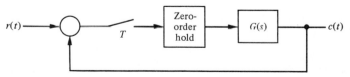

Figure P11.7

Determine the response of the compensated system and compare with the uncompensated system.

REFERENCES

1. G. F. Franklin and J. D. Powell, *Digital Control of Dynamic Systems,* Addison-Wesley, Reading, Mass., 1980.

2. E. J. Kompass, "Fitting Computers to the Control Task," *Control Engineering,* June 1978; pp. 50–51.

3. R. C. Dorf, *Introduction to Computers and Computer Science,* 2nd ed., Boyd and Fraser Publishing Co., San Francisco, 1977.

4. T. J. Harrison, *Minicomputers in Industrial Control,* Instrument Society of America, Pittsburgh, Pa., 1978.

5. P. H. Garrett, *Analog Systems for Microprocessors and Minicomputers,* Reston Publishing Co., Reston, Va., 1978.

6. D. M. Auslander et al., "Direct Digital Process Control," *Proceedings of the IEEE,* February 1978; pp. 199–207.

7. R. F. Stengel, "Digital Controllers for VTOL Aircraft," *IEEE Transactions on Aerospace and Electronic Systems,* January 1978; pp. 54–62.

8. H. Amrehn, "Computer Control in the Polymerization Industry," *Automatica,* **13,** 1977; pp. 533–545.

9. B. Harding, "Valve Control with a Microcomputer," *Instruments and Control Systems,* October 1978; pp. 73–77.

10. D. Bursky, "Micros Getting Faster and Denser," *Electronic Design,* January 4, 1979; pp. 62–66.

11. W. S. Blashke and J. McGill, *The Control of Industrial Processes by Digital Techniques,* Elsevier, New York, 1976.

12. G. Flynn, "Applying Electronics to Automobiles," *Product Engineering,* September 1978; pp. 63–65.

13. C. R. Walker, *Toward the Automatic Factory,* Greenwood Press, Westport, Conn., 1977.

14. J. L. Nevins and D. E. Whitney, "Computer-Controlled Assembly," *Scientific American,* February 1978; pp. 62–74.

15. N. P. Ruzie, "The Automated Factory–A Dream Coming True?" *Control Engineering,* April 1978; pp. 58–62.

16. K. Capek, *Rossum's Universal Robots,* English version by P. Selver and N. Playfair, Doubleday, Page and Co., New York, 1923.

17. R. Malone, *The Robot Book,* Harcourt, Brace, Jovanovich, New York, 1978.

18. J. S. Albus and J. M. Evans, "Robot Systems," *Scientific American,* October 1976; pp. 77–86.

19. T. Loofbourrow, *How to Build a Computer-Controlled Robot,* Hayden Book Co., Rochelle Park, N.J., 1978.

20. E. Heer, "New Luster for Space Robots and Automation," *Astronautics and Aeronautics,* September 1978; pp. 48–60.

21. R. Allan, "The Microcomputer Invades the Production Line," *IEEE Spectrum,* January 1979; pp. 53–57.

22. Y. Koren and J. G. Bollinger, "Design Parameters for Sampled-Data Drives for CNC Machine Tools," *IEEE Transactions on Industry Applications,* June 1978; pp. 255–263.

23. R. C. Dorf, *Time Domain Analysis and Design of Control Systems,* Addison-Wesley, Reading, Mass., 1965; Ch 9.

APPENDIXES

APPENDIX A / LaPlace Transform Pairs

Table A.1

$F(s)$	$f(t)$, $t \geqslant 0$
1. 1	$\delta(t_0)$, unit impulse at $t = t_0$
2. $1/s$	1, unit step
3. $\dfrac{n!}{s^{n+1}}$	t^n
4. $\dfrac{1}{(s + a)}$	e^{-at}
5. $\dfrac{1}{(s + a)^n}$	$\dfrac{1}{(n - 1)!} t^{n-1} e^{-at}$
6. $\dfrac{a}{s(s + a)}$	$1 - e^{-at}$
7. $\dfrac{1}{(s + a)(s + b)}$	$\dfrac{1}{(b - a)} (e^{-at} - e^{-bt})$
8. $\dfrac{s + \alpha}{(s + a)(s + b)}$	$\dfrac{1}{(b - a)} [(\alpha - a)e^{-at} - (\alpha - b)e^{-bt}]$
9. $\dfrac{ab}{s(s + a)(s + b)}$	$1 - \dfrac{b}{(b - a)} e^{-at} + \dfrac{a}{(b - a)} e^{-bt}$
10. $\dfrac{1}{(s + a)(s + b)(s + c)}$	$\dfrac{e^{-at}}{(b - a)(c - a)} + \dfrac{e^{-bt}}{(c - a)(a - b)} + \dfrac{e^{-ct}}{(a - c)(b - c)}$
11. $\dfrac{s + \alpha}{(s + a)(s + b)(s + c)}$	$\dfrac{(\alpha - a)e^{-at}}{(b - a)(c - a)} + \dfrac{(\alpha - b)e^{-bt}}{(c - b)(a - b)} + \dfrac{(\alpha - c)e^{-ct}}{(a - c)(b - c)}$
12. $\dfrac{ab(s + \alpha)}{s(s + a)(s + b)}$	$\alpha - \dfrac{b(\alpha - a)}{(b - a)} e^{-at} + \dfrac{a(\alpha - b)}{(b - a)} e^{-bt}$
13. $\dfrac{\omega}{s^2 + \omega^2}$	$\sin \omega t$
14. $\dfrac{s}{s^2 + \omega^2}$	$\cos \omega t$

Table A.1 (Cont.)

$F(s)$	$f(t),\ t \geqslant 0$
15. $\dfrac{s + \alpha}{s^2 + \omega^2}$	$\dfrac{\sqrt{\alpha^2 + \omega^2}}{\omega}\ \sin(\omega t + \phi),\quad \phi = \tan^{-1} \omega/\alpha$
16. $\dfrac{\omega}{(s + a)^2 + \omega^2}$	$e^{-at} \sin \omega t$
17. $\dfrac{(s + a)}{(s + a)^2 + \omega^2}$	$e^{-at} \cos \omega t$
18. $\dfrac{s + \alpha}{(s + a)^2 + \omega^2}$	$\dfrac{1}{\omega}[(\alpha - a)^2 + \omega^2]^{1/2} e^{-at} \sin(\omega t + \phi),\quad \phi = \tan^{-1} \dfrac{\omega}{\alpha - a}$
19. $\dfrac{\omega_n^2}{s^2 + 2\zeta\omega_n s + \omega_n^2}$	$\dfrac{\omega_n}{\sqrt{1 - \zeta^2}} e^{-\zeta\omega_n t} \sin \omega_n \sqrt{1 - \zeta^2}\ t,\ \zeta < 1$
20. $\dfrac{1}{s[(s + a)^2 + \omega^2]}$	$\dfrac{1}{a^2 + \omega^2} + \dfrac{1}{\omega\sqrt{a^2 + \omega^2}} e^{-at} \sin(\omega t - \phi),$ $\phi = \tan^{-1} \omega/-a$
21. $\dfrac{\omega_n^2}{s(s^2 + 2\zeta\omega_n s + \omega_n^2)}$	$1 - \dfrac{1}{\sqrt{1 - \zeta^2}} e^{-\zeta\omega_n t} \sin(\omega_n\sqrt{1 - \zeta^2}\ t + \phi),\ \phi = \cos^{-1} \zeta,$ $\zeta < 1$
22. $\dfrac{(s + \alpha)}{s[(s + a)^2 + \omega^2]}$	$\dfrac{\alpha}{a^2 + \omega^2} + \dfrac{1}{\omega}\left[\dfrac{(\alpha - a)^2 + \omega^2}{a^2 + \omega^2}\right]^{1/2} e^{-at} \sin(\omega t + \phi),$ $\phi = \tan^{-1} \dfrac{\omega}{\alpha - a} - \tan^{-1} \dfrac{\omega}{-a}$
23. $\dfrac{1}{(s + c)[(s + a)^2 + \omega^2]}$	$\dfrac{e^{-ct}}{(c - a)^2 + \omega^2} + \dfrac{e^{-at} \sin(\omega t - \phi)}{\omega[(c - a)^2 + \omega^2]^{1/2}},\quad \phi = \tan^{-1} \dfrac{\omega}{c - a}$

APPENDIX B

Table B.1 Symbols and Units

Parameter or variable name	Symbol	SI	English
Acceleration, angular	$\alpha(t)$	rad/sec^2	rad/sec^2
Acceleration, translational	$a(t)$	m/sec^2	ft/sec^2
Friction, rotational	f	$\dfrac{\text{n-m}}{\text{rad/sec}}$	$\dfrac{\text{ft-lb}}{\text{rad/sec}}$
Friction, translational	f	$\dfrac{\text{n}}{\text{m/sec}}$	$\dfrac{\text{lb}}{\text{ft/sec}}$
Inertia, rotational	J	$\dfrac{\text{n-m}}{\text{rad/sec}^2}$	$\dfrac{\text{ft-lb}}{\text{rad/sec}^2}$
Mass	M	kg	slugs
Position, rotational	$\Theta(t)$	rad	rad
Position, translational	$x(t)$	m	ft
Speed, rotational	$\omega(t)$	rad/sec	rad/sec
Speed, translational	$v(t)$	m/sec	ft/sec
Torque	$T(t)$	n-m	ft-lb

Table B.2 Conversion Factors

To convert	Into	Multiply by	To convert	Into	Multiply by
Btu	ft-lb	778.3	kw	Btu/min	56.92
Btu	joules	1,054.8	kw	ft-lb/min	4.426×10^4
Btu/hr	ft-lb/sec	0.2162	kw	hp	1.341
Btu/hr=	watts	0.2931			
Btu/min	hp	0.02356	miles		
Btu/min	kw	0.01757	(statute)	ft	5,280
Btu/min	watts	17.57	mph	ft/min	88
			mph	ft/sec	1.467
cal	joules	4.182	mph	m/sec	0.44704
cm	ft	3.281×10^{-2}	mils	cms	2.540×10^{-3}
cm	in	0.3937	mils	in	0.001
cm^3	ft^3	3.531×10^{-5}	min (angles)	deg	0.01667
			min (angles)	rad	2.909×10^{-4}
deg (angle)	rad	0.01745			
deg/sec	rpm	0.1667			
dynes	gm	1.020×10^{-3}	n-m	ft-lb	0.73756
dynes	lb	2.248×10^{-6}	n-m	dyne-cm	10^7
dynes	newtons	10^{-5}	n-m-sec	watt	1.0
ft/sec	miles/hr	0.6818	oz	gm	28.349527
ft/sec	miles/min	0.01136	oz-in	dyne-cm	70,615.7
ft-lb	gm-cm	1.383×10^4	$oz-in^2$	$gm-cm^2$	1.829×10^2
ft-lb	oz-in	192	oz-in	ft-lb	5.208×10^{-3}
ft-lb/min	Btu/min	1.286×10^{-3}	oz-in	gm-cm	72.01
ft-lb/sec	hp	1.818×10^{-3}			
ft-lb/sec	kw	1.356×10^{-3}	lb (force)	newtons	4.4482
$\dfrac{\text{ft-lb}}{\text{rad/sec}}$	$\dfrac{\text{oz-in}}{\text{rpm}}$	20.11	lb/ft^3	gm/cm^3	0.01602
			lb.-ft-sec^2	$oz-in^2$	7.419×10^4
gm	dynes	980.7	rad	deg	57.30
gm	lb	2.205×10^{-3}	rad	min	3,438
$gm-cm^2$	$oz-in^2$	5.468×10^{-3}	rad	sec	2.063×10^5
gm-cm	oz-in	1.389×10^{-2}	rad/sec	deg/sec	57.30
gm-cm	ft-lb	1.235×10^{-5}	rad/sec	rpm	9.549
			rad/sec	rps	0.1592
hp	Btu/min	42.44	rpm	deg/sec	6.0
hp	ft-lb/min	33,000	rpm	rad/sec	0.1047
hp	ft-lb/sec	550.0			
hp	watts	745.7	sec (angle)	deg	2.778×10^{-4}
			sec (angle)	rad	4.848×10^{-6}
in	meters	2.540×10^{-2}	slugs (mass)	kg	14.594
in	cm	2.540	slug-ft^2	km^2	1.3558
joules	Btu	9.480×10^{-4}			
joules	ergs	10^7	watts	Btu/hr	3.413
joules	ft-lb	0.7376	watts	Btu/min	0.05688
joules	watt-hr	2.778×10^{-4}	watts	ft-lb/min	44.27
			watts	hp	1.341×10^{-3}
kg	lb	2.205	watts	n-m/sec	1.0
kg	slugs	6.852×10^{-2}	watt-hr	Btu	3.413

APPENDIX C / An Introduction to Matrix Algebra

C.1 DEFINITIONS

There are many situations in which we have to deal with rectangular arrays of numbers or functions. The rectangular array of numbers (or functions)

$$\mathbf{A} = \begin{bmatrix} a_{11} & a_{12} & \cdots & a_{1n} \\ a_{21} & a_{22} & \cdots & a_{2n} \\ \vdots & \vdots & & \vdots \\ a_{m1} & a_{m2} & \cdots & a_{mn} \end{bmatrix} \tag{C.1}$$

is known as a *matrix*. The numbers a_{ij} are called *elements* of the matrix, with the subscript i denoting the row and the subscript j denoting the column.

A matrix with m rows and n columns is said to be a matrix of *order* (m, n) or alternatively called an $m \times n$ (m by n) matrix. When the number of the columns equals the number of rows, $m = n$, the matrix is called a *square matrix* of order n. It is common to use bold-faced capital letters to denote an $m \times n$ matrix.

A matrix which is comprised of only one column, that is an $m \times 1$ matrix, is known as a column matrix or more commonly a *column vector*. We shall represent a column vector with bold-faced lower-case letters as

$$\mathbf{y} = \begin{bmatrix} y_1 \\ y_2 \\ \vdots \\ y_m \end{bmatrix}. \tag{C.2}$$

Analogously, a *row vector* is an ordered collection of numbers written in a row, that is, a $1 \times n$ matrix. We will use bold-faced lower-case letters to represent vectors, and therefore a row vector will be written as

$$\mathbf{z} = [z_1, z_2, \ldots, z_n] \tag{C.3}$$

with n elements.

A few matrices with distinctive characteristics are given special names. A square matrix in which all the elements are zero except those on the principal diagonal, a_{11}, a_{22}, . . . , a_{nn}, is called a *diagonal matrix*. Then, for example, a 3×3 diagonal matrix would be

$$\mathbf{B} = \begin{bmatrix} b_{11} & 0 & 0 \\ 0 & b_{22} & 0 \\ 0 & 0 & b_{33} \end{bmatrix}. \tag{C.4}$$

If all the elements of a diagonal matrix have the value 1, then the matrix is known as the *identity matrix* **I**, which is written as

$$\mathbf{I} = \begin{bmatrix} 1 & 0 & \cdots & 0 \\ 0 & 1 & \cdots & 0 \\ \vdots & \vdots & \ddots & \vdots \\ 0 & 0 & \cdots & 1 \end{bmatrix}. \tag{C.5}$$

When all the elements of a matrix are equal to zero, the matrix is called the *zero* or *null matrix*. When the elements of a matrix have a special relationship so that $a_{ij} = a_{ji}$, it is called a *symmetrical* matrix. Thus, for example, the matrix

$$\mathbf{H} = \begin{bmatrix} 3 & -2 & 1 \\ -2 & 6 & 4 \\ 1 & 4 & 8 \end{bmatrix} \tag{C.6}$$

is a symmetrical matrix of order (3, 3).

C.2 ADDITION AND SUBTRACTION OF MATRICES

The addition of two matrices is possible only for matrices of the same order. The sum of two matrices is obtained by adding the corresponding elements. Thus if the elements of **A** are a_{ij} and of **B** are b_{ij} and if

$$\mathbf{C} = \mathbf{A} + \mathbf{B}, \tag{C.7}$$

then the elements of **C** which are c_{ij} are obtained as

$$c_{ij} = a_{ij} + b_{ij}. \tag{C.8}$$

Then, for example, the matrix addition for two 3×3 matrices is as follows:

$$\mathbf{C} = \begin{bmatrix} 2 & 1 & 0 \\ 1 & -1 & 3 \\ 0 & 6 & 2 \end{bmatrix} + \begin{bmatrix} 8 & 2 & 1 \\ 1 & 3 & 0 \\ 4 & 2 & 1 \end{bmatrix} = \begin{bmatrix} 10 & 3 & 1 \\ 2 & 2 & 3 \\ 4 & 8 & 3 \end{bmatrix}. \tag{C.9}$$

From the operation used for performing the operation of addition, we note that the process is commutative, that is

$$\mathbf{A} + \mathbf{B} = \mathbf{B} + \mathbf{A}. \tag{C.10}$$

Also, we note that the addition operation is associative, so that

$$(\mathbf{A} + \mathbf{B}) + \mathbf{C} = \mathbf{A} + (\mathbf{B} + \mathbf{C}). \tag{C.11}$$

In order to perform the operation of subtraction we note that if a matrix $\mathbf{A}$ is multiplied by a constant α, then every element of the matrix is multiplied by this constant. Therefore, we may write

$$\alpha\mathbf{A} = \begin{bmatrix} \alpha a_{11} & \alpha a_{12} & \cdots & \alpha a_{1n} \\ \alpha a_{12} & \alpha a_{22} & \cdots & \alpha a_{2n} \\ \vdots & \vdots & & \vdots \\ \alpha a_{m1} & \alpha a_{m2} & \cdots & \alpha a_{mn} \end{bmatrix}. \tag{C.12}$$

Then, in order to carry out a subtraction operation, we use $\alpha = -1$, and $-\mathbf{A}$ is obtained by multiplying each element of $\mathbf{A}$ by -1. Then, for example,

$$\mathbf{C} = \mathbf{B} - \mathbf{A} = \begin{bmatrix} 2 & 1 \\ 4 & 2 \end{bmatrix} - \begin{bmatrix} 6 & 1 \\ 3 & 1 \end{bmatrix} = \begin{bmatrix} -4 & 0 \\ 1 & 1 \end{bmatrix}. \tag{C.13}$$

C.3 MULTIPLICATION OF MATRICES

Matrix multiplication is defined in such a way as to assist in the solution of simultaneous linear equations. The multiplication of two matrices $\mathbf{AB}$ requires that the number of columns of $\mathbf{A}$ is equal to the number of rows of $\mathbf{B}$. Thus, if $\mathbf{A}$ is of order $m \times n$ and $\mathbf{B}$ is of order $n \times q$, then the product is of order $m \times q$. The elements of a product

$$\mathbf{C} = \mathbf{AB} \tag{C.14}$$

are found by multiplying the ith row of $\mathbf{A}$ and the jth column of $\mathbf{B}$ and summing these products to give the element c_{ij}. That is,

$$c_{ij} = a_{i1}b_{1j} + a_{i2}b_{2j} + \cdots + a_{iq}b_{qj} = \sum_{k=1}^{q} a_{ik}b_{kj}. \tag{C.15}$$

Thus we obtain c_{11}, the first element of $\mathbf{C}$, by multiplying the first row of $\mathbf{A}$ by the first column of $\mathbf{B}$ and summing the products of the elements. We should note that, in general, matrix multiplication is not commutative, that is

$$\mathbf{AB} \neq \mathbf{BA}. \tag{C.16}$$

Also, we will note that the multiplication of a matrix of $m \times n$ by a column vector (order $n \times 1$) results in a column vector of order $m \times 1$.

A specific example of a multiplication of a column vector by a matrix is

$$\mathbf{x} = \mathbf{Ay} = \begin{bmatrix} a_{11} & a_{12} & a_{13} \\ a_{21} & a_{22} & a_{23} \end{bmatrix} \begin{bmatrix} y_1 \\ y_2 \\ y_3 \end{bmatrix}$$

$$= \begin{bmatrix} (a_{11}y_1 + a_{12}y_2 + a_{13}y_3) \\ (a_{21}y_1 + a_{22}y_2 + a_{23}y_3) \end{bmatrix}. \tag{C.17}$$

Note that $\mathbf{A}$ is of order 2×3 and $\mathbf{y}$ is of order 3×1. Therefore, the resulting matrix $\mathbf{x}$ is of order 2×1 which is a column vector with two rows. There are two elements of $\mathbf{x}$, and

$$x_1 = (a_{11}y_1 + a_{12}y_2 + a_{13}y_3) \tag{C.18}$$

is the first element obtained by multiplying the first row of $\mathbf{A}$ by the first (and only) column of $\mathbf{y}$.

Another example, which the reader should verify, is

$$\mathbf{C} = \mathbf{AB} = \begin{bmatrix} 2 & -1 \\ -1 & 2 \end{bmatrix} \begin{bmatrix} 3 & 2 \\ -1 & -2 \end{bmatrix} = \begin{bmatrix} 7 & 6 \\ -5 & -6 \end{bmatrix}. \tag{C.19}$$

For example, the element c_{22} is obtained as $c_{22} = -1(2) + 2(-2) = -6$.

Now we are able to use this definition of multiplication in representing a set of simultaneous linear algebraic equations by a matrix equation. Consider the following set of algebraic equations:

$$\begin{aligned} 3x_1 + 2x_2 + x_3 &= u_1, \\ 2x_1 + x_2 + 6x_3 &= u_2, \\ 4x_1 - x_2 + 2x_3 &= u_3. \end{aligned} \tag{C.20}$$

We may identify two column vectors as

$$\mathbf{x} = \begin{bmatrix} x_1 \\ x_2 \\ x_3 \end{bmatrix} \quad \text{and} \quad \mathbf{u} = \begin{bmatrix} u_1 \\ u_2 \\ u_3 \end{bmatrix}. \tag{C.21}$$

Then we may write the matrix equation

$$\mathbf{Ax} = \mathbf{u}, \tag{C.22}$$

where

$$\mathbf{A} = \begin{bmatrix} 3 & 2 & 1 \\ 2 & 1 & 6 \\ 4 & -1 & 2 \end{bmatrix}.$$

One immediately notes the utility of the matrix equation as a compact form of a set of simultaneous equations.

The multiplication of a row vector and a column vector may be written as

$$\mathbf{xy} = [x_1, x_2, \cdots, x_n] \begin{bmatrix} y_1 \\ y_2 \\ \vdots \\ y_n \end{bmatrix} = x_1y_1 + x_2y_2 + \cdots + x_ny_n. \tag{C.23}$$

Thus we note that the multiplication of a row vector and a column vector results in a number which is a sum of a product of specific elements of each vector.

As a final item in this section, we note that the multiplication of any matrix by the identity matrix results in the original matrix, that is $\mathbf{AI} = \mathbf{A}$.

C.4 OTHER USEFUL MATRIX OPERATIONS AND DEFINITIONS

The *transpose* of a matrix $\mathbf{A}$ is denoted in this text as $\mathbf{A}^T$. One will often find the notation $\mathbf{A}'$ for $\mathbf{A}^T$ in the literature. The transpose of a matrix $\mathbf{A}$ is obtained by interchanging the rows and columns of $\mathbf{A}$. Then, for example, if

$$\mathbf{A} = \begin{bmatrix} 6 & 0 & 2 \\ 1 & 4 & 1 \\ -2 & 3 & -1 \end{bmatrix},$$

then

$$\mathbf{A}^T = \begin{bmatrix} 6 & 1 & -2 \\ 0 & 4 & 3 \\ 2 & 1 & -1 \end{bmatrix}. \tag{C.24}$$

Therefore, we are able to denote a row vector as the transpose of a column vector and write

$$\mathbf{x}^T = [x_1, x_2, \ldots, x_n]. \tag{C.25}$$

Since $\mathbf{x}^T$ is a row vector, we obtain a matrix multiplication of $\mathbf{x}^T$ by $\mathbf{x}$ as follows:

$$\mathbf{x}^T\mathbf{x} = [x_1, x_2, \ldots, x_n] \begin{bmatrix} x_1 \\ x_2 \\ \vdots \\ x_n \end{bmatrix} = x_1^2 + x_2^2 + \cdots + x_n^2. \tag{C.26}$$

Thus the multiplication $\mathbf{x}^T\mathbf{x}$ results in the sum of the squares of each element of $\mathbf{x}$.

The transpose of the product of two matrices is the product in reverse order of their transposes, so that

$$(\mathbf{AB})^T = \mathbf{B}^T\mathbf{A}^T. \tag{C.27}$$

The sum of the main diagonal elements of a square matrix $\mathbf{A}$ is called the *trace* of $\mathbf{A}$ written as

$$\text{tr } \mathbf{A} = a_{11} + a_{22} + \cdots + a_{nn}. \tag{C.28}$$

The *determinant* of a square matrix is obtained by enclosing the elements of the matrix $\mathbf{A}$ within vertical bars as, for example,

$$\det \mathbf{A} = \begin{vmatrix} a_{11} & a_{12} \\ a_{21} & a_{22} \end{vmatrix}. \tag{C.29}$$

If the determinant of $\mathbf{A}$ is equal to zero, then the determinant is said to be singular. The value of a determinant is determined by obtaining the minors and cofactors of the determinants. The *minor* of an element a_{ij} of a determinant of order n is a determinant of order $(n - 1)$ obtained by removing the row i and the column j of the original determinant. The cofactor of a given element of a determinant is the minor of the element with either a plus or minus sign attached; hence

$$\text{cofactor of } a_{ij} = \alpha_{ij} = (-1)^{i+j}M_{ij},$$

where M_{ij} is the minor of a_{ij}. For example, the cofactor of the element a_{23} of

$$\det \mathbf{A} = \begin{vmatrix} a_{11} & a_{12} & a_{13} \\ a_{21} & a_{22} & a_{23} \\ a_{31} & a_{32} & a_{33} \end{vmatrix} \tag{C.30}$$

is

$$\alpha_{23} = (-1)^5 M_{23} = -\begin{vmatrix} a_{11} & a_{12} \\ a_{31} & a_{32} \end{vmatrix}. \tag{C.31}$$

The value of a determinant of second-order (2×2) is

$$\begin{vmatrix} a_{11} & a_{12} \\ a_{21} & a_{22} \end{vmatrix} = (a_{11}a_{22} - a_{21}a_{12}). \tag{C.32}$$

The general nth-order determinant has a value given by

$$\det \mathbf{A} = \sum_{j=1}^{n} a_{ij}\alpha_{ij} \qquad \text{with } i \text{ chosen for one row, or} \tag{C.33}$$

$$\det \mathbf{A} = \sum_{i=1}^{n} a_{ij}\alpha_{ij} \qquad \text{with } j \text{ chosen for one column.}$$

That is, the elements a_{ij} are chosen for a specific row (or column) and that entire row (or column) is expanded according to Eq. (C.33). For example, the value of a specific 3×3 determinant is

$$\det \mathbf{A} = \det \begin{bmatrix} 2 & 3 & 5 \\ 1 & 0 & 1 \\ 2 & 1 & 0 \end{bmatrix} = 2\begin{vmatrix} 0 & 1 \\ 1 & 0 \end{vmatrix} - 1\begin{vmatrix} 3 & 5 \\ 1 & 0 \end{vmatrix} + 2\begin{vmatrix} 3 & 5 \\ 0 & 1 \end{vmatrix}$$
$$= 2(-1) - (-5) + 2(3) = 9, \tag{C.34}$$

where we have expanded in the first column.

The *adjoint matrix* of a square matrix $\mathbf{A}$ is formed by replacing each element a_{ij} by the cofactor α_{ij} and transposing. Therefore

$$\text{Adjoint } \mathbf{A} = \begin{bmatrix} \alpha_{11} & \alpha_{12} & \cdots & \alpha_{1n} \\ \alpha_{21} & \alpha_{22} & \cdots & \alpha_{2n} \\ \cdot & \cdot & & \cdot \\ \cdot & \cdot & & \cdot \\ \alpha_{n1} & \alpha_{n2} & \cdots & \alpha_{nn} \end{bmatrix}^T = \begin{bmatrix} \alpha_{11} & \alpha_{21} & \cdots & \alpha_{n1} \\ \alpha_{12} & \alpha_{22} & \cdots & \alpha_{n2} \\ \cdot & \cdot & & \cdot \\ \cdot & \cdot & & \cdot \\ \alpha_{1n} & \alpha_{2n} & \cdots & \alpha_{nn} \end{bmatrix}. \tag{C.35}$$

C.5 MATRIX INVERSION

The inverse of a square matrix A is written as A^{-1} and is defined as satisfying the relationship

$$A^{-1}A = AA^{-1} = I. \tag{C.36}$$

The inverse of a matrix A is

$$A^{-1} = \frac{\text{adjoint of } A}{\det A} \tag{C.37}$$

when the det A is not equal to zero. For a 2×2 matrix we have the adjoint matrix

$$\text{adjoint } A = \begin{bmatrix} a_{22} & -a_{12} \\ -a_{21} & a_{11} \end{bmatrix} \tag{C.38}$$

and the det $A = a_{11}a_{22} - a_{12}a_{21}$. Consider the matrix

$$A = \begin{bmatrix} 1 & 2 & 3 \\ 2 & -1 & 4 \\ 0 & -1 & 1 \end{bmatrix}. \tag{C.39}$$

The determinant has a value det $A = -7$. The cofactor α_{11} is

$$\alpha_{11} = (-1)^2 \begin{vmatrix} -1 & 4 \\ -1 & 1 \end{vmatrix} = 3. \tag{C.40}$$

In a similar manner we obtain

$$A^{-1} = \frac{\text{adjoint } A}{\det A} = \left(-\frac{1}{7}\right) \begin{bmatrix} 3 & -5 & 11 \\ -2 & 1 & 2 \\ -2 & 1 & -5 \end{bmatrix}. \tag{C.41}$$

C.6 MATRICES AND CHARACTERISTIC ROOTS

A set of simultaneous linear algebraic equations may be represented by the matrix equation

$$y = Ax, \tag{C.42}$$

where the y vector may be considered as a transformation of the vector x. The question may be asked whether or not it may happen that a vector y may be a scalar multiple of x. Trying $y = \lambda x$, where λ is a scalar we have

$$\lambda x = Ax. \tag{C.43}$$

Alternatively, Eq. (C.43) may be written as

$$\lambda x - Ax = (\lambda I - A)x = 0, \tag{C.44}$$

where $\mathbf{I}$ = identity matrix. Thus the solution for $\mathbf{x}$ exists if and only if

$$\det (\lambda \mathbf{I} - \mathbf{A}) = 0. \tag{C.45}$$

This determinant is called the characteristic determinant of $\mathbf{A}$. Expansion of the determinant of Eq. (C.45) results in the *characteristic equation*. The characteristic equation is an nth-order polynomial in λ. The n roots of this characteristic equation are called the *characteristic roots*. For every possible value λ_i $(i = 1, 2, \ldots, n)$ of the nth-order characteristic equation, we may write

$$(\lambda_i \mathbf{I} - \mathbf{A})\mathbf{x}_i = 0. \tag{C.46}$$

The vector $\mathbf{x}_i$ is the *characteristic vector* for the ith root. Let us consider the matrix

$$\mathbf{A} = \begin{bmatrix} 2 & 1 & 1 \\ 2 & 3 & 4 \\ -1 & -1 & -2 \end{bmatrix}. \tag{C.47}$$

The characteristic equation is found as follows:

$$\det \begin{bmatrix} (\lambda - 2) & -1 & -1 \\ -2 & (\lambda - 3) & -4 \\ 1 & 1 & (\lambda + 2) \end{bmatrix} = (-\lambda^3 + 3\lambda^2 + \lambda - 3) = 0. \tag{C.48}$$

The roots of the characteristic equation are $\lambda_1 = 1$, $\lambda_2 = -1$, $\lambda_3 = 3$. When $\lambda = \lambda_1 = 1$, we find the first characteristic vector from the equation

$$\mathbf{A}\mathbf{x}_1 = \lambda_1 \mathbf{x}_1, \tag{C.49}$$

and we have $\mathbf{x}_1^T = k[1, -1, 0]$, where k is an arbitrary constant usually chosen equal to one. Similarly, one finds

$$\mathbf{x}_2^T = [0, 1, -1]$$

and

$$\mathbf{x}_3^T = [2, 3, -1]. \tag{C.50}$$

C.7 THE CALCULUS OF MATRICES

The derivative of a matrix $\mathbf{A} = \mathbf{A}(t)$ is defined as

$$\frac{d}{dt}[\mathbf{A}(t)] = \begin{bmatrix} da_{11}(t)/dt & da_{12}(t)/dt & \cdots & da_{1n}(t)/dt \\ \vdots & \vdots & & \vdots \\ da_{n1}(t)/dt & da_{n2}(t)/dt & \cdots & da_{nn}(t)/dt \end{bmatrix}. \tag{C.51}$$

That is, the derivative of a matrix is simply the derivative of each element $a_{ij}(t)$ of the matrix.

The *matrix exponential function* is defined as the power series

$$\exp [\mathbf{A}] = e^{\mathbf{A}} = \mathbf{I} + \frac{\mathbf{A}}{1!} + \frac{\mathbf{A}^2}{2!} + \cdots + \frac{\mathbf{A}^k}{k!} + \cdots = \sum_{k=0}^{\infty} \frac{\mathbf{A}^k}{k!}, \tag{C.52}$$

where $\mathbf{A}^2 = \mathbf{A}\mathbf{A}$ and, similarly, $\mathbf{A}^k$ implies $\mathbf{A}$ multiplied k times. This series may be shown to be convergent for all square matrices. Also a matrix exponential which is a function of time is defined as

$$e^{\mathbf{A}t} = \sum_{k=0}^{\infty} \frac{\mathbf{A}^k t^k}{k!}. \tag{C.53}$$

If one differentiates with respect to time, then

$$\frac{d}{dt} (e^{\mathbf{A}t}) = \mathbf{A} e^{\mathbf{A}t}. \tag{C.54}$$

Therefore, for a differential equation

$$\frac{d\mathbf{x}}{dt} = \mathbf{A}\mathbf{x}, \tag{C.55}$$

one might postulate a solution $\mathbf{x} = e^{\mathbf{A}t}\mathbf{c} = \boldsymbol{\phi}\mathbf{c}$, where the matrix $\boldsymbol{\phi}$ is $\boldsymbol{\phi}$ and $e^{\mathbf{A}t}$ and $\mathbf{c}$ is an unknown column vector. Then

$$\frac{d\mathbf{x}}{dt} = \mathbf{A}\mathbf{x} \tag{C.56}$$

or

$$\mathbf{A} e^{\mathbf{A}t} = \mathbf{A} e^{\mathbf{A}t}, \tag{C.57}$$

and we have in fact satisfied the relationship, Eq. (C.55). Then, the value of $\mathbf{c}$ is simply $\mathbf{x}(0)$, the initial value of $\mathbf{x}$ since, when $t = 0$, we have $\mathbf{x}(0) = \mathbf{c}$. Therefore the solution to Eq. (C.55) is

$$\mathbf{x}(t) = e^{\mathbf{A}t}\mathbf{x}(0). \tag{C.58}$$

REFERENCES

1. R. C. Dorf, *Matrix Algebra—A Programmed Introduction*, Wiley, New York, 1969.
2. C. R. Wylie, Jr., *Advanced Engineering Mathematics*, 4th ed., McGraw-Hill, New York, 1975.

APPENDIX D / The Evaluation of the Transition Matrix of a Linear Time-Invariant System by Means of a Computational Algorithm

The state variable differential equation of a linear, time-invariant system may be written as

$$\dot{x} = Ax + Bu, \tag{D.1}$$

as shown in Section 9.2. The solution to this equation is

$$x(t) = \boldsymbol{\phi}(t)x(0) + \int_0^t \boldsymbol{\phi}(t - \tau)Bu(\tau) \, d\tau, \tag{D.2}$$

where $\boldsymbol{\phi}(t)$ is the transition matrix of the system. An efficient computer algorithm for the evaluation of $\boldsymbol{\phi}(t)$ has been developed by Faddeev and utilized by Morgan [1, 2].

In order to develop the algorithm for $\boldsymbol{\phi}(t)$, let us consider the unforced form of Eq. (D.1) which is

$$\dot{x} = Ax. \tag{D.3}$$

Taking the Laplace transform of Eq. (D.3), we have

$$s\,X(s) - x(0) = AX(s)$$

or

$$(s\,I - A)X(s) = x(0). \tag{D.4}$$

Multiplying both sides of Eq. (D.4) by the inverse of $(s\,I - A)$, we have

$$X(s) = (s\,I - A)^{-1}x(0). \tag{D.5}$$

Therefore, reconsidering Eq. (D.2), we recognize that

$$\boldsymbol{\phi}(s) = (s\,I - A)^{-1}, \tag{D.6}$$

where $\boldsymbol{\phi}(s) = \mathcal{L}\{\boldsymbol{\phi}(t)\}$. Therefore, in order to obtain the transition matrix we will endeavor to find $\phi(s)$ and then simply take the inverse Laplace transform in order to obtain $\phi(t)$. Since $\phi(s)$ is a rational function of s for linear time-invariant systems, we may write Eq. (D.6) as

$$(s\mathbf{I} - \mathbf{A})^{-1} = \frac{\mathbf{B}(s)}{\det(s\mathbf{I} - \mathbf{A})} \qquad (D.7)$$

by utilizing the definition of matrix inversion (see Appendix C).

Now, we recognize $\det(s\mathbf{I} - \mathbf{A})$ as the characteristic equation and therefore

$$\det(s\mathbf{I} - \mathbf{A}) = d(s) = s^n - d_1 s^{n-1} - d_2 s^{n-2} - \cdots - d_n, \qquad (D.8)$$

where we recall that d_1 equals the sum of the characteristic roots of $\mathbf{A}$ (see Section 5.2). The matrix function $\mathbf{B}(s)$ is written as a polynomial as

$$\mathbf{B}(s) = \mathbf{B}_0 s^{n-1} + \mathbf{B}_1 s^{n-2} + \cdots + \mathbf{B}_{n-2} s + \mathbf{B}_{n-1}. \qquad (D.9)$$

Rewriting Eq. (D.7) as

$$\mathbf{I}[\det(s\mathbf{I} - \mathbf{A})] = (s\mathbf{I} - \mathbf{A})\mathbf{B}(s) \qquad (D.10)$$

and substituting Eqs. (D.8) and (D.9), we have

$$\begin{aligned} \mathbf{I}(s^n - d_1 s^{n-1} &- d_2 s^{n-2} - \cdots - d_n) \\ &= (s\mathbf{I} - \mathbf{A})(\mathbf{B}_0 s^{n-1} + \mathbf{B}_1 s^{n-2} + \cdots + \mathbf{B}_{n-1}) \end{aligned} \qquad (D.11)$$

Equating equal powers of s, we have

$$\begin{aligned} \mathbf{B}_0 &= \mathbf{I}, \\ \mathbf{B}_1 &= \mathbf{A} - d_1\mathbf{I}, \\ \mathbf{B}_2 &= \mathbf{A}\mathbf{B}_1 - d_2\mathbf{I}, \\ &\vdots \\ \mathbf{B}_k &= \mathbf{A}\mathbf{B}_{k-1} - d_k\mathbf{I}. \end{aligned} \qquad (D.12)$$

The coefficient d_1 is equal to the sum of the roots of $\mathbf{A}$ and may be shown to be equal to the trace of $\mathbf{A}$ [1]. That is,

$$d_1 = \operatorname{tr} \mathbf{A} = a_{11} + a_{22} + \cdots + a_{nn}. \qquad (D.13)$$

In a similar manner, we find that

$$d_2 = \frac{1}{2} \operatorname{tr} \mathbf{A}\mathbf{B}_1,$$

$$\vdots \qquad\qquad\qquad (D.14)$$

$$d_k = \frac{1}{k} \operatorname{tr} \mathbf{A}\mathbf{B}_{k-1}.$$

The algorithm represented by Eqs. (D.12) and (D.14) is readily programmed for digital computer calculation. Basically, this method is a convenient computer technique for the inversion of the matrix $(s\mathbf{I} - \mathbf{A})$. In order to illustrate the procedure let us consider the example of Section 9.6.

Example D.1. The flow graph of the system considered in Example 9.7 is shown in Fig. 9.16. The matrix differential equation of this system is

$$\dot{x} = \mathbf{A}x, \tag{D.15}$$

where

$$\mathbf{A} = \begin{bmatrix} 0 & -2 \\ 1 & -3 \end{bmatrix}.$$

The characteristic equation of this system is

$$\begin{aligned}
\det(s\mathbf{I} - \mathbf{A}) = d(s) &= s^2 + 3s + 2 \\
&= (s + 1)(s + 2) \\
&= s^2 - d_1 s - d_2.
\end{aligned} \tag{D.16}$$

The coefficient d_1 is

$$d_1 = \text{tr } \mathbf{A} = -3, \tag{D.17}$$

which is in fact the sum of the roots of Eq. (D.16). Then we have

$$\mathbf{B}_1 = \mathbf{A} - d_1\mathbf{I} = \begin{bmatrix} 0 & -2 \\ 1 & -3 \end{bmatrix} - (-3)\begin{bmatrix} 1 & 0 \\ 0 & 1 \end{bmatrix} = \begin{bmatrix} 3 & -2 \\ 1 & 0 \end{bmatrix}. \tag{D.18}$$

Continuing, we obtain

$$\mathbf{AB}_1 = \begin{bmatrix} -2 & 0 \\ 0 & -2 \end{bmatrix}. \tag{D.19}$$

Finally, we find that

$$d_2 = 1/2 \text{ tr } \mathbf{AB}_1 = 1/2(-4) = -2. \tag{D.20}$$

and $\mathbf{B}_2 = \mathbf{0}$, the null matrix. Then, considering Eq. (D.7), we obtain

$$\begin{aligned}
\boldsymbol{\phi}(s) = \frac{\mathbf{B}(s)}{\det(s\mathbf{I} - \mathbf{A})} &= \frac{\mathbf{I}s + \mathbf{B}_1}{(s^2 + 3s + 2)} \\
&= \frac{\begin{bmatrix} (s+3) & -2 \\ 1 & s \end{bmatrix}}{(s^2 + 3s + 2)},
\end{aligned} \tag{D.21}$$

which is identical to the solution we obtained as Eq. (9.89) in Section 9.7. Therefore,

$$\boldsymbol{\phi}(t) = \mathcal{L}^{-1}\{\boldsymbol{\phi}(s)\} = \begin{bmatrix} (2e^{-t} - e^{-2t}) & (-2e^{-t} + 2e^{-2t}) \\ (e^{-t} - e^{-2t}) & (-e^{-t} + 2e^{-2t}) \end{bmatrix}. \tag{D.22}$$

Equation (D.22) can also be directly obtained from Eq. (D.7), after $\mathbf{B}(s)$ is calculated as

$$\boldsymbol{\phi}(t) = \sum_{i=0}^{n} (\operatorname{tr} \mathbf{B}(s_i))^{-1} \mathbf{B}(s_i) e^{s_i t},$$

where s_i are distinct roots of the characteristic equation [2].

REFERENCES

1. D. K. Faddeev and V. N. Faddeeva, *Computational Methods of Linear Algebra*. Freeman, San Francisco, 1963; pp. 260–265.

2. B. S. Morgan, Jr., "Computational Procedures for Sensitivity Coefficients in Time-Invariant Multivariable Systems," *Proceedings of the Allerton Conference on Circuit and System Theory*, 1965; pp. 252—258.

3. B. S. Morgan, Jr., "Sensitivity Analysis and Synthesis of Multivariable Systems," *IEEE Transactions on Automatic Control*, **11**, 3, July, 1966, pp. 506–512.

APPENDIX E / Program Transition

Program transition is a digital computer program which may be utilized to obtain the transition matrix, $\phi(t)$, of a linear time-invariant system. The program also determines the values of the state variables at the time intervals selected by the user.

The program uses a truncated matrix series to determine

$$\phi(t) = e^{Ft} \sum_{k=0}^{N} \frac{(Ft)^k}{k!}, \tag{E.1}$$

where $N + 1$ is the number of terms used. Of course, when N equals infinity, the exponential series is exactly equal to $\phi(t)$. A truncated series where N is equal to ten or twenty terms is often quite sufficient for normal accuracy requirements. However, in each case the user of this program may select the number of terms N to be used.

The program considers the linear vector differential equation

$$\dot{x} = Ax + Bu, \tag{E.2}$$

where x = state vector and u = input signal. A new vector y is defined as

$$y = \begin{bmatrix} x \\ m \end{bmatrix}, \tag{E.3}$$

where m is used to generate the input signal $u(t)$, and the state vector and input vector are joined together as one vector. Then the new vector differential equation is

$$\dot{y} = Fy. \tag{E.4}$$

The solution to Eq. (E.4) is then

$$\begin{aligned} y(t) &= \phi(t)y(0) \\ &= e^{Ft}y(0). \end{aligned} \tag{E.5}$$

In order to evaluate the state variables at specific time intervals we let $t = T$, and we have

$$\phi(t) = e^{FT}. \tag{E.6}$$

The truncated series uses $N + 1$ terms, denoted as IPROX terms in the program, so that

$$e^{FT} \simeq I + FT + \frac{(FT)^2}{2} + \cdots + \frac{(FT)^{\text{IPROX}}}{(\text{IPROX})!}. \tag{E.7}$$

The number of terms necessary for the approximation is a function of the accuracy requirements. However, one notes that $1/10! = 0.276 \times 10^{-6}$ and $1/20! = 0.41 \times 10^{-18}$. Usually, 15 or 20 terms will suffice and a simple way to check the accuracy is to calculate $\phi(T)$ for both $N = 15$ and $N = 20$. The approximation is accurate for a smaller number of terms when T is chosen so that $f_{ij}T \leqslant 1$ for all the elements of F.

The input vector m generates the input signals for all polynomials or sine wave inputs. Thus, a step or ramp input may be readily used. The input vector is used to generate the system input $u(t)$ so that for a unit ramp input, for example, we have

$$\dot{m} = \begin{bmatrix} \dot{m}_1 \\ \dot{m}_2 \end{bmatrix} = \begin{bmatrix} 0 & 1 \\ 0 & 0 \end{bmatrix} \begin{bmatrix} m_1 \\ m_2 \end{bmatrix} \tag{E.8}$$

and

$$m(0) = [0, 1]^T.$$

In general we use the input vector $m = [m_1, m_2, \ldots m_r]^T$ of order r.

This program is written in the Fortran language. The maximum value that $(n + r)$ may have is equal to 13, where n is the order of the system (the number of state variables).

Three sets of data cards are used. The first lists the time interval T, the value of $(n + r) = \text{LIM}$, and the number of terms of the truncated series IPROX. This list should be according to the format F5.3, 2I2. The second set of data lists the elements of the F matrix. The elements should be listed according to the format 5E14.7 in the following order: $f_{11}, f_{12}, \ldots, f_{1\text{LIM}}, f_{21}, f_{22}, \ldots, f_{2\text{LIM}}, \ldots, f_{\text{LIM1}}, f_{\text{LIM2}}, \ldots, f_{\text{LIMLIM}}$. The set of data lists the initial value of the augmented state vector, $y(0)$. These are listed according to the format 5E14.7 in the order

$$x_1(0), x_2(0), \ldots, x_n(0), m_1(0), m_2(0), \ldots, m_r(0).$$

The output of the program is punched cards. The program is shown on the last page of this appendix.

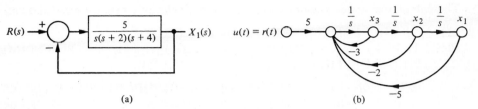

(a)   (b)

Figure E.1

Example E.1. Let us consider the system shown in Fig. E.1(a), where the state variables are identified in Fig. E.1(b). The state vector differential equation is

$$\dot{x} = \begin{bmatrix} 0 & 1 & 0 \\ 0 & 0 & 1 \\ -5 & -2 & -3 \end{bmatrix} x + \begin{bmatrix} 0 \\ 0 \\ 5 \end{bmatrix} r(t). \tag{E.9}$$

Consider this system with a unit step input so that $m_1(t) = r(t)$, $\dot{m}_1 = 0$, and $m_1(0) = 1$. Then the augmented system matrix equation is

$$\dot{y} = \frac{d}{dt}\begin{bmatrix} x_1 \\ x_2 \\ x_3 \\ m_1 \end{bmatrix} = \begin{bmatrix} 0 & 1 & 0 & 0 \\ 0 & 0 & 1 & 0 \\ -5 & -2 & -3 & 5 \\ 0 & 0 & 0 & 0 \end{bmatrix}\begin{bmatrix} x_1 \\ x_2 \\ x_3 \\ m_1 \end{bmatrix} \tag{E.10}$$

and

$$y^T(0) = [0, 0, 0, 1].$$

Initially, let us take $T = 0.5$ and $N + 1 = \text{IPROX} = 10$ terms. Also note that LIM $= (n + r) = 4$. Then the first data card is

$$\underbrace{0.5 \quad 0 \quad 0 \quad 0 \quad 0}_{T} \quad \underbrace{4}_{\text{LIM}} \quad \underbrace{0 \quad 1 \quad 0}_{\text{IPROX}}$$

The elements of the **F** matrix are:

0.0 E+00	1.0 E+00	0.0 E+00	0.0 E+00	0.0 E+00
0.0 E+00	1.0 E+00	0.0 E+00	−5.0 E+00	−2.0 E+00
−3.0 E+00	5.0 E+00	0.0 E+00	0.0 E+00	0.0 E+00
0.0 E+00				

The initial conditions on **y**(t) are:

$$0.0 \quad \text{E}+00 \quad 0.0 \quad \text{E}+00 \quad 0.0 \quad \text{E}+00 \quad 1.0 \quad \text{E}+00$$

The calculated values of $\phi(T)$ and the state variables at $t = kT$ are as follows:

PHI MATRIX ELEMENTS PHI(I,J) IN THE ORDER
PHI(1,1), PHI(1,2), . . . , PHI(1,LIM), PHI(2,1), PHI(2,2), . . . , PHI(2,LIM),
. . . , PHI(LIM,1), PHI(LIM,2), . . . , PHI(LIM,LIM).

9275.5310E−04	4612.7664E−04	7661.1473E−05	7244.6941E−05
−3830.5737E−04	7743.3010E−04	2314.4224E−04	3830.5737E−04
−1157.2111E−03	−8459.4183E−04	8000.3500E−05	1157.2111E−03
0000.0000E−99	0000.0000E−99	0000.0000E−99	1000.0000E−03

STATE VECTOR ELEMENTS IN THE ORDER
X(1), X(2), . . . , X(N), M(1), M(2), . . . , M(R).

7244.6941E−05	3830.5737E−04	1157.2111E−03	1000.0000E−03
4049.9637E−04	9197.4641E−04	8419.1140E−04	1000.0000E−03
9368.6017E−04	1134.9617E−03	−2215.1300E−05	1000.0000E−03
1463.2687E−03	8978.9444E−04	−8888.1740E−04	1000.0000E−03
1775.7904E−03	3120.9832E−04	−1366.7745E−03	1000.0000E−03
1758.8397E−03	−3718.3446E−04	−1271.1169E−03	1000.0000E−03
1434.9635E−03	−8727.9190E−04	−6652.8120E−04	1000.0000E−03
9498.8507E−04	−9964.1910E−04	1817.6180E−04	1000.0000E−03
5078.1596E−04	−7102.9310E−04	9154.4770E−04	1000.0000E−03
2859.6538E−04	−1495.9333E−04	1243.6665E−03	1000.0000E−03
3639.7023E−04	4455.1858E−04	1052.3338E−03	1000.0000E−03
6961.7680E−04	8321.6883E−04	4433.2830E−04	1000.0000E−03
1136.0118E−03	8633.5998E−04	−3169.1100E−04	1000.0000E−03
1500.1268E−03	5430.7870E−04	−9131.0050E−04	1000.0000E−03
1644.4497E−03	1761.4910E−05	−1111.2163E−03	1000.0000E−03
1520.7547E−03	−4904.0383E−04	−8495.6650E−04	1000.0000E−03
1191.7293E−03	−7758.3900E−04	−2557.3820E−04	1000.0000E−03
⋮	⋮	⋮	⋮
9132.1203E−04	−3852.7640E−05	1531.1250E−04	1000.0000E−03
9134.5785E−04	3884.8360E−05	1452.7370E−04	1000.0000E−03
9487.7707E−04	9685.4530E−05	7890.6600E−05	1000.0000E−03
1003.2099E−03	1128.8102E−04	−1634.4600E−05	1000.0000E−03
1053.7945E−03	8239.4770E−05	−1005.1280E−04	1000.0000E−03

When IPROX was increased to 15, the following output was obtained:

PHI MATRIX ELEMENTS PHI(I,J) IN THE ORDER
PHI(1,1), PHI(1,2), . . . , PHI(1,LIM), PHI(2,1), PHI(2,2), . . . , PHI(2,LIM),
. . . , PHI(LIM,1), PHI(LIM,2), . . . , PHI(LIM,LIM).

9275.5290E−04	4612.7663E−04	7661.1336E−05	7244.7177E−05
−3830.5669E−04	7743.3020E−04	2314.4263E−04	3830.5669E−04

$-1157.2131E-03$	$-8459.4194E-04$	$8000.2300E-05$	$1157.2131E-03$
$0000.0000E-99$	$0000.0000E-99$	$0000.0000E-99$	$1000.0000E-03$

STATE VECTOR ELEMENTS IN THE ORDER
$X(1), X(2), \ldots, X(N), M(1), M(2), \ldots, M(R).$

$7244.7177E-05$	$3830.5669E-04$	$1157.2131E-03$	$1000.0000E-03$
$4049.9648E-04$	$9197.4612E-04$	$8419.1230E-04$	$1000.0000E-03$
$9368.6023E-04$	$1134.9616E-03$	$-2215.1000E-05$	$1000.0000E-03$
$\vdots$	$\vdots$	$\vdots$	$\vdots$
$9487.7665E-04$	$9685.4350E-05$	$7890.7300E-05$	$1000.0000E-03$
$1003.2095E-03$	$1128.8120E-04$	$-1634.4000E-05$	$1000.0000E-03$
$1053.7943E-03$	$8239.5210E-05$	$-1005.1240E-04$	$1000.0000E-03$

When IPROX was increased to 20, the following PHI matrix was obtained:

$9275.5290E-04$	$4612.7663E-04$	$7661.1336E-05$	$7244.7177E-05$
$-3830.5669E-04$	$7743.3020E-04$	$2314.4263E-04$	$3830.5669E-04$
$-1157.2131E-03$	$-8459.4194E-04$	$8000.2300E-05$	$1157.2131E-03$
$0000.0000E-99$	$0000.0000E-99$	$0000.0000E-99$	$1000.0000E-03$

Evidently, ten terms is sufficient in this case and 15 terms would provide five place accuracy. Clearly, the program is easy to use and the number of terms is readily changed.

```
PROGRAM TRANSITION
      DIMENSION A(13,13), B(13,13), C(13,13), D(13,13), X(13), Y(13)
   77 READ 202,T,LIM,IPROX
  202 FORMAT(F5.3,2I3)
      READ 201,((A(I,J),J = 1,LIM),I = 1,LIM)
      READ 201,(Y(I),I = 1,LIM)
  201 FORMAT(5E14.7)
      DO 1 I = 1,LIM
      DO 1 J = 1,LIM
    1 A(I,J) = T*A(I,J)
      DO 2 I = 1,LIM
      DO 2 J = 1,LIM
      C(I,J) = 0.0
      D(I,J) = A(I,J)
      B(I,J) = A(I,J)
    2 CONTINUE
      DO 5 IK = 2,IPROX
      DEN = IK
      DO 4 I = 1,LIM
      DO 4 J = 1,LIM
```

```
      DO 4 K = 1,LIM
      C(I,J) = C(I,J) + A(I,K)*B(K,J)/DEN
   4  CONTINUE
      DO 6 I = 1,LIM
      DO 6 J = 1,LIM
      B(I,J) = C(I,J)
   6  D(I,J) = D(I,J) + C(I,J)
      DO 7 I = 1,LIM
      DO 7 J = 1,LIM
   7  C(I,J) = 0.0
   5  CONTINUE
      DO 10 I = 1,LIM
  10  C(I,I) = 1.0
      DO 20 I = 1,LIM
      DO 20 J = 1,LIM
  20  D(I,J) = D(I,J) + C(I,J)
      PUNCH 29
  29  FORMAT(/41HPHI MATRIX ELEMENTS PHI(I,J) IN THE ORDER
      ,/43HPHI(1,1), PHI(1,2), . . . , PHI(1,LIM), PHI(2,1), 26H, PHI(2,2),
      . . . , PHI(2,LIM),/,44H . . . , PHI(LIM,1), PHI(LIM,2), . . . ,
      PHI(LIM,LIM).)
      PUNCH 30,((D(I,J), J = 1,LIM), I = 1,LIM)
  30  FORMAT(4(4X,E14.4))
      PUNCH 203
 203  FORMAT(///34HSTATE VECTOR ELEMENTS IN THE ORDER
      /42HX(1), X(2), ... , X(N), M(1), M(2), ... , M(R). /)
      DO 400 L = 1,100
      DO 101 I = 1,LIM
 101  X(I) = 0.0
      DO 100 I = 1,LIM
      DO 100 J = 1,LIM
 100  X(I) = X(I) + D(I,J)*Y(J)
      DO 200 K = 1,LIM
 200  Y(K) = X(K)
      PUNCH 30,(Y(I),I = 1,LIM)
 400  CONTINUE
      GO TO 77
      END
```

APPENDIX F / Decibel Conversion

M	0	1	2	3	4	5	6	7	8	9
0.0	$m =$	-40.00	-33.98	-30.46	-27.96	-26.02	-24.44	-23.10	-21.94	-20.92
0.1	-20.00	-19.17	-18.42	-17.72	-17.08	-16.48	-15.92	-15.39	-14.89	-14.42
0.2	-13.98	-13.56	-13.15	-12.77	-12.40	-12.04	-11.70	-11.37	-11.06	-10.75
0.3	-10.46	-10.17	-9.90	-9.63	-9.37	-9.12	-8.87	-8.64	-8.40	-8.18
0.4	-7.96	-7.74	-7.54	-7.33	-7.13	-6.94	-6.74	-6.56	-6.38	-6.20
0.5	-6.02	-5.85	-5.68	-5.51	-5.35	-5.19	-5.04	-4.88	-4.73	-4.58
0.6	-4.44	-4.29	-4.15	-4.01	-3.88	-3.74	-3.61	-3.48	-3.35	-3.22
0.7	-3.10	-2.97	-2.85	-2.73	-2.62	-2.50	-2.38	-2.27	-2.16	-2.05
0.8	-1.94	-1.83	-1.72	-1.62	-1.51	-1.41	-1.31	-1.21	-1.11	-1.01
0.9	-0.92	-0.82	-0.72	-0.63	-0.54	-0.45	-0.35	-0.26	-0.18	-0.09
1.0	0.00	0.09	0.17	0.26	0.34	0.42	0.51	0.59	0.67	0.75
1.1	0.83	0.91	0.98	1.06	1.14	1.21	1.29	1.36	1.44	1.51
1.2	1.58	1.66	1.73	1.80	1.87	1.94	2.01	2.08	2.14	2.21
1.3	2.28	2.35	2.41	2.48	2.54	2.61	2.67	2.73	2.80	2.86
1.4	2.92	2.98	3.05	3.11	3.17	3.23	3.29	3.35	3.41	3.46
1.5	3.52	3.58	3.64	3.69	3.75	3.81	3.86	3.92	3.97	4.03
1.6	4.08	4.14	4.19	4.24	4.30	4.35	4.40	4.45	4.51	4.56
1.7	4.61	4.66	4.71	4.76	4.81	4.86	4.91	4.96	5.01	5.06
1.8	5.11	5.15	5.20	5.25	5.30	5.34	5.39	5.44	5.48	5.53
1.9	5.58	5.62	5.67	5.71	5.76	5.80	5.85	5.89	5.93	5.98
2.	6.02	6.44	6.85	7.23	7.60	7.96	8.30	8.63	8.94	9.25
3.	9.54	9.83	10.10	10.37	10.63	10.88	11.13	11.36	11.60	11.82
4.	12.04	12.26	12.46	12.67	12.87	13.06	13.26	13.44	13.62	13.80
5.	13.98	14.15	14.32	14.49	14.65	14.81	14.96	15.12	15.27	15.42
6.	15.56	15.71	15.85	15.99	16.12	16.26	16.39	16.52	16.65	16.78
7.	16.90	17.03	17.15	17.27	17.38	17.50	17.62	17.73	17.84	17.95
8.	18.06	18.17	18.28	18.38	18.49	18.59	18.69	18.79	18.89	18.99
9.	19.08	19.18	19.28	19.37	19.46	19.55	19.65	19.74	19.82	19.91
	0.	1.	2.	3.	4.	5.	6.	7.	8.	9.

Decibels $= 20 \log_{10} M$

ANSWERS

Answers to Selected Problems

Chapter 1

1.1

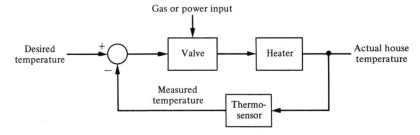

1.7   The feedback is positive.
Time lost per day = 5/3 minutes.
Total error after 15 days = 25 minutes.

Chapter 2

2.1 $R_1 i_1 + \dfrac{1}{C_1} \displaystyle\int i_1 dt + L_1 \dfrac{d(i_1 - i_2)}{dt} + R_2(i_1 - i_2) = v(t)$ loop 1.

$L_2 \dfrac{di_2}{dt} + \dfrac{1}{C_2} \displaystyle\int i_2 dt + R_2(i_2 - i_1) + L_1 \dfrac{d(i_2 - i_1)}{dt} = 0$ loop 2.

2.4   a) $v_0 = \dfrac{v_{in}}{2}$ for $-.5 \leqslant v_{in} \leqslant .5$.

b) $v_0 = 2v_{in} - 1$ for $0.5 \leqslant v_{in} \leqslant 1.5$.

2.5   a) $\delta Q = k\delta P$; $k = \dfrac{K}{2(\Delta P_0)^{1/2}}$;   and $\Delta P = P_1 - P_2$.

2.7 $T(s) = \dfrac{s + 1/R_1 C}{s + (R_1 + R_2)/R_1 R_2 C}$.

2.8 $T(s) = \dfrac{s^2 + 4s + 8}{s^2 + 8s + 8}$.

479

2.14 $\dfrac{\theta(s)}{V_f(s)} = \dfrac{0.0278}{s(s + 1.39)}$.

2.16 $x_1 = 2; x_2 = 3$.

2.18 $T(s) = \dfrac{V_2(s)}{V_1(s)} = \dfrac{Y_1 Z_2 Y_3 Z_4}{1 + Y_1 Z_2 + Y_3 Z_2 + Y_3 Z_4 + Y_1 Z_2 Z_4 Y_3}$.

2.20   a) $\dfrac{e_0}{e_{in}} = \dfrac{g_m R_s}{1 + g_m R_s}$.

     b) $\dfrac{e_0}{e_{in}} = \dfrac{20}{21}$.

2.25 $C(s) = G(s)R(s)$.

With the effect of the disturbance eliminated.

Chapter 3

3.1   a) Open loop $S_R^T = \dfrac{1}{RCs + 1}$.

     Closed loop $S_R^T = \dfrac{1}{RCs + 1 + KR}$

     where

$$T(s) = \dfrac{G_1(s)}{1 + KG_1(s)}.$$

3.3   a) Closed loop $T(s) = \dfrac{KG_1(s)}{1 + KK_t G_1(s)}$; $G_1(s) = \dfrac{1}{\tau s + 1}$; $S_K^T = \dfrac{1}{1 + KK_t G_1(s)}$.

     b) $\mathfrak{J}(s) = \dfrac{G_1(s)}{1 + KK_t G_1(s)}$ $\mathfrak{J}_e(s) \approx \dfrac{\mathfrak{J}_e(s)}{KK_t}$.

     c) $e_{ss} = A/1 + KK_t$ closed loop error.

3.11   b) $T(s) = \dfrac{1}{s - k}$.

     $c(t) = c(0)e^{kt}$ when $t \geqslant 0$.

     After one year $c(1) = 1000e^{.05} = \$1051.30$.

     Total after five years $c = \$5823.70$.

3.12   c) $S_{K_1}^{T_1} = 0.01$; $S_{K_1}^{T_2} = 0.1$.

Chapter 4

4.1   a) $E(s) = \dfrac{R(s)}{(1 + K_a K_m K_t)/(s\tau_m + 1)}$

     b) $K_a K_m \geqslant 24$ $K_a K_m \geqslant 39$.

4.4   b) Compromise: Percent overshoot $= 11\%$; $T_p = 2.19$ seconds; $K = 2.98$; $\zeta = 0.58$; $\omega_n = 1.725$.

4.5   a)   Set $K_2 = 1$ so step response has $e_{ss} = 0$.

$$T(s) = \frac{2}{s^2 + 2s + 2}; \qquad \zeta = 1/\sqrt{2}.$$

     c)   ITAE for step input

$$T(s) = \frac{\omega_n^2}{s^2 + 1.4\omega_n s + \omega_n^2}; \qquad \zeta = 1/\sqrt{2}$$

$$= \frac{2}{s^2 + 2s + 2}.$$

4.7   a)   A type one system, $K_p = \infty$, $K_v = 1/K_3$.

     b)   Set $\zeta = 0.6$, $\omega_n = 120$ and $K_1 K_2 = 36 \times 10^4$.

4.9 $K_v = 2.05$.

4.12 $T(s) = \dfrac{3.25\omega_n^2 s + \omega_n^3}{s^3 + 1.75\omega_n s^2 + 3.25\omega_n^2 s + \omega_n^3}; \qquad \omega_n = 4$

$$= \frac{52s + 64}{s^3 + 7s^2 + 52s + 64}.$$

4.15 $T_s = 0.195$ seconds.

Chapter 5

5.1   a)   Stable

     b)   Stable

     c)   Unstable

     d)   Unstable

     e)   Unstable

5.2   a) $-1 \le k_a \le 2.64$.

     b) $k_a = 0.8$ A reasonable approximation, since real pole is 4.5 times the real part of the complex roots.

5.4   a) $K \le 600$.

     b) $s = \pm j28.3$.

5.7   a)   Stable for all values of $K_v > 0$.

     b) $K_v = 5750$.

5.10 $0 \le K \le 28.1$.

5.13   a) $0 < K < 10.69$.

Chapter 6

6.1   a) $\phi_A = +60°, -60°, -180°$.

        $\sigma_A = -\frac{2}{3}$.

Locus crosses imaginary axis at $K = 2$ and $s = \pm j1$.

Breakaway from real axis at $s = -\frac{1}{3}$.

6.5 a) $K_2 = 1.6$; complex roots: $s = -3.83 \pm j3.88$.

real roots: $s = -1.33, -0.045$.

6.8 Require a zero degree locus, $\underline{/GH} = 0°, \pm 360°, \ldots$

6.10 b) $K_2 = 10.1$.

6.14 Stable for $320 \leqslant K \leqslant 27,000$.

6.26 b) Stable for $K \geqslant 2$.

c) $K = 2$, $s = \pm j1.4$.

d) Complex roots do not dominate.

6.29 $K = 7.35$.

6.30 $S_R^{r_1} = \frac{5}{6}$; $S_R^{r_2} = -\frac{10}{3}$.

Chapter 7

7.1 a)

ω	0	1	5	∞
$\lvert GH \rvert$	1	.4	.037	0
ϕ	$0°$	$-90°$	$-153°$	$-180°$

7.2 a)

ω	.5	1	2	8
db	-3.27	-8	-15.3	-36.4
ϕ	$-59°$	$-90°$	$-121°$	$-162°$

7.3 b) Evaluate at $\omega = 1.1\omega_n$.

Twin T: $\lvert G \rvert = 0.05$.

Bridged T: $\lvert G \rvert = 0.707$; A narrower band filter.

7.6 a) $GH(s) = \dfrac{0.8(5s + 1)}{s(0.25s + 1)^2}$.

7.14 $G(s) = \dfrac{809.7}{s(s^2 + 6.35s + 161.3)}$.

Chapter 8

8.5 c) Phase margin $= +7°$ $\omega_c = 1.5$.

Gain margin $= +2$ db when $k = 13.3$.

8.6 a) Phase margin $= +20°$.

Gain margin $= +6$db.

b) $Mp_\omega = 10.5$ db at $\omega_c = .0075$.

8.7   b)   Phase margin = 88°.

Gain margin is infinite, since stable for all $K_1 > 0$.

8.8   a)   Phase margin = $-9°$,   unstable.

$\omega_c = 49$.

8.9   a) $\omega_c = 8$ and phase margin = 83°.

8.12   a)   Phase margin = $+7°$, stable.

b)   Phase margin = 30.7°, unstable.

8.17 $K = 4$ yields phase margin = 30°.

8.18   a) $K_1 = 95$.

Chapter 9

9.1   c) $\dot{\mathbf{x}} = \begin{bmatrix} -R/L & -1/L \\ 1/C & 0 \end{bmatrix} \mathbf{x} + \begin{bmatrix} 1/L \\ 0 \end{bmatrix} u(t)$,   where $x_1 = i,\ x_2 = v_c$.

9.4   b) $\dot{\mathbf{x}} = \begin{bmatrix} 0 & 1 & 0 \\ 0 & 0 & 1 \\ -1 & -3 & -2 \end{bmatrix} \mathbf{x} + \begin{bmatrix} 0 \\ 0 \\ 1 \end{bmatrix} r(t)$, $c(t) = [3,3,1]\ \mathbf{x}$.

9.5   c) $\dot{\mathbf{x}} = \begin{bmatrix} 0 & 1 \\ -1 & 1 \end{bmatrix} \mathbf{x} + \begin{bmatrix} 0 \\ 1 \end{bmatrix} r(t)$, $c(t) = x_1(t)$.

9.6   a)   for stability: $h > k,\ ab > kh$.

b)   When $h > k$, then rabbits grow in number.

9.7   c) $\det[sI - A] = s^3 + 10^{-4}s^2 + 10^{-2}s + 4 \times 10^{-6}$
Unstable system

9.10   b)   Unstable system

9.11   a) $\phi(t) = \begin{bmatrix} (2t + 1)e^{-t} & -2te^{-t} \\ 2te^{-t} & (-2t + 1)e^{-t} \end{bmatrix}$.

9.11   b) $x_1(t) = x_2(t) = 10e^{-t}$.

9.14   a)   System is unstable.

b)   Add $r(t) = -Kx_2(t)$; then stable for $K > 1$.

Chapter 10

10.1   a)   Lead network compensation:

$$G(s) = \frac{20}{s^2}.$$

$$G_c(s) = \frac{(s/2.8) + 1}{(s/22.4) + 1}.$$

Phase margin $\approx 50°$.

   b) Root locus, choose $\zeta = .5$, $\omega_n = 4.5$.
$$G_c(s)G(s) = \frac{40(s + 2.28)}{s^2(s + 9.0)}.$$

10.5 $G_c(s) = K_2 + \dfrac{K_3}{s} = \dfrac{K_2 s + K_3}{s}$.

$e_{ss} = 0$.

One solution: Let $K_3/K_2 = .1$ and cancel one pole of $G(s)$,
                set $\zeta = .6$, $\zeta\omega_n = 5$, $\omega_n = 8.333$,
                roots $s = -5 \pm j6.67$.

10.6 Set phase margin $= 60°$.
$$G_c(s)G(s) = \frac{1000(s/14 + 1)}{(s/70 + 1)(.1s + 1)(10s + 1)}.$$

Simulation yields: Percent overshoot $= 22\%$, $T_p = 0.05$ sec.

10.9 Desire $K_v = 100$, use root locus.

Desire complex roots at $s = -3.2 \pm j2.4$ $(\zeta = 0.8)$.
$$G_c(s)G(s) = \frac{207(s + .2)}{s(s + 10)^2(s + .004)}.$$

10.11 $J = p_{11} = \dfrac{1}{2(2k - 3)}$,   so desire k large.

10.12 a) $J = \dfrac{(1 + \lambda k^2)}{2(2k - 3)}$.

      b) $\lambda = 1$; $k = 3.3$.

10.13 $J = 1/(2k - 1)$.

10.15 $k = 0.9$ for minimum J.

10.16 c) For Problem 10.15: $k = 0.9$.
          Roots $s = -.45 \pm j.835$; $\zeta = .47$.

10.17 $k_2 = \sqrt{20}$ for minimum J; $J = \dfrac{k_2^2 - 20}{8k_2}$.

10.18 Set zero at $s = -20$; then find pole at $s = -60$. Then $K_v = 21$.

Chapter 11

11.3 $G(z) = \dfrac{z}{z - e^{-T}}$, $R(z) = \dfrac{Tz}{(z - 1)^2}$,

    $C(z) = Tz^{-1} + T(2 + e^{-T})z^{-2} + T(2 + e^{-T})(1 + e^{-T})z^{-3} + \cdots$

11.5 $C(z) = \dfrac{0.632z}{(z - 1)(z + 0.264)}$,

    $C(z) = 0.632z^{-1} + 0.465z^{-2} + 0.511z^{-3} + \cdots$

11.6 Initial value $c(0) = 0$.
Final value $c(\infty) = 0.5$.

11.7 Select $T/\tau = 0.5$ and overshoot $= 0.3$.
Then $T = 0.25$ sec, $K = 2.2$, $e_{ss} = 0.7$.
If $T/\tau = 0.25$, $T = .125$ sec.
Then $K = 2.8$, $e_{ss} = 0.6$.

11.8 $D(z) = \dfrac{1.582(z - 0.3678)}{z + 0.418}$.

Index

Index